Progress in Molecular and Subcellular Biology

Series Editors Werner E.G. Müller
Philippe Jeanteur, Robert E. Rhoads, Đurđica Ugarković,
Márcio Reis Custódio

53

Volumes Published in the Series

Progress in Molecular
and Subcellular Biology

Subseries:
Marine Molecular Biotechnology

Volume 36
Viruses and Apoptosis
C. Alonso (Ed.)

Volume 38
Epigenetics and Chromatin
Ph. Jeanteur (Ed.)

Volume 40
Developmental Biology of Neoplastic Growth
A. Macieira-Coelho (Ed.)

Volume 41
Molecular Basis of Symbiosis
J. Overmann (Ed.)

Volume 44
Alternative Splicing and Disease
Ph. Jeanlevr (Ed.)

Volume 45
Asymmetric Cell Division
A. Macieira Coelho (Ed.)

Volume 48
Centromere
Đurđica Ugarković (Ed.)

Volume 49
Aestivation
C.A. Navas and J.E. Carvalho (Eds.)

Volume 50
miRNA Regulation of the Translational Machinery
R.E. Rhoads (Ed.)

Volume 51
Long Non-Coding RNAs
Đurđica Ugarkovic (Ed.)

Volume 37
Sponges (Porifera)
W.E.G. Müller (Ed.)

Volume 39
Echinodermata
V. Matranga (Ed.)

Volume 42
Antifouling Compounds
N. Fusetani and A.S. Clare (Eds.)

Volume 43
Molluscs
G. Cimino and M. Gavagnin (Eds.)

Volume 46
Marine Toxins as Research Tools
N. Fusetani and W. Kem (Eds.)

Volume 47
Biosilica in Evolution, Morphogenesis, and Nanobiotechnology
W.E.G. Müller and M.A. Grachev (Eds.)

Volume 52
Molecular Biomineralization
W.E.G. Müller (Ed.)

Volume 53
Biology of Marine Fungi
C. Raghukumar (Ed.)

Chandralata Raghukumar
Editor

Biology of Marine Fungi

Springer

Editor
Dr. Chandralata Raghukumar
National Institute of Oceanography
Marine Biotechnology Laboratory
Dona Paula
403004 Panjim
India
lata_raghukumar@rediffmail.com

ISSN 0079-6484
ISBN 978-3-642-23341-8 e-ISBN 978-3-642-23342-5
DOI 10.1007/978-3-642-23342-5
Springer Heidelberg Dordrecht London New York

Library of Congress Control Number: 2011943185

Printed on acid-free paper

Springer is part of Springer Science+Business Media (www.springer.com)

Preface

The late Dr. Stephen Thomas Moss of Portsmouth Polytechnic, UK, edited the book "Biology of Marine Fungi" in the year 1986. The present book, appearing nearly 25 years later, is dedicated to the memory of Dr. Moss, the eminent mycologist who left us prematurely. His expertise covered marine, as well as the trichomycetous fungi and electron microscopy of fungi. None of the authors who have contributed in the present book were authors in the 1986 book, indicating that a new crop of marine mycologists have appeared on the scene, bringing new ideas, techniques and approaches, and having been groomed and guided by the earlier generation in their quest to understand the biology of marine fungi.

There have been quite a few books on marine fungi. Johnson and Sparrow's pioneering monograph in 1961, the book on higher marine fungi by Kohlmeyer and Kohlmeyer in 1979 and in recent times, reviews on diverse topics in marine mycology published as "Fungi in marine environment" by Kevin Hyde in 2002 as a special issue of the journal *Fungal Diversity* are examples. Several books published as proceedings of the marine mycology symposia containing original research papers have appeared regularly. The present book is aimed to present marine fungal biology in its broad perspective, with several reviews and extensive literature coverage that will help the students and teachers working in this field.

With increasing realization of the importance of oceans in our daily lives, the diversity of marine microbes, flora and fauna has attracted the attention of the scientific community. Several exciting papers on marine fungi have appeared in the last two decades. Thus, fungi have been reported from extreme marine environments, such as hydrothermal vents, deep-sea trenches, cold methane seeps, deep-sea subsurface, hypersaline, anoxic and suboxic waters. Culture-dependent as well as culture-independent approaches are helping in assessing marine fungal diversity. Several new phylotypes have been reported based on these techniques, which may contribute to the hidden diversity of fungi on this planet. This should pave way for developing methods to culture the uncultured fungi.

Unraveling the functional diversity of fungi using metagenomics approach is slowly gathering momentum, the first report being that of fungi in coral holobiont.

This approach emphasizes the ecological role of fungi present in any niche. With increasing availability of molecular tools, the metagenomics of fungi in various marine environments will become a popular approach in future.

In this book, I have tried to include reviews on some of the recent advances made in different topics. The role of unicellular thraustochytrids, although not strictly fungi, but eukaryotic osmoheterotrophs, in marine food web is discussed by Bongiorni in the first chapter. The oomycetous pathogens of marine fisheries and shell fisheries with their descriptions and symptoms are illustrated by Hattai. The chapter on fungal endosymbionts of algae highlights their diversity and metabolites. The diversity and ecology of planktonic fungi in the world's oceans and their significance in ocean carbon and nutrient cycling is reviewed by Wang. Fungi associated with sponges and corals and their secondary metabolites have enthused interest of pharma companies. In light of this, one chapter discusses the ecology of fungi in corals and their role in coral ecosystem. Dead Sea with its extremely high salt content is an extremophilic environment. A chapter is devoted to fungi in this environment. Another chapter discusses diversity and adaptation of fungi growing in hypersaline environment of salterns. Sumathi et al. have reviewed molecular diversity of fungi in oxygen-deficient environment from world oceans. Nagano and Nagahama discuss the cultured and uncultured fungal diversity in deep-sea sediments. This chapter highlights the presence of novel fungal lineages, their adaptations and ecological role in extreme environments. Pang et al. discuss fine tuning of identification of marine ascomycetes by classical morphological taxonomy using several new criteria. Sridhar et al. provide new data on diversity of marine fungi from beaches of Portugal. A chapter is devoted to describe marine genera of Xylariaceae of the phylum Ascomycota. Sarma has compiled information on diversity of marine fungi occurring on the mangrove plant *Rhizophora* in 18 different sites world over. Finally, two chapters describe biotechnological potentials of marine fungi in various fields of industrial enzymes, secondary metabolites and bioremediation.

The untapped potential of oceanic resources is being increasingly realized. These resources are in the field of bioactive molecules, enzymes, nutraceuticals, antibiotics, energy, aquaculture, source of gene pool and biodiversity. Most papers contributed in this book have tried to address the current global understanding of the biology, role and biotechnological potential of fungi in the oceans.

I thank all the authors who readily agreed to contribute to this. I acknowledge various reviewers for critically reading the manuscripts. My special thanks to Dr. VVR Sarma who persuaded me to edit this book. I am thankful to the Director of NIO, Dr. Shetye for the support and to Council for Scientific and Industrial Research (CSIR), New Delhi, for the financial assistance.

Dr. Steve Moss: An Appreciation of an Outstanding Mycologist and Friend

Dr. Steve Moss (General Secretary and President Elect) and Professor Tony Whalley (Honorary Treasurer) preparing for the AGM of the British Mycological Society in 2000

Stephen (Steve) Thomas Moss was born in Buxton, Derbyshire, but was very much a Norfolk man. In his youth, he was an outstanding swimmer at county level, a fact which he kept modestly to himself, and this is not surprising because those who knew Steve will remember his gentle and unassuming nature. I first met Steve in Florida, USA, at the International Mycological Congress (IMC2) in 1977 and it was clear then that he was a very talented individual but was also warm and friendly. In future years, we became close friends through the British Mycological Society and working together in Malaysia, China, and Thailand. Steve was an outstanding communicator with endless patience as a teacher and at the same time inspirational. Many will remember his operation of a scanning electron microscope at the 100th anniversary exhibition of the British Mycological Society in London in September 1996. He captivated school children, members of the public, and members of parliament over the 3-day event and did a wonderful job in bringing mycology alive. His ability to enthuse his audience was exceptional and his appearance on the popular TV show, Big Breakfast, was educational, entertaining, and memorable.

As a mycologist, he was exceptionally talented with strong international recognition for his research on thraustochytrids and marine fungi. He did not achieve this easy way having studied part-time for his M.Sc. at Birkbeck College, University of London, funding his studies as a part-time lecturer or demonstrator at a range of technical colleges in the London area. Later these teaching experiences in the technical colleges provided him with the foundation to become a teacher and communicator par excellence. At Birkbeck, he was introduced to fungi by one of the best-known mycologists at that time, Terence Ingold, and this is when his interest and love of fungi started. On completion of his M.Sc. he moved to the University of Reading to undertake his PhD under the guidance of Michael Dick. It was here that he became a proficient electron microscopist. This was followed by an NSF-funded Research Associateship with Robert Lichtwardt at the University of Kansas (USA) which resulted in many publications on the Trichomycetes and established him as a world authority. He then returned to the UK taking up a position at the University of Portsmouth where he formed a very productive partnership with Gareth Jones working mainly on marine ascomycetes and biodeterioration of submerged woods. Electron microscopes played an important part of his studies and many of his publications demonstrate his considerable skill in producing unrivaled images of appendages of trichospores and attachment mechanisms of marine ascomycete spores. He could easily have succeeded as an electron microscopist.

Perhaps it was the electron microscope which was responsible for the detailed and meticulous way Steve tackled his academic career and through his many contributions to the British Mycological Society. He was a devoted and outstanding general secretary and many council members and officers remarked on the emails sent in the early hours! I was particularly fortunate to have the support of Steve during my presidential year and when he assumed office as president he was alive with ideas and an exciting vision for the future of the society. Sadly much of this was lost with his untimely and tragic death at such an early age.

We worked together in Malaysia, China, and Thailand which Jan could sometimes join and I have many fond memories of these trips. He ignored a cobra in the Gombak Reserve in Malaysia to collect a *Xylaria* specimen I had requested. He spent many late nights sorting out problems with the electron microscopes at the University of Malaya and training staff and students. Steve even sorted out the plumbing in the university guest house in Wuhan, China, when we were there running a fungal workshop. It was perhaps his encouragement and interaction with students both in the UK and in the abroad which stands as a testimony to Steve. He was an inspirational teacher and always found time to help to resolve problems. It was hardly surprising that his lectures and workshops were highly sought after. He passionately and expertly promoted mycology with the general public, academics, politicians, and the media at large. He was also a dedicated family man and when time permitted would take his dinghy out on the Solent with his daughter Becky as crew. His wife Jan strongly supported him in his research and later BMS activities and visits with them to the Jolly Roger near their home in Gosport were highly recommended.

Two measures which indicate the respect and love for Steve were demonstrated by the presence of an exceptional number of mourners at his funeral with former students and colleagues traveling from far corners of the world and the frequent references to Steve when I am with our friends and former students in Thailand, Malaysia, and China. Steve's memory lives on!

Tony Whalley

Contents

Contributors

Mohamed T. Abdelghany Department of Microbiology, Al-Azhar University, Cairo, Egypt

Ayman Farrag Ahmed Department of Microbiology, Al-Azhar University, Cairo, Egypt

Lucia Bongiorni IMAR-Department of Oceanography and Fisheries (DOP), University of the Azores, Horta, Azores 9901-862, Portugal, lbongiorni@uac.pt

F. Cássio Centre of Molecular and Environmental Biology (CBMA), Department of Biology, University of Minho, Campus de Gualtar, Braga 4710-057, Portugal

Sukanyanee Chareprasert Program in Biotechnology, Chulalongkorn University, Bangkok 10330, Thailand

Wai-Lun Chiang Department of Biology and Chemistry, City University of Hong Kong, 83 Tat Chee Avenue, Kowloon Tong, Hong Kong, China

Raymond K.K. Chow Department of Biology and Chemistry, City University of Hong Kong, 83 Tat Chee Avenue, Kowloon Tong, Hong Kong SAR, PR China

Samir Damare Marine Biotechnology Laboratory, CSIR-National Institute of Oceanography, Dona Paula, Goa 403004, India, samir@nio.org

Hussain H. El-sheikh Department of Microbiology, Al-Azhar University, Cairo, Egypt

Dominik Forster Faculty of Biology, University of Kaiserslautern, Erwin Schrödinger Str., Kaiserslautern 67663, Germany

Ji-Dong Gu School of Biological Sciences, The University of Hong Kong, Pokfulam Road, Hong Kong SAR, PR China

Nina Gunde-Cimerman Biology Department, University of Ljubljana, Večna pot 111, Ljubljana SI-1000, Slovenia

Centre of Excellence for Integrated Approaches in Chemistry and Biology of Proteins (CIPKeBiP), Jamova 39, Ljubljana SI-1000, Slovenia, nina.gunde-cimerman@bf.uni-lj.si

Kishio Hatai Nippon Veterinary and Life Science University, Tokyo, Japan, khatai0111@nvlu.ac.jp

Ravindran J. National Institute of Oceanography, Regional Centre, Kochi, Kerala 682 018, India

Cathrine Sumathi Jebaraj National Institute of Oceanography, Council of Scientific and Industrial Research, Dona Paula 403 004, Goa India, cathrine@nio.org

Jen-Sheng Jheng Institute of Marine Biology, National Taiwan Ocean University, 2 Pei-Ning Road, Keelung 20224, Taiwan, ROC

K.S. Karamchand Department of Biosciences, Mangalore University, Mangalore, Karnataka India

Frank Kauff Faculty of Biology, University of Kaiserslautern, Erwin Schrödinger Str., Kaiserslautern 67663, Germany

Ahmed M.A. Khalil School of Pharmacy and Biomolecular Sciences, Liverpool John Moores University, Byrom Street, Liverpool L3 3AF, UK

Qian Li Department of Oceanography, University of Hawaii at Manoa, Honolulu, HI 96822 USA

Xianhua Liu School of Environmental Science and Engineering, Tianjin University, Tianjin 300071, China

Zhu-Hua Luo Key Laboratory of Marine Biogenetic Resources, Third Institute of Oceanography, State Oceanic Administration, 178 Daxue Road, Xiamen 361005, PR China

Takahiko Nagahama Department of Food and Nutrition, Higashi-Chikushi Junior College, 5-1-1 Shimoitozu, Kokurakita-ku, Kitakyusyu, Fukuoka 800-0351, Japan

Institute of Biogeosciences, Japan Agency for Marine-Earth Science and Technology (JAMSTEC), 2-15, Natsushima-cho, Yokosuka 237-0061, Japan, jamstec@gmail.com

Yuriko Nagano Institute of Biogeosciences, Japan Agency for Marine-Earth Science and Technology (JAMSTEC), 2-15, Natsushima-cho, Yokosuka 237-0061, Japan

Aharon Oren Department of Plant and Environmental Sciences, The Institute of Life Sciences, The Hebrew University of Jerusalem, Jerusalem 91904, Israel, orena@cc.huji.ac.il

Ka-Lai Pang Institute of Marine Biology, National Taiwan Ocean University, 2 Pei-Ning Road, Keelung, 20224, Taiwan, ROC, klpang@ntou.edu.tw

C. Pascoal Centre of Molecular and Environmental Biology (CBMA), Department of Biology, University of Minho, Campus de Gualtar, Braga 4710-057, Portugal

Ana Plemenitaš Biology Department, University of Ljubljana, Vrazov, Trg 2, Ljubljana SI-1000, Slovenia

Seshagiri Raghukumar Myko Tech Private Limited, 313 Vainguinnim Valley, Dona Paula, Goa 403004, India

Chandralata Raghukumar National Institute of Oceanography, (Council for Scientific and Industrial Research), Dona Paula, Goa 403 004, India, lata_raghukumar@rediffmail.com

Vemuri Venkateswara Sarma Department of Biotechnology, School of Life Sciences, Pondicherry University, Kalapet, Pondicherry 605 014, India, sarmavv@yahoo.com

George P. Sharples School of Pharmacy and Biomolecular Sciences, Liverpool John Moores University, Byrom Street, Liverpool L3 3AF, UK

Prakitsin Sihanonth Department of Microbiology, Chulalongkorn University, Bangkok 30330, Thailand

Purnima Singh Marine Biotechnology Laboratory, CSIR-National Institute of Oceanography, Dona Paula, Goa 403004, India

Hamdy G. Soliman Department of Microbiology, Al-Azhar University, Cairo, Egypt

K.R. Sridhar Department of Biosciences, Mangalore University, Mangalore, Karnataka India, sirikr@yahoo.com

Thorsten Stoeck Faculty of Biology, University of Kaiserslautern, Erwin Schrödinger Str., Kaiserslautern 67663, Germany

T.S. Suryanarayanan Vivekananda Institute of Tropical Mycology (VINSTROM), Ramakrishna Mission Vidyapith, Chennai 600004, India, t_sury2002@yahoo.com

Nuttika Suwannasai Department of Biology, Srinakharinwirot University, Bangkok 10110, Thailand

L.L.P. Vrijmoed Department of Biology and Chemistry, City University of Hong Kong, 83 Tat Chee Avenue, Kowloon Tong, Hong Kong SAR, PR China, bhlilian@cityu.edu.hk

Guangyi Wang School of Environment and Energy, Peking University Shenzhen Graduate School, Shenzhen 518055, China

Department of Oceanography, University of Hawaii at Manoa, Honolulu, HI 96822, USA, guangyi@hawaii.edu, gywang@pkusz.edu.cn

Xin Wang Department of Oceanography, University of Hawaii at Manoa, Honolulu, HI 96822, USA

Anthony J.S. Whalley School of Pharmacy and Biomolecular Sciences, Liverpool John Moores University, Byrom Street, Liverpool L3 3AF, UK, A.J.Whalley@ljmu.ac.uk

Margaret A. Whalley School of Pharmacy and Biomolecular Sciences, Liverpool John Moores University, Byrom Street, Liverpool, L3 3AF, UK

Yi-Rui Wu Department of Civil and Environmental Engineering, National University of Singapore, 21 Lower Kent Ridge Road, Kent Ridge 119077, Singapore

Janja Zajc Biology Department, University of Ljubljana, Večna pot 111, Ljubljana, SI-1000, Slovenia

Polona Zalar Biology Department, University of Ljubljana, Večna pot 111, Ljubljana SI-1000, Slovenia

Chapter 1
Thraustochytrids, a Neglected Component of Organic Matter Decomposition and Food Webs in Marine Sediments

Lucia Bongiorni

Contents

Abstract Decomposition of organic matter in marine sediments is a critical step influencing oxygen and carbon fluxes. In addition to heterotrophic bacteria and fungi, osmoheterotrophic protists may contribute to this process, but the extent of their role as decomposers is still unknown. Among saprophytic protists, the thraustochytrids have been isolated from different habitats and substrates. Recently, they have been reported to be particularly abundant in marine sediments characterized by the presence of recalcitrant organic matter such as seagrass and mangrove detritus where they can reach biomass comparable to those of other protists and bacteria. In addition, their capacity to produce a wide spectrum of enzymes suggests a substantial role of thraustochytrids in sedimentary organic decomposition. Moreover, thraustochytrids may represent a food source for several benthic microorganisms and animals and may be involved in the upgrading of nutrient-poor organic detritus. This chapter presents an overview on studies of thraustochytrids in benthic ecosystems and discusses future prospectives and possible methods to quantify their role in benthic food webs.

L. Bongiorni (✉)
IMAR-Department of Oceanography and Fisheries (DOP), University of the Azores, 9901-862 Horta, Azores, Portugal
e-mail: lbongiorni@uac.pt

C. Raghukumar (ed.), *Biology of Marine Fungi*,
Progress in Molecular and Subcellular Biology 53,
DOI 10.1007/978-3-642-23342-5_1,

1.1 Introduction

Marine sediments are the largest habitat on our planet in terms of areal coverage. These habitats are greatly dependent upon the supply of organic material falling from the water column, most of which enters and accumulates as polymeric organic compounds. The balance between organic matter degradation and preservation in these habitats has immense consequences for the global carbon and oxygen cycles (Berner 1989; Hedges and Keil 1995). Part of this organic matter is labile, i.e., immediately degraded, while another part (such as humic and fulvic acids, structural carbohydrates, etc.) is recalcitrant to microbial decomposition and accumulate easily in marine sediments (Mayer et al. 1995; Middelburg et al. 1999). The enzymatic hydrolysis of high molecular weight and recalcitrant organic polymers is usually the rate-limiting step in mineralization in natural environments (Sigee 2005). While prokaryotes have been for a long time considered the main players of organic matter decomposition (Sime-Ngando and Colombet 2009; Weinbauer and Rassoulzadegan 2004), it is known that apart from fungi, several protists (e.g., heterotrophic nanoflagellates and photosynthetic dinoflagellates) can also make important and unique contributions in this process (Mohapatra and Fukami 2004; Stoecker and Gustafson 2003).

Among saprophytic marine protists, the thraustochytrids (stramenopilan fungi) have been found to be abundant in coastal sediments (Bongiorni and Dini 2002; Bongiorni et al. 2005a, b; Santangelo et al. 2000). Thraustochytrids, together with heterotrophic protozoa, are highly adapted to solid substrate pervasion and are involved in the breakdown, scavenging, and mineralization of highly refractory organic matter through extracellular digestion (Fig. 1.1, Bongiorni and Dini 2002; Bongiorni et al. 2005a, b; Raghukumar 2002; Santangelo et al. 2000). Due to these peculiar characteristics, thraustochytrids likely represent an important component of the food webs in benthic environments characterized by high levels of organic matter such as seagrass beds and mangrove areas (Raghukumar and Raghukumar 1999; Raghukumar et al. 1994). However, till now, knowledge on the quantitative relevance of thraustochytrids within the benthic microbial loop is poor especially

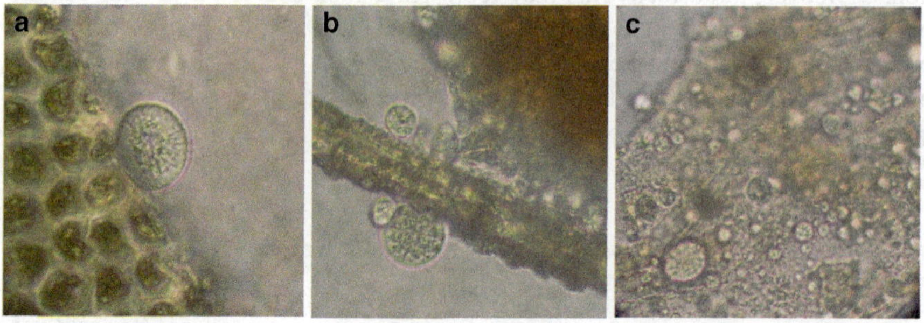

Fig. 1.1 Thraustochytrid cells colonizing dead algae (**a**, **b**) and crustacean exoskeleton (**c**) substrates

due to the lack of appropriate methodologies to measure their abundance, biomass and production in marine sediments. This chapter summarizes the state of the art regarding the study of thraustochytrids in marine sediments and discusses possible research developments with focus on novel methods and approaches that might provide new insights into their ecological role.

1.2 Methodologies for Detecting and Counting Thraustochytrids in Sediments

Knowledge of abundance and distribution of microorganisms are indispensable for understanding their ecological roles. It is therefore evident that the development of methods for accurate determination of their numbers and biomass is a crucial step. Although thraustochytrids abundance has been investigated in the pelagic realm of several geographic locations (see Raghukumar 2002, 2008 and literature cited), studies on sediments have been for a long time hampered, like for other protists, by the lack of adequate counting methodologies. This is especially due to difficulties in separating cells from sediments. Numbers of benthic thraustochytrids were estimated for the first time using the Most Probable Number (MPN) technique modified by Gaertner (1968). The modification consisted in using pine pollen to bait fresh sediment samples. This method, later applied on Mediterranean Sea shores' samples by Bongiorni and Dini (2002) resulted to be efficient in detecting relative abundance and seasonal fluctuation of local thraustochytrid populations. However, this technique presents several limitations. Direct counts might result in underestimating densities as not all thraustochytrid species can grow on pine pollen (Raghukumar 2002). Moreover, labour intensive analysis of fresh material and cultures imposes the examination of small number of replicated samples, thus limiting the extent of *in situ* experimental designs (Santangelo et al. 2000). In the last decades, counting methodologies have been largely improved by staining bacteria and protists with fluorochromes and observing them under the epifluorescence microscope (Sherr et al. 1993). The acriflavine direct detection method (AfDD) elaborated by Raghukumar and Schaumann (1993) specifically for thraustochytrids has allowed enormous advances in the study of these organisms. Acriflavine is a fluorochrome that stains the nucleic DNA and the cell wall of thraustochytrids (made of sulphated polysaccharides) making them fluorescing yellow/green and orange/red, respectively. The uniqueness of the method consists in the fact that autotrophic cells can be discriminated by chlorophyll-autofluorescence, and most other heterotrophic protists can be recognized by their shape or by the lack of a red-fluorescencing cell wall. However, the problem in using fluorochrome for detection of benthic microorganisms has always been the background fluorescence due to sediments. This problem has been partially solved by dividing cells from sediments on the base of their densities mainly by centrifuging samples in a nonlinear silica-gel gradient (Alongi 1993; Epstein 1995, 1997a, b;

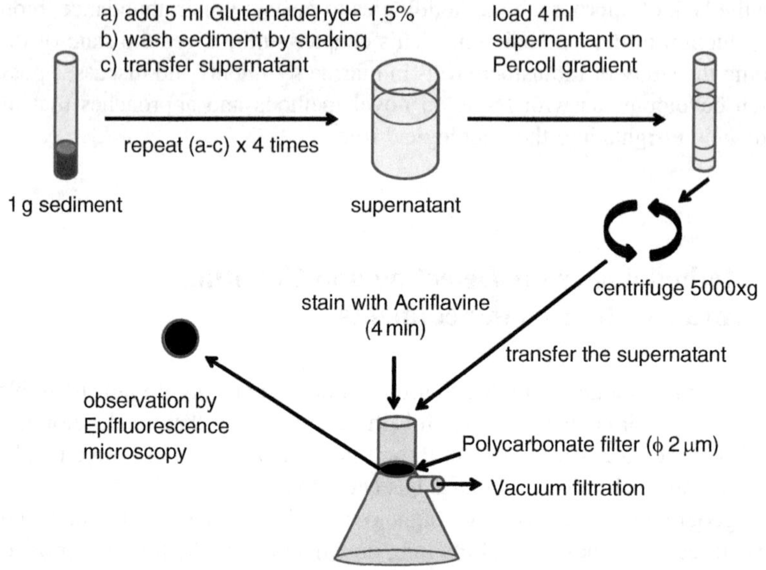

Fig. 1.2 Main steps of the method to enumerate thraustochytrids in sediments as described in Santangelo et al. (2000). Sediment samples are fixed and washed for four times by shaking with 1.5% glutaraldehyde-seawater solution. From each washing step, the supernatant is collected (total ca. 20 ml). Successively an aliquot (4 ml) of the supernatant is loaded on a Percoll gradient. The supernatant and the gradient are centrifuged, concentrated on a polycarbonate 2 μm pore-membrane and stained with acriflavine before observation under an epifluorescence microscope

Starink et al. 1994; Tao and Taghon 1997). A similar method was successfully applied by Santangelo et al. (2000) to simultaneously enumerate the thraustochytrid and ciliate components of benthic protistan community (Fig. 1.2). This protocol combined with the one developed by Raghukumar and Schaumann (1993) for thraustochytrids and that by Epstein (1995) for nano, micro- and meiobenthos consists in separating cells from sediments as described before and then staining them in sequence with acriflavine for thraustochytrids, diamino-2-phenylindole hydrochloride (DAPI) and fluorescein isothiocyanate (FITC) for ciliates (Santangelo et al. 2000). The same method was also used for counting on same samples different component of protistan community (thraustochytrids, ciliates and heterotrophic nanoflagellates, Bongiorni et al. 2005a). However, even the acriflavine method might underestimate the true abundances of thraustochytrids as small cells have a tiny wall and most thraustochytrid zoospores lack cell walls. Because of such problems, the developments of new methods able to specifically and precisely detect thraustochytrid cells are becoming essential (Takao et al. 2007).

Molecular techniques allow direct investigation of the community structure, diversity and phylogeny of microorganisms in almost any environment and quantification of the individual types of microorganisms or entire microbial communities. These techniques can be applied in conjunction with microscopic and biochemical methods to study diversity and functioning of bacterial and fungal communities.

The use of fluorescently labelled oligonucleotide probes complementary to unique target sites on the rRNA (Fluorescent *In Situ* Hybridization – FISH) allows microscopic examination of individual cells in complex microbial assemblages (Theron and Cloete 2000). These techniques have been widely used for analysing population composition of natural bacterial communities (DeLong et al. 1989; Giovannoni et al. 1988; Glöckner et al. 1996; Ishii et al. 2004; Llobet-Brossa et al. 1998). Recently, Takao et al. (2007) established a FISH protocol, and its efficiency, targeting thraustochytrid phylogenetic group (TPG). Another molecular method employing the use of real-time quantitative PCR (qPCR) assay has been successfully applied to detect QPX, a thraustochytrid which parasitized hard clams (Liu et al. 2009; Lyons et al. 2006). This technique is highly sensitive for measuring copy numbers of target nucleic acid sequences. It has been used to detect and quantify prokaryotes as well as eukaryotes, with the power of identification at genus and species levels, from a large number of habitats (Galluzzi et al. 2004; Skovhus et al. 2004; Takai and Horikoshi 2000). Although the detection of thraustochytrids through molecular techniques seems to be promising, further improvements are needed before they will become routine tools for monitoring thraustochytrid abundances in natural environments. Next developments will have to include the design of primers specific for this group of organisms, their validation on a wide representative of thraustochytrid strains and their optimization for application in different types of natural samples, including sediments and mud.

1.3 Roles of Thraustochytrids in Benthic Food Webs

The roles played by fungi and thraustochytrids in benthic food webs have so far received little attention (Raghukumar 2002). Thraustochytrids densities in marine sediments and benthic detritus are the highest reported in literature for that group (Table 1.1) exceeding values measured in planktonic environments (typically of few hundred cells per ml of water). In benthic habitats, protists are known to be sensitive to environmental conditions and generally display a significant numerical response to organic enrichment (Danovaro 2000; Manini et al. 2003). Organic matter represents an important food source for thraustochytrids as suggested by the positive correlation found between thraustochytrid densities and amounts of total sediment organic matter in Mediterranean beaches and intertidal areas (Bongiorni and Dini 2002; Bongiorni et al. 2004; Santangelo et al. 2000). Although thraustochytrids might often represent a negligible fraction of microbial abundance and a minor fraction of the total microbial biomass (Bongiorni and Dini 2002; Bongiorni et al. 2005b), they can reach extremely high densities in the presence of large amounts of organic matter recalcitrant to enzymatic decomposition. Thraustochytrids are found to be particularly abundant in marine sediments underlying seagrass (*Posidonia oceanica*) beds and mangrove detritus (Table 1.1, Bongiorni et al. 2005a; Wong et al. 2005). Moreover, they were reported in high densities in polluted harbour and in sediments characterized by high protein and lipid contents

Table 1.1 Average thraustochytrid abundance in sediments and organic detritus of different benthic habitats

Area	Habitat type	Abundance	Method	References
North Sea (Germany)	Fladen Ground	18–73 cells ml^{-1}[a]	Modified MPN	Gaertner and Raghukumar (1980)
				Raghukumar and Gaertner (1980)
Southern-North Sea	–	10–18.5 cells ml^{-1}[a]	Modified MPN	Gaertner (1967)
NE-Atlantic Ocean (Africa)	Upwelling region	$<$0.001–22 cells ml^{-1}[a]	Modified MPN	Gaertner (1982)
W- Atlantic Ocean	–	$<$0.001–65.7 cells ml^{-1}[a]	Modified MPN	Gaertner (1982)
NW-Mediterranean Sea (Italy)	Surf zone sediments	44 $\pm$ 40 cells ml^{-1}	Modified MPN	Bongiorni and Dini (2002)
NW-Mediterranean Sea (Italy)	Subtidal sediments	61 $\pm$ 53 cells ml^{-1}	Modified MPN	Bongiorni and Dini (2002)
NW-Mediterranean Sea (Italy)	Subtidal sediments	42.4 $\pm$ 35.2 cells ml^{-1}	Percoll + AfDD	Santangelo et al. (2000)
S-Mediterranean Sea (Italy)	Fish farm, sediments	3.35 $\pm$ 0.88 $\times$ 10^3 cells g^{-1}	Percoll + AfDD	Bongiorni et al. (2005a)
S-Mediterranean Sea (Italy)	Subtidal sediments	0.96 $\pm$ 0.36 $\times$ 10^3 cells g^{-1}	Percoll + AfDD	Bongiorni et al. (2005a)
S-Mediterranean Sea (Italy)	Fish farm (*Posidonia* bed)	13.45 $\pm$ 3.87 $\times$ 10^3 cells g^{-1}	Percoll + AfDD	Bongiorni et al. (2005a)
S-Mediterranean Sea (Italy)	Sediments (*Posidonia* bed)	1.47 $\pm$ 0.76 $\times$ 10^3 cells g^{-1}	Percoll + AfDD	Bongiorni et al. (2005a)
E-Mediterranean Sea (Cyprus)	Fish farm, sediments	17.49 $\pm$ 5.71 $\times$ 10^3 cells g^{-1}	Percoll + AfDD	Bongiorni et al. (unpublished)
E-Mediterranean Sea (Cyprus)	Subtidal sediments	11.66 $\pm$ 1.18 $\times$ 10^3 cells g^{-1}	Percoll + AfDD	Bongiorni et al. (unpublished)
E-Mediterranean Sea (Cyprus)	Fish farm (*Posidonia* bed)	15.03 $\pm$ 4.91 $\times$ 10^3 cells g^{-1}	Percoll + AfDD	Bongiorni et al. (unpublished)
E-Mediterranean Sea (Cyprus)	Sediments (*Posidonia* bed)	7.79 $\pm$ 0.78 $\times$ 10^3 cells g^{-1}	Percoll + AfDD	Bongiorni et al. (unpublished)
Central Adriatic Sea (Italy)	Surf zone sediments	6.71 $\pm$ 1.04 $\times$ 10^3 cells g^{-1}	Percoll + AfDD	Bongiorni et al. (2005b)
Central Adriatic Sea (Italy)	Port, subtidal sand-mud	7.34 $\pm$ 1.91 $\times$ 10^3 cells g^{-1}	Percoll + AfDD	Bongiorni et al. (2005b)
N-Adriatic Sea (Goro lagoon, Italy)	Mussel farm, mud	34.59 $\pm$ 2.71 $\times$ 10^3 cells g^{-1}	Percoll + AfDD	Bongiorni et al. (unpublished)
Arabian Sea (India)	Anoxic sediments	64–2622 CFU g^{-1}[b]	Agar plating	Sumathi and Raghukumar (2009)
China Sea (Hong Kong)	Leaves detritus (Mangrove area)	4.8–560 $\times$ 10^3 CFU g^{-1}[a]	Agar plating	Wong et al. (2005)
China Sea (Hong Kong)	Sediments (Mangrove area)	0.1–1.6 $\times$ 10^3 CFU g^{-1}[a]	Agar plating	Wong et al. (2005)

[a]Density range

[b]Total fungal abundance (thraustochytrids were the next most abundant fungi after *Aspergillus* spp. and *Humicola* sp.)

due to fish-farm biodeposition (Table 1.1, Bongiorni et al. 2005a, b, Bongiorni et al. unpublished). In *P. oceanica* beds characterized by the presence of a highly refractory organic matter composition, dominated by structural carbohydrates (up to 70% of the biopolymeric C fraction), thraustochytrids largely dominated over heterotrophic nanoflagellates biomass (0.124 vs. 0.044 μg C g^{-1}), reaching up to 50% of protist total abundance (as sum of thraustochytrids, nanoflagellates and ciliates, Bongiorni et al. 2005a). Experiments carried on pattern of microbial colonization of dead mangrove leaves and detritus showed that thraustochytrid density and biomass increase in an advanced state of the decomposition process (Fell and Findlay 1988; Raghukumar et al. 1994, 1995). In decaying brown algae, thraustochytrid biomass peaked only after 21 days of the experiment almost equalling bacterial biomass at that time (0.07 vs. 1.1% detrital carbon, Sathe-Pathak et al. 1993). A similar phenomenon was also observed in decaying kelp (Miller and Jones 1983). All these results point to a key role of thraustochytrids in the degradation of the most refractory portion of sedimentary organic matter.

Production of enzymes has been reported sporadically for thraustochytrids (Bahnweg 1979a, b; Bremer and Talbot 1995; Raghukumar and Raghukumar 1999; Raghukumar et al. 1994; Sharma et al. 1994). Enzymes such as cellulase, amylase and xylanase were observed to be produced by several thraustochytrids strains (Bremer and Talbot 1995; Nagano et al. 2011; Raghukumar et al. 1994; Sharma et al. 1994). A later study on the enzymatic profiles of 11 benthic and epiphytic thraustochytrid strains showed that these organisms can produce a wide pool of enzymes involved in the hydrolysis of all classes of organic compounds (Bongiorni et al. 2005b). Benthic thraustochytrids exhibited a good production of lipase, a selection of protease and a poor presence of enzymes degrading carbohydrates. The dominance of enzymatic activities such as alkaline phosphatase and aminopeptidase suggested that thraustochytrids might be implicated in sediment N and P cycling. Moreover, 8 out of 11 strains produced high amount of N-acetyl-β-glucosaminidase, a chitin degrading enzyme (Bongiorni et al. 2005b), thus confirming previous observation (Heiland and Ulken 1989). Chitin occurs commonly as an exo- and endoskeletal material in many marine organisms, and represents a dominant fraction of particulate carbon and nitrogen in the marine sediments (Gooday 1990; Place 1996). However, more studies are needed to determine and quantify the importance of thraustochytrids in the degradation of these refractory polymers. It is also worth of note that a rough estimate of thraustochytrid contribution to the total extracellular enzymatic activities in two types of sediments showed that these organisms may contribute up to 4% of the total β-D-glucosidase, 1% of aminopeptidase and 1.4% of alkaline phosphatase activities (Bongiorni et al. 2005b). However, these data could be underestimates since they were based only on culturable thraustochytrids. If it is postulated that a certain portion of thraustochytrids could be either unculturable or not culturable with the available nutrient media, their contribution to total enzymatic activities in sediments would become even more consistent.

Thraustochytrids are potentially involved in different processes along benthic food webs. Estimates of thraustochytrids biomass in organic-rich sediments suggest that thraustochytrids could likely represent a significant food source in the microbial loop (Bongiorni et al. 2004, 2005a; Fenchel 1980; Raghukumar and Balasubramanian 1991; Santangelo et al. 2000). Different studies observed ciliates and amoebae feeding on thraustochytrid cells (Raghukumar and Balasubramanian 1991; Ulken 1981) and the association found between thraustochytrids and ciliates abundance in Mediterranean sediments seems to support this hypothesis (Bongiorni et al. 2004). Probably, another important role of these stramenopilan fungi is to supply essential nutrients to detritivores. Thraustochytrids represent a source of valuable vitamins, amino acids (e.g., lysine and methionine), sterols and polyunsaturated fatty acids (PUFAs) that cannot be de novo synthesized by detritus-feeding animals (Arts et al. 2001; Huang et al. 2001; Phillips 1984; Unagul et al. 2005; Zhukova and Kharlamenko 1999). Thraustochytrids might be also implicated, as other heterotrophic protists, in the improvement of the biochemical constituents of poor quality detritus for subsequent use by higher trophic levels, a phenomenon dubbed "trophic upgrading" (Tang and Taal 2005). Several detritivores ingest detritus as a way to feed upon living microorganisms attached to the particles. It has not been clearly demonstrated whether thraustochytrids by colonizing organic particles and detritus can function as trophic upgraders in detrital food webs. Although the role of thraustochytrids in all these energetic pathways remains to be studied in detail, the little clues available suggest that thraustochytrids may represent important energy links within the benthic food webs.

1.4 Conclusions and Future Perspectives in the Study of Thraustochytrids in Marine Sediments

The high thraustochytrid abundances and biomass constantly found in marine sediments suggest their prominent role in benthic habitats. Benthic thraustochytrids produce a wide spectrum of enzymatic activities which are likely to be important in the degradation of the recalcitrant portion of organic matter and in sediment biogeochemical cycles. Moreover, thraustochytrids may be implicated in the transformation of poor nutritive detritus into a more palatable food for detritivores. However, studies on such mechanisms are still poor and more benthic habitats need to be explored in regard to the occurrence of thraustochytrids. The deep-sea is largely unexplored and contains unique benthic features such as canyons, seamounts, deep-water coral reefs, hydrothermal vents and fluid seepages that support unique microbial and faunal communities. Thraustochytrids have been detected or isolated in deep-waters and only occasionally in deep-sea sediments including hydrothermal vents (Damare and Raghukumar 2008; Edgecomb et al. 2002; Raghukumar and Raghukumar 1999; Raghukumar et al. 2001). Ability of these organisms to produce active proteases at high pressure has been documented, thus suggesting a contribution to decomposition processes in deep-sea conditions (Raghukumar and Raghukumar 1999). Thraustochytrids have been also reported to inhabit anoxic sediments (Sumathi and Raghukumar 2009) but no studies have been attempted to understand their role in these habitats.

The lack of information on thraustochytrids (and in particular in marine sediments) is mainly due to methodological difficulties in quantifying their abundance, biomass and production. Development of new techniques for assessing thraustochytrid abundance and improvements of existing ones has been described. A special effort should be also specifically directed towards the development of techniques for measurements of thraustochytrid biomass, productivity and role in different sedimentary food webs. There are presently no biochemical methods to measure the biomass of thraustochytrids, as a distinctive signature compound such as ergosterol for eumycota fungi has not been found yet. Thraustochytrid biomass is usually estimated by measuring cell biovolume under epifluorescent microscopy and then multiplying the value obtained by the conversion factor 310 fg C μm^{-3} (Kimura et al. 1999). However, since this factor was obtained by coastal planktonic thraustochytrids, in order to have more realistic measurements of biomass in sediment, more C/N measurements from different thraustochytrid strains isolated from representative benthic habitats are urgently needed. Fatty acid signatures (22:5ω6 and 22:5ω3 fatty acids) have been used as alternative method to measure thraustochytrid biomass in mangrove detritus (Fell and Findlay 1988). Thraustochytrid-derived fatty acids, 14:0, 16:0 22:6ω3 and 22:5ω6, were also used to track thraustochytrid diet of krill in controlled conditions (Alonzo et al. 2005). The use of stable isotope (^{13}C)-labelled substrates in combination with biomarker analysis represents a new approach in microbial ecology enabling direct identification of microbes involved in specific processes and also allows for the incorporation of microorganism into food web studies (Boschker and Middelburg 2002). Another recent technique, the stable-isotope probing (SIP), offers a powerful tool for identifying microorganisms that are actively involved in specific metabolic processes. Following this technique ^{13}C-DNA, produced during the growth of metabolically distinct microbial groups on a ^{13}C-enriched carbon source, can be resolved from ^{12}C-DNA by density-gradient centrifugation. DNA isolated from the target group of microorganisms can be characterized taxonomically and functionally by gene probing and sequence analysis (Radajewski et al. 2000). It is likely that the use of multidisciplinary approaches such as combining molecular, biochemical and microscopic tools with isotopic tracers' experiments may provide considerable insights into the dynamics of thraustochytrids in benthic environments.

References

Alongi DM (1993) Extraction of protists in aquatic sediments via density gradient centrifugation. In: Kemp PF, Sherr BF, Sherr EB, Cole JJ (eds) Aquatic microbial ecology. Lewis Publishers, Boca Raton

Alonzo F, Virtue P, Nicol S, Nichols PD (2005) Lipids as trophic markers in Antarctic krill. III. Temporal changes in digestive gland lipid composition of *Euphausia superba* in controlled conditions. Mar Ecol Prog Ser 296:81–91

Arts MT, Ackman RG, Holub BJ (2001) "Essential fatty acids" in aquatic ecosystems: a crucial link between diet and human health and evolution. Can J Fish Aquat Sci 58(1):122–137

Bahnweg G (1979a) Studies on the physiology of Thraustochytriales I. Growth requirements and nitrogen nutrition of *Thraustochytrium* spp., *Schizochytrium* sp., *Japonochytrium* sp., *Ulkenia* spp., and *Labyrinthuloides* spp. Veröff Inst Meeresforsch Bremerh 17:245–268

Bahnweg G (1979b) Studies on the physiology of Thraustochytriales II. Carbon nutrition of *Thraustochytrium* spp., *Schizochytrium* sp., *Japonochytrium* sp., *Ulkenia* spp., and *Labyrinthuloides* spp. Veröff Inst Meeresforsch Bremerh 17:269–273

Berner RA (1989) Biogeochemical cycles of carbon and sulphur and their effects on atmospheric oxygen over Phanerozoic time. Palaeogeogr Palaeoclimatol Palaeoecol 73:97–112

Bongiorni L, Dini F (2002) Distribution and abundance of thraustochytrids in different Mediterranean coastal habitats. Aquat Microb Ecol 30:49–56

Bongiorni L, Pignataro L, Santangelo G (2004) Thraustochytrids (fungoid protists): an unexplored component of marine sediment microbiota. Sci Mar 68(1):43–48

Bongiorni L, Mirto S, Pusceddu A, Danovaro R (2005a) Response of benthic protozoa and thraustochytrid protists to fish-farm impact in seagrass (*Posidonia oceanica*) and soft bottom sediments. Microb Ecol 50:268–276

Bongiorni L, Pusceddu A, Danovaro R (2005b) Enzymatic activities of epiphytic and benthic thraustochytrids involved in organic matter degradation. Aquat Microb Ecol 41:299–305

Boschker HTS, Middelburg JJ (2002) Stable isotopes and biomarkers in microbial ecology. FEMS Microbiol Ecol 40:85–95

Bremer GB, Talbot G (1995) Cellulolytic enzyme activity in the marine protist *Schizochytrium aggregatum*. Bot Mar 38:37–41

Damare V, Raghukumar S (2008) Abundance of thraustochytrids and bacteria in the equatorial Indian Ocean, with relation to transparent exopolymeric particles (TEPs). FEMS Microbiol Ecol 65(1):40–49

Danovaro R (2000) Benthic microbial loop and meiofaunal response to oil induced disturbance in coastal sediments: a review. Int J Environ Pollut 13:380–391

DeLong EF, Wickham GS, Pace NR (1989) Phylogenetic stains: ribosomal RNA-based probes for the identification of single cells. Science 243:1360–1363

Edgecomb VP, Kysela DT, Teske A, Gomez A, Sogin ML (2002) Benthic eukaryotic diversity in the Guaymas basin hydrothermal vent environment. Proc Natl Acad Sci U S A 99:7658–7662

Epstein SS (1995) Simultaneous enumeration of protozoa and micrometazoa from marine sandy sediments. Aquat Microb Ecol 9:219–227

Epstein SS (1997a) Microbial food webs in marine sediments. I. Trophic interactions and grazing rates in two tidal flat communities. Microb Ecol 34:188–198

Epstein SS (1997b) Microbial food webs in marine sediments. II. Seasonal changes in trophic interactions in sandy tidal flat communities. Microb Ecol 34:188–198

Fell JW, Findlay RH (1988) Biochemical indicators of microbial decomposition process in coastal and oceanic environments. In: Thompson MF, Tirmizi N (eds) Marine science of the Arabian Sea. American Institute of Sciences, Washington

Fenchel T (1980) Relation between particle size selection and clearance in suspension feeding ciliates. Limnol Oceanogr 25:733–738

Gaertner A (1967) Marine niedere Pilze in Nordsee und Nordatlantik. Ber Dtsch Bot Ges 82:287–306

Gaertner A (1968) Eine Methode des quantitativen Nachweises niederer mit Pollen koederbarer Pilze im Meerwasser und im Sediment. Veröff Inst Meeresforsch Bremerh 3:75–92

Gaertner A (1982) Lower marine fungi from the Northwest African upwelling areas and from the Atlantic off Portugal. Meteor Forsch Ergebn D 34:9–30

Gaertner A, Raghukumar S (1980) Ecology of thraustochytrids (lower marine fungi) in the Fladen Ground and other parts of the North Sea. I. Meteor Forsch Ergebn A 22:165–185

Galluzzi L, Penna A, Bertozzini E, Vila M, Garces E, Magnani M (2004) Development of a real-time PCR assay for rapid detection and quantification of *Alexandrium minutum* (a dinoflagellate). Appl Environ Microbiol 70:1199–1206

Giovannoni SJ, DeLong EF, Olsen GJ, Pace NR (1988) Phylogenetic group-specific oligodeoxynucleotide probes for identification of single microbial cells. J Bacteriol 170:720–726

Glöckner FO, Amann R, Alfreider A et al (1996) An *in situ* hybridization protocol for detection and identification of planktonic bacteria. Syst Appl Microbiol 19:403–406

Gooday GW (1990) The ecology of chitin degradation. Adv Microb Ecol 11:387–430

Hedges JI, Keil RG (1995) Sedimentary organic matter preservation: an assessment and speculative synthesis. Mar Chem 49:81–115

Heiland R, Ulken A (1989) Untersuchungen zum Chitinabbau von niederen Pilzen. Nova Hedwigia 48:495–504

Huang J, Aki T, Hachida K, Yokochi T, Kawamoto S, Shigeta S, Ono K, Suzuki O (2001) Profile of polyunsaturated fatty acids produced by *Thraustochytrium* sp. KK17-3. J Am Oil Chem Soc 78:605–610

Ishii K, Mussmann M, MacGregor BJ, Amann R (2004) An improved fluorescence in situ hybridization protocol for the identification of bacteria and archaea in marine sediment. FEMS Microbiol Ecol 50:203–212

Kimura H, Fukuba T, Naganuma T (1999) Biomass of thraustochytrid protists in coastal water. Mar Ecol Prog Ser 189:27–33

Liu Q, Allam B, Collier JL (2009) Quantitative real-time PCR assay for QPX (Thraustochytriidae), a parasite of the hard clam (*Mercenaria mercenaria*). Appl Environ Microbiol 75(14):4913–4918

Llobet-Brossa E, Rosselló-Mora R, Amann R (1998) Microbial community composition of Wadden sea sediments as revealed by fluorescence in situ hybridization. Appl Environ Microbiol 64:2691–2696

Lyons MM, Smolowitz R, Dungan CF, Roberts SB (2006) Development of a real time quantitative PCR assay for the hard clam pathogen Quahog Parasite Unknown (QPX). Dis Aquat Organ 72:45–52

Manini E, Fiordelmondo C, Gambi C, Pusceddu A, Danovaro R (2003) Benthic microbial loop functioning in coastal lagoons: a comparative approach. Oceanol Acta 26:27–38

Mayer LM, Schick LL, Sawyer T, Plante CJ, Jumars PA, Self RL (1995) Bioavailable amino acids in sediments: a biomimetic, kinetics-based approach. Limnol Oceanogr 40:511–520

Middelburg JJ, Nieuwenhuize J, Van-Breugel P (1999) Black carbon in marine sediments. Mar Chem 65:245–252

Miller JD, Jones EBG (1983) Observations on the association of thraustochytrids marine fungi with decaying seaweed. Bot Mar 26:345–351

Mohapatra BR, Fukami K (2004) Production of aminopeptidase by marine heterotrophic nanoflagellates. Aquat Microb Ecol 34:129–137

Nagano N, Matsui S, Kuramura T, Taoka Y, Honda D, Hayashi M (2011) The distribution of extracellular cellulase activity in marine eukaryotes, thraustochytrids. Mar Biotechnol 13:133–136

Phillips NW (1984) Role of different microbes and substrates as potential supplies of specific essential nutrients to marine detritivores. Bull Mar Sci 35:283–298

Place AR (1996) The biochemical basis and ecological significance of chitin digestion. In: Muzzarelli RAA (ed) Chitin enzymology, vol 2. Atec Edizioni, Grottammare

Radajewski S, Ineson P, Parekh NP, Murrell JC (2000) Stable-isotope probing as a tool in microbial ecology. Nature 403:646–649

Raghukumar S (2002) Ecology of the marine protists, the Labyrinthulomycetes (thraustochytrids and labyrinthulids). Eur J Protistol 38:127–145

Raghukumar S (2008) Thraustochytrid marine protists: production of PUFAs and other emerging technologies. Mar Biotechnol 10(6):631–640

Raghukumar S, Balasubramanian R (1991) Occurrence of thraustochytrids fungi in coral and coral mucus. Indian J Mar Sci 20:176–181

Raghukumar S, Gaertner A (1980) Ecology of the thraustochytrids (lower marine fungi) in the Falden Ground and other parts of the North Sea II. Veröff Inst Meeresforsh Bremerh 18:289–308

Raghukumar S, Raghukumar C (1999) Thraustochytrid fungoid protists in faecal pellets of the tunicate *Pagea confoederata*, their tolerance to deep-sea conditions and implication in degradation processes. Mar Ecol Prog Ser 190:133–140

Raghukumar S, Schaumann K (1993) An epifluorescence microscope method for direct detection of and enumeration of the fungilike marine protists: the thraustochytrids. Limnol Oceanogr 38 (1):182–187

Raghukumar S, Sharma S, Raghukumar C, Sathe-Pathak V (1994) Thraustochytrid and fungal component of marine detritus. IV. Laboratory studies on decomposition of the leaves of the mangrove *Rhizophora apiculata* Blume. J Exp Mar Biol Ecol 183:113–131

Raghukumar S, Sathe-Pathak V, Sharma S, Raghukumar C (1995) Thraustochytrid and fungal component of marine detritus. III. Field studies on decomposition of the mangrove *Rhizophora apiculata*. Aquat Microb Ecol 9:117–125

Raghukumar S, Ramaiah N, Raghukumar C (2001) Dynamics of thraustochytrid protists in the water column of the Arabian Sea. Aquat Microb Ecol 24:175–186

Santangelo G, Bongiorni L, Pignataro L (2000) Abundance of thraustochytrids and ciliated protozoans in a Mediterranean sandy shore determined by an improved, direct method. Aquat Microb Ecol 23:55–61

Sathe-Pathak V, Raghukumar S, Raghukumar C, Sharma S (1993) Thraustochytrid and fungal component of marine detritus I-Field studies of decomposition of the brown algae *Sargassium cinereum*. Ind J Mar Sci 22:159–169

Sharma S, Raghukumar C, Raghukumar S, Sathe-Pathak V, Chandramohan D (1994) Thraustochytrid and fungal components of marine detritus. IV. Laboratory studies on decomposition of the brown alga *Sargassium cinereum*. J Exp Mar Biol Ecol 175:217–242

Sherr EB, Caron DA, Sherr BF (1993) Staining of heterotrophic protists for visualization via epifluorescence microscopy. In: Kemp PF, Sherr BF, Sherr EB, Cole JJ (eds) Aquatic microbial ecology. Lewis Publishers, Boca Raton

Sigee DC (2005) Freshwater microbiology. Wiley, Chichester, England

Sime-Ngando T, Colombet J (2009) Virus et prophages dans les écosystèmes aquatiques. Can J Microbiol 55:95–109

Skovhus TL, Ramsing NB, Holmstro C, Kjelleberg S, Dahllo I (2004) Real-time quantitative PCR for assessment of abundance of *Pseudoalteromonas* species in marine samples. Appl Environ Microbiol 70:2373–2382

Starink M, Bar-Gilissen MJ, Bak RPM, Cappenberg TE (1994) Quantitative centrifugation to extract benthic protozoa from freshwater sediments. Appl Environ Microbiol 60:167–173

Stoecker DK, Gustafson DE (2003) Cell-surface proteolytic activity of photosynthetic dinoflagellates. Aquat Microb Ecol 30:175–183

Sumathi JC, Raghukumar C (2009) Anaerobic denitrification in fungi from the coastal marine sediments off Goa, India. Mycol Res 113:100–109

Takai K, Horikoshi K (2000) Rapid detection and quantification of members of archaeal community by quantitative PCR using fluorogenic probes. Appl Environ Microbiol 66(11):5066–5072

Takao Y, Tomaru Y, Nagasaki K, Sasakura Y, Yokoama R, Honda D (2007) Fluorescence in situ hybridization using reformatted targeted probe for specific detection of thraustochytrids (Labyrinthulomycetes). Plank Benthos Res 2(2):91–97

Tang KW, Taal M (2005) Trophic modification of food quality by heterotrophic protists: species-specific effects on copepod egg production and egg hatching. J Exp Mar Biol Ecol 318:85–98

Tao SF, Taghon GL (1997) Enumeration of protozoa and bacteria in muddy sediment. Microb Ecol 33:144–148

Theron J, Cloete TE (2000) Molecular techniques for determining microbial diversity and community structure in natural environment. Crit Rev Microbiol 26(1):37–57

Ulken A (1981) On the role of phycomycetes in the food web of different mangrove swamps with brackish waters and waters of high salinity. Kieler Meeresforsh Sonderh 5:425–428

Unagul P, Assantachai C, Phadungruengluij S, Suphantharika M, Verduyn C (2005) Properties of the docosahexaenoic acid-producer *Schizochytrium mangrovei* Sk-02: effects of glucose, temperature and salinity and their interaction. Bot Mar 48:387–394

Weinbauer MG, Rassoulzadegan F (2004) Are viruses driving microbial diversification and diversity? Environ Microbiol 6:1–11

Wong MKM, Vrijmoed LLP, Au DWT (2005) Abundance of thraustochytrids on fallen decaying leaves of *Kandelia candel* and mangrove sediments in Futian National Nature Reserve, China. Bot Mar 48:374–378

Zhukova NV, Kharlamenko VI (1999) Sources of essential fatty acids in the marine microbial loop. Aquat Microb Ecol 17:153–157

Chapter 2
Diseases of Fish and Shellfish Caused by Marine Fungi

Kishio Hatai

Contents

Abstract Fungal diseases are problematic in cultured fish and shellfish, their seeds, and sometimes wild marine animals. In this chapter fungal diseases found in marine animals, especially in Japan, are described. Pathogens in the fungal diseases are divided into two groups. One of them is marine Oomycetes, which cause fungal diseases in marine shellfish and abalones. The diseases caused by the fungi of this group and the fungal characteristics are introduced. The pathogens include members of the genera *Lagenidium, Haliphthoros, Halocrusticida, Halioticida, Atkinsiella*, and *Pythium*. On the other hand, some fungal diseases caused by mitosporic fungi are also known in marine fish and shellfish. The diseases caused

K. Hatai (✉)
Nippon Veterinary and Life Science University, Tokyo, Japan
e-mail: khatai0111@nvlu.ac.jp

C. Raghukumar (ed.), *Biology of Marine Fungi*,
Progress in Molecular and Subcellular Biology 53,
DOI 10.1007/978-3-642-23342-5_2,

by these fungi and the fungal characteristics are described. The pathogens include members of the genera *Fusarium, Ochroconis, Exophiala, Scytalidium, Plectosporium*, and *Acremonium.*

2.1 Introduction

Many diseases of fish and shellfish have been observed in studies at monitoring sites in the oceans around the world. The impact of these diseases on population sizes in marine ecosystems in general is poorly understood. Marine fishes and prawns are very popular as seafood especially in Japan, because the Japanese like to eat them raw as "sushi" or "sashimi." Therefore, the culture of fish and shellfish and their seed production are important industries in the sea around Japan. The yield from these industries is gradually increasing along with demand. However, the industry is facing serious problems with infectious diseases. These diseases include fungal infections. No strategies are currently available for effectively controlling fungal diseases with antifungal substances. Therefore, some of these diseases cause high mortality rates, which results in significant economic losses.

In this chapter some of the most economically important diseases caused by species of Oomycetes and mitosporic fungi, found mainly in marine fish and shellfish in Japan, are described. Procedures for identification of these pathogens are also included. Accurate identification of pathogens is necessary in studies designed to improve production rates.

2.2 Fungal Diseases of Shellfish Caused by Oomycetes

Five fungal genera have been reported in Japan as pathogenic Oomycetes of marine shellfish including abalone. For classification of the fungi, an observation on the mode of zoospore production is essential and important. All fungi of the marine Oomycetes were isolated from the lesions using PYGS agar (peptone, 1.25 g; yeast extract, 1.25 g; glucose, 3.00 g; agar, 12–15 g; sea water, 1,000 mL). For inhibition of the most bacterial growth, an addition of ampicillin and streptomycin sulfate in the medium is required. After fungal colonies develop on the agar plates, each one is transferred onto fresh PYGS agar to make a pure culture. The fungi are maintained at 20–25°C and subcultured on PYGS agar approximately at monthly intervals. For morphological observations, the isolates were inoculated into PYGS broth and incubated at 25°C for 3–5 days. Small colonies in PYGS broth are rinsed twice with sterilized artificial seawater and transferred into Petri dishes containing 25 mL of sterilized artificial seawater, and then incubated at 25°C to induce zoospore production. The fungi of the five genera are illustrated in Fig. 2.1.

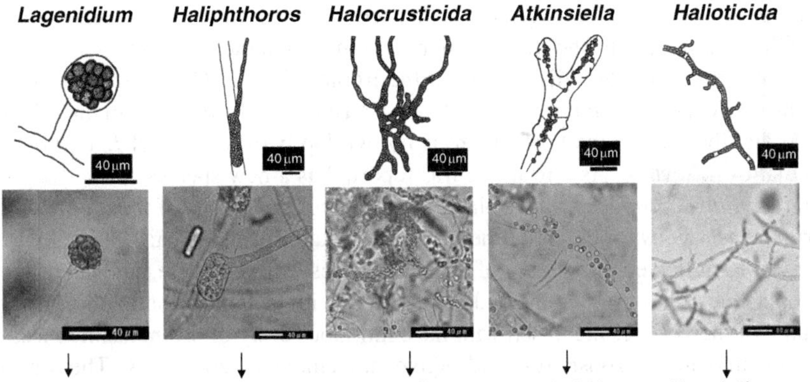

Fig. 2.1 Mode of zoospore production in the fungi of the five genera reported as pathogenic Oomycetes of marine shellfish including abalone in Japan

2.2.1 Lagenidium *Infection*

Species of *Lagenidium* have been found on various hosts from both freshwater and marine habitats (Sparrow 1960, 1973a). As parasites of animals, *L. callinectes* has been described from ova of the blue crab, *Callinectes sapidus* (Couch 1942; Rogers-Talbert 1948; Bland and Amerson 1973) and ova of the barnacle, *Chelonibia patula* (Johnson and Bonner 1960); *L. chthamalophilum* has been isolated from ova of the barnacle, *Chthamalus fragilis* (Johnson 1958), while *L. giganteum* has been reported in mosquito larvae, *Daphne* and copepods (Couch 1935). *Lagenidium callinectes* has been reported from certain marine algae (Fuller et al. 1964). In addition, unidentified species of *Lagenidium* were isolated from cultivated crustaceans, e.g., white shrimp, *Penaeus setiferus* (Lightner and Fontaine 1973), the Dungeness crab, *Cancer magister* (Armstrong et al. 1976) and the American lobster, *Homarus americanus* (Nilson et al. 1976).

A new fungus, *Lagenidium scyllae*, was isolated from ova and larvae of the mangrove crab, *Scylla serrata*, in Philippine (Bian et al. 1979). This fungus was very similar to *L. callinectes*. However, there were some differences between the two. Discharge tubes of *L. scyllae* were longer than those of *L. callinectes*. In the vesicle of *L. callinectes*, a gelatinous envelope is obvious and protoplasmic material never fills more than half of the inside vesicle (Couch 1942), whereas, in that of this fungus, the gelatinous envelope was not seen and the protoplasmic material nearly fills in the whole inside vesicle. According to Couch (1942) and Bland and Amerson (1973), zoospores of *L. callinectes* were discharged by the bursting of vesicle, which was rather persistent for several hours after the spores had emerged. In *L. scyllae*, however, all the spores were simultaneously discharged by rapid deliquescence of the vesicle or the spores were liberated one by one through the opening of vesicle. Moreover, the collapsed vesicles were never persistent. It seemed to be related to the thin and non-gelatinous characters of the vesicle. The mangrove crab is widely distributed in the tropical Pacific Ocean and the Indian Ocean, while the blue crab is in the Atlantic Coast of North America. In the relation to the distributions of the hosts, there is no overlap between that of *L. callinectes* and that of *L. scyllae*. As a result, it was reported as a new species.

Lagenidium callinectes was isolated from the eggs and zoea of the marine crab, *Portunus pelagicus* for the first time in Japan (Nakamura and Hatai 1995a). Masses of protoplasm flowed into the tip of discharge tubes, where vesicles appeared. Each protoplasmic mass was connected in a chain with a protoplasmic thread. The volume of the vesicles increased with the continuous entry of protoplasmic masses, division into initial zoospores, and active movement of zoospores. The way of zoospore liberation varied: sometimes they were released simultaneously by rupture of the vesicle, sometimes singly through a hole in the vesicle wall. When zoospores were discharged singly, vesicles usually persisted for a few minutes (Fig. 2.2).

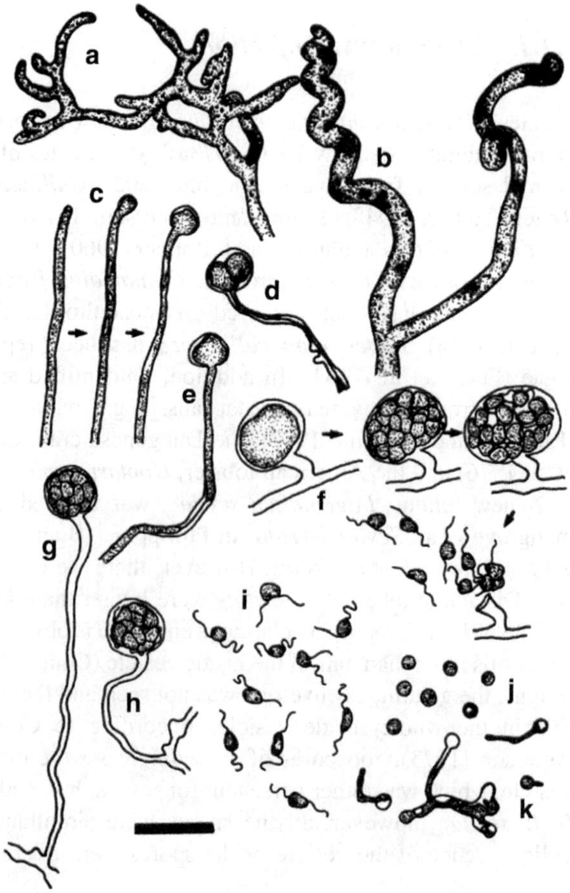

Fig. 2.2 Morphological characteristics of *L. callinectes* isolated from an egg of *P. pelagicus*. Scale = 50 μm. (**a**) Irregularly branched hyphae with numerous shiny rod granules; (**b**) Coiled hyphae in PYGS broth; (**c**) Vesicle formation; (**d, e**) Protoplasmic masses flow into the vesicle with a protoplasmic thread; (**f**) Division into initial zoospores and zoospores liberation; (**g, h**) Mature vesicles; (**i**) Zoospores; (**j**) Encysted zoospores; (**k**) Germination (Nakamura and Hatai 1995a)

Lagenidium callinectes was also isolated from larvae of mangrove crab, *Scylla serrata*, in Bali, Indonesia (Hatai et al. 2000).

A fungal infection occurred in the eggs and larvae of mangrove crab, *Scylla serrata*, affecting the seed production in Bali, Indonesia. The fungus isolated in August 1993 was a new species in the genus *Lagenidium*, and named *L. thermophilum*, because of its rapid and thermotolerant growth and unique discharge process. Masses of protoplasm occupied nearly all of the vesicles and divided into individual zoospores with two flagellae. The envelopes of the vesicles were not apparent. Zoospore liberation occurred after the vesicles separated from the discharge tubes, namely the vesicles left the discharge tubes before the zoospores were released (Nakamura et al. 1995). The manner of zoospore discharge varied: either zoospores were all discharged simultaneously when the vesicles burst, or they were released in ones or twos through opening in the vesicles. Generally, the former was observed among the bigger vesicle and the latter among the smaller ones. Collapsed vesicles were not persistent. The isolate grew at 15–45°C with an optimum at 30–40°C. This species differed from *L. callinectes* in its salt requirements. As *L. callinectes* grew on media containing seawater or 1–2.5% (w/w) NaCl, it seems to be a marine fungus. However, as *L. thermophilum* also grew on media without seawater, it is obvious that it is not exclusively marine.

L. thermophilum was also found in eggs and larvae of black tiger shrimp, *Penaeus monodon*, at a hatchery in August 2000, Thailand (Muraosa et al. 2006). The characteristics feature of asexual reproduction of the fungus was that zoospores swam away in seawater after the vesicle separated from the discharge tube (Fig. 2.3). This was the first report of *L. thermophilum* infection in black tiger shrimp in Thailand.

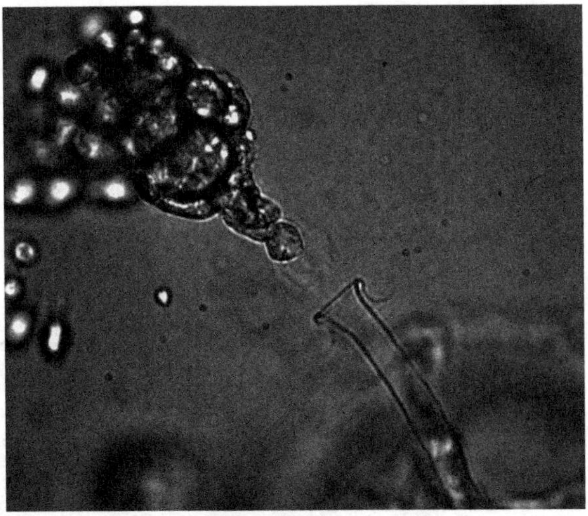

Fig. 2.3 Zoospores swim away in seawater after the vesicle separated from the discharge tube

2.2.2 Haliphthoros *Infection*

The genus *Haliphthoros* was a monotypic genus erected by Vishniac (1958) as the type of the family Haliphthraceae (Saprolegniales). *Haliphthoros milfordensis*, the type species of the genus, has been reported as an endoparasite of eggs of the oyster drill, *Urosalpinx cinerea* (Vishniac 1958). Since then it has been isolated from juveniles of the American lobster, *Homarus americanus* (Fisher et al. 1975), adults of the white shrimp, *Penaeus setiferus* (Tharp and Bland 1977), and a few marine algae (Fuller et al. 1964).

A new species, *H. philippinensis*, was isolated from larvae of the jumbo tiger prawn, *Penaeus monodon* in Philippines (Hatai et al. 1980). The hyphae were stout, branched, irregular, non-septate, developing within the bodies of larvae of the prawn, and it was holocarpic. In pure cultures, the hyphae were homotrichous, at first somewhat uniform, sometimes highly vacuolated, 10–37.5 μm in diameter, becoming fragmentary by means of cytoplasmic constriction with age (Fig. 2.4). Fragments with a dense cytoplasm were variable in size and shape, globose, elongate or tubular, often with protuberances, up to 190 × 100 μm, not disarticulated, connected in bead-like chains, functioning as sporangia and developing discharge tubes which were straight, wavy or coiled, up to 620 × 7.5–12.5 μm. Zoospores were polyplanetic. Encysted spores were spherical, 5–7.5(−12.5) μm in diameter, producing a delicate germ tube. Germ tube was simple, sometimes once branched and up to 250 μm in length. Sexual reproduction was not observed. The fungus showed a close resemblance to *H. milfordensis*, but it differed from the latter in a number of features as described below.

When the fragment with protuberance on the medium was transferred into sea water, the protuberance might again constrict and transform into another sporangium, or might extend and serve as a part of the discharge tube. The sporangia of

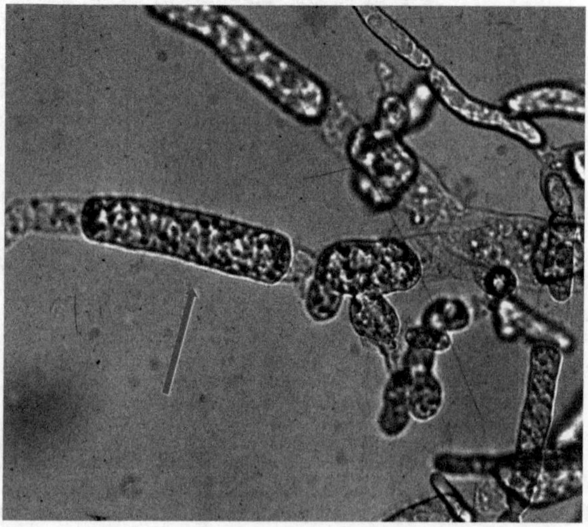

Fig. 2.4 Fragment (*arrow*) formation of genus *Haliphthoros*

this fungus are very variable in shape and often with protuberances: globose, elongate, tubular, or irregular-shaped. These are distinctive for the fungus. In this fungus, discharge of zoospores is also peculiar and diverse from that of *Haliphthoros* reported previously. The zoospores were released not only through the orifice of discharge tube but also through the opening of the sporangium. According to Vishniac (1958) and Fuller et al. (1964), zoospores of *H. milfordensis* were monoplanetic and monomorphic. This fungus, however, has polyplanetic and polymorphic zoospores including primary and secondary types. *H. philippinensis* has a wide range of temperature requirement for its growth. Possibly owing to its tropical habitation, it could be tolerant even up to 36°C. This feature was different from that of *H. milfordensis* which could not grow at 35°C (Vishniac 1958).

Haliphthoros milfoldensis infection was found in abalone, *Haliotis sieboldii*, temporarily held in aquaria with circulating sea water adjusted to 15°C by a cooling system in Japan (Hatai 1982). The typical external symptom of diseased abalones was flat or tubercle-like swelling formed on mantle, epipode and dorsal surface on foot (Fig. 2.5). The mycelium was always observed in the lesions. Zoospores of a fungus, which was isolated from the lesion, formed within the fragment were liberated through the orifice of discharge tube. Encysted spores were spherical, usually 7 μm in diameter. The fungus grew at a temperature range of 4.9–26.5°C, with optimum of 11.9–24.2°C.

In June 1994, fungal diseases occurred in the eggs and zoeae of crab, *Portunus pelagicus*, in Japan. Fragments of the isolated fungus were clearly constructed of concentrated masses of protoplasm in the hyphae, tuberculate, saccate or irregular, and quite variable in size and shape. They changed into zoosporangia-producing discharge tubes. Many vacuoles appeared in the zoosporangia and the extending

Fig. 2.5 The typical external symptom of diseased abalones was flat or tubercle-like swelling (*arrow*) formed on mantle

discharge tubes, and were also observed in the active mycelia. It was identified as *Haliphthoros milfordens* (Nakamura and Hatai 1995a).

In July 1997, *Haliphthoros milfordens* infection occurred in the eggs and zoeae of the mangrove crab, *Scylla serrata*, in Bali, Indonesia. The mortality rate reached almost 100% in the larvae (Hatai et al. 2000). The colonies on PYGS agar were whitish and reached a diameter of 20–25 mm after 5 days at 25°C. Hyphae in PYGS broth were stout, aseptate, branched with numerous shiny spherical granules, and sometimes concentrated masses of protoplasm were observed in the hyphae. In artificial seawater, fungal fragments were clearly observed to be concentrated masses of protoplasm in the hyphae. They changed into zoosporangia-producing discharge tubes. One discharge tube was usually formed on the lateral side of each zoosporangium. Division of the protoplasm started in the sporangia and continued in the discharge tubes just before zoospore liberation (Fig. 2.6).

In March 2001, *Haliphthoros milfordens* was isolated from larvae of the black tiger prawn, *Penaeus monodon* in Nha Trang, Vietnam (Chukanhom et al. 2003). This was the first report of disease in the black tiger prawn in Vietnam.

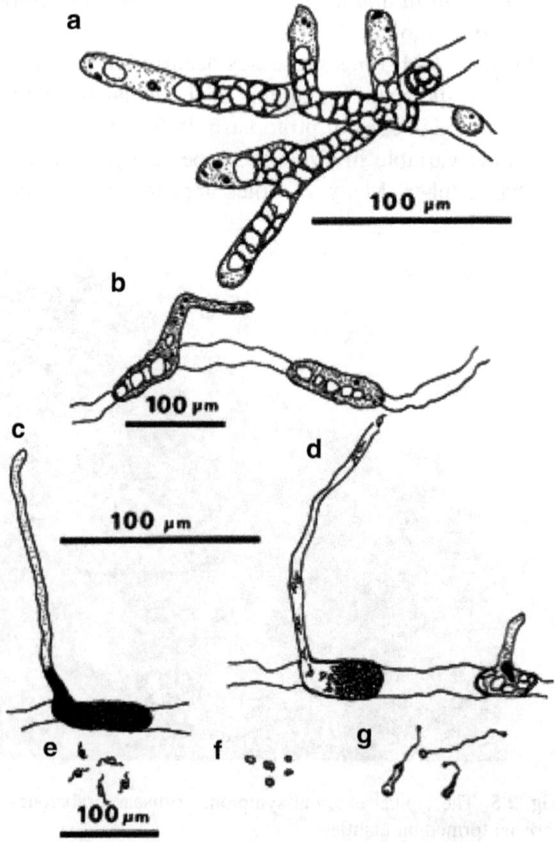

Fig. 2.6 Morphological characteristics of *Haliphthoros milfordensis* isolated from a zoea of *S. serrata*. (**a**) Hyphae in PYGS broth; (**b**) Fragments. Discharge tube formation on the left fragment; (**c**) Zoospore formation; (**d**) Zoospore liberation; (**e**) Zoospores; (**f**) Encysted zoospores; (**g**) Germination (Hatai et al. 2000)

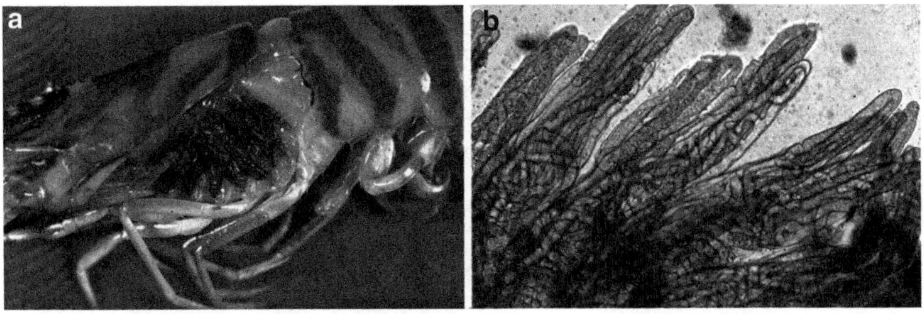

Fig. 2.7 *H. milfordensis* infection in juvenile kuruma prawns, *Penaeus japonicus*. (**a**) Showing *black* gills. (**b**) Hyphae found in gills. Hyphae are only growing within gills

H. milfordensis was also isolated from gill lesions of juvenile kuruma prawns, *Penaeus japonicus*, with black gill disease (Fig. 2.7) at a private farm in August 1989, Japan (Hatai et al. 1992). *H. milfordensis* has been known as a pathogen of various marine organisms. Especially, the fungus has been well known to be an important fungal pathogen of eggs and larvae of crustaceans. However, this was the first case of *H. milfordensis* infection in juvenile kuruma prawn. This condition has previously been known in kuruma prawn as the typical clinical sign of *Fusarium solani* infection.

2.2.3 Halocrusticida *Infection*

A new genus *Halocrusticida* gen. nov. (Lagenidiales, Haliphthoraceae) was proposed for the six species formerly reported as the fungi in the genus *Atkinsiella* except *A. dubia* (Nakamura and Hatai 1995b). These six species of *Atkinsiella* (Table 2.1) were reported from various aquatic animals (Martin 1977; Bian and Egusa 1980; Nakamura and Hatai 1994, 1995a; Kitancharoen et al. 1994; Kitancharoen and Hatai 1995). A key to the species of *Halocrusticida* is described in Table 2.2. Mycelia contained granular clusters without oil droplets and vacuoles on *A. dubia*, but many vacuoles and numerous shiny granules were found on the others. Central protoplasmic masses supported by several protoplasmic threads in the process of zoospore production were observed on *A. dubia*, but not on the others. The most apparent difference between *A. dubia* and the other six species of *Atkinsiella* was the behavior of zoospores in the first motile stage. Zoospores encysted within zoosporangia and discharge tubes following the first motile stage in *A. dubia*, while zoospores in the first motile stage were released from zoosporangia in the other six species.

The definition of the genus *Halocrusticida* is as follows. Thallus is endobiotic, holocarpic, stout, and branched. Zoosporangia are the same in size and shape as thalli. Discharge tubes develop one to several per sporangium. Zoospores in the first motile stage emerge from the zoosporangia. Zoospores are monoplanetic or

Table 2.1 Six species reported previously as *Atkinsiella*[a]

Species	References	Host	Locality
A. entomophaga	Martin (1977)	Insect eggs	USA
A. hamanaensis	Bian and Egusa (1980)	Eggs and larvae of *Scylla serrata*	Japan
A. parasitica	Nakamura and Hatai (1994)	Rotifer (*Brachionus plicatilis*)	Japan
A. awabi	Kitancharoen et al. (1994)	Abalone (*Haliotis sieboldii*)	Japan
A. okinawaensis	Nakamura and Hatai (1995a)	Zoea of the crab (*Portunus pelagicus*)	Japan
A. panulirata	Kitancharoen and Hatai (1995)	Philozoma of spiny lobster (*Panulirus japonicus*)	Japan

[a]Nakamura and Hatai (1995b)

Table 2.2 Key to species of *Halocrusticida*

1 Colonies filamentous, less than two tubes produced from each sporangium	*H. awabi*
1 Colonies lobed, bulbous................	2
2 Encysted spores more than 9 μm, parasitic on insect eggs	*H. entomophaga*
2 Encysted spores less than 9 μm, parasitic on crustaceans	3
3 Branched discharge tubes present	4
3 Branched discharge tubes absent	5
4 Zoospores generally formed two or more deep in the discharge tubes.....	*H. okinawaensis*
4 Zoospores generally formed in a single row in the discharge tubes.........	*H. parasitica*
5 Pigmentation from gray to light brown, optimum temperature for growth 30–32°C	*H. hamanaensis*
5 No pigmentation, optimum temperature for growth 25°C..................	*H. panulirata*

diplanetic, isokont, laterally biflagellate. Germinating zoospore has a slender germ tube. Sexual reproduction is absent. It is parasitic on aquatic animals, especially marine crustaceans.

Halocrusticida hamanaensis was originally reported as *Atkinsiella hamanaensis* (Bian and Egusa 1980). The fungus was isolated from ova of mangrove crab, *Scylla serrata* in Japan. The swollen hyphal tips up to 150 μm in diameter contained dense cytoplasm. Each sporangium was formed through the formation of septa and several lateral or terminal discharge tubes. The discharge tubes were straight or wavy, measuring 40–1,150 × 5–15 μm. Zoospores measured 6.3 (5–10) × 4.5 (3.8–5) μm in size, were pyriform or slipper-shaped, with two lateral flagella. The encysted spores were 5 (4.5–7.5) μm in diameter, spherical, subglobose, or angular. The encysted spore in sterile sea water developed a hair-like filament, 10–270 μm in length.

Halocrusticida awabi was originally reported as *Atkinsiella awabi* (Kitancharoen et al. 1994). The fungus was isolated from diseased abalone, *Haliotis sieboldii* in Japan. It showed external signs of infection of tubercle-like swelling on the mantle and melanized lesions on the peduncle. The hyphae were stout, irregular, branched, 16–140 μm diameter. Sporangia were formed through the formation of septa and lateral or terminal discharge tubes which were wavy or coiled. Zoospores were pyriform, biflagellate, and diplanetic. The encysted spore generally developed a hair-like filament with globular enlarged tip in PYGS broth. Direct germination without filament formation also occurred occasionally. The fungus was exclusively marine and grew best in shrimp extract medium at 25°C.

Halocrusticida parasitica was originally reported as *Atkinsiella parasitica* (Nakamura and Hatai 1994). In May 1992 the rotifer, *Brachionus plicatilis* did not increase in number when it was bred in a concrete tank as food supply for seed production of crustaceans and fishes. Because protozoa were observed microscopically on the surface of rotifers, a bath treatment with 25 ppm formalin was first conducted to solve the problem in the tank. However, no increase in the number of rotifers in the tank was found following the treatment. Further detailed microscopical observation revealed thick, non-septate hyphae measuring about 10 μm diameter in the eggs and bodies of many rotifers examined. Discharge tubes were extended outside the bodies (Fig. 2.8), and zoospores with lateral biflagella were released into the seawater through the tubes. Vesicles were not formed at the tip of discharge tubes (Nakamura and Hatai 1994; Nakamura et al. 1994a). The fungus isolated from the rotifer was characterized by producing monoplanetic, lateral biflagellate zoospores, and infrequently branched discharge tubes. The zoosporogenesis of the species is shown in Fig. 2.9.

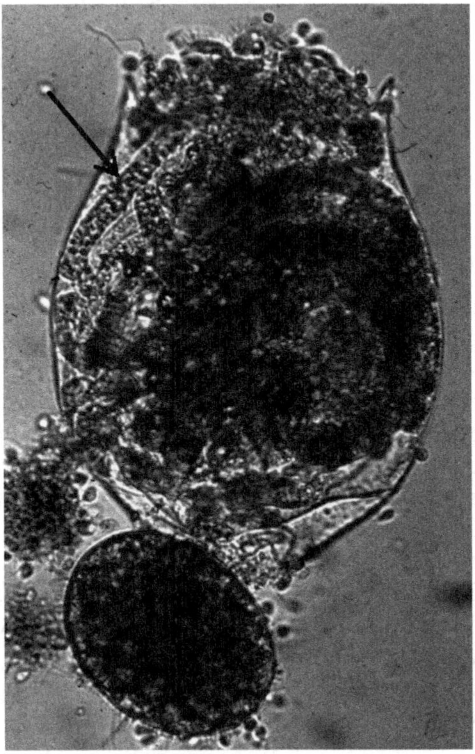

Fig. 2.8 Hyphae in bodies of a rotifer (*arrow*)

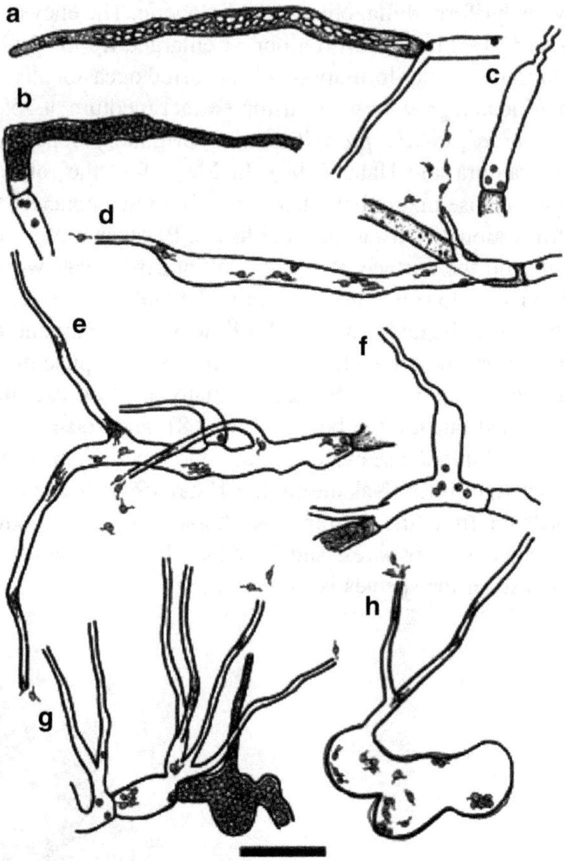

Fig. 2.9 Zoosporogenesis of *Halocrusticida* (*Atkinsiella*) *parasitica*. Scale bar = 50 μm. (**a**) Numerous large vacuoles appeared at an early stage of zoosporogenesis, and later discharge tubes developed. (**b**) Zoospore formation in a zoosporangium and a discharge tube at the final stage of zoosporogenesis. (**c**–**f**) One to three discharge tubes formed from a zoosporangium. (**g**, **h**) Branched discharge tubes

Halocrusticida panulirata was originally reported as *Atkinsiella panulirata* (Kitancharoen and Hatai 1995). This species was isolated from philozoma of the diseased spiny lobster, *Panulirus japonicus* in Japan. The fungus exhibited slow growth, occasionally submerged, with a creamy white, raised moist colony. Hyphae were stout, arranged in radiating pattern, irregularly branched, 10–22 μm diameter, occasionally separated by cross walls into subthalli. Thalli occasionally consisted of swollen features. Sporangia formed from the subthalli had one to three or partly coiled discharge tubes at the terminal or subterminal area. Zoospores were pyriform or reniform, biflagellate, isokont, and diplanetic. Encysted spores germinated as a hair-like filament with a globular enlarged tip in sterilized synthetic seawater, and directly as stout initial hyphae in PYGS broth. Gemmae spontaneously occurred in 3-day-old culture in PYGS broth at 25°C (Fig. 2.10). They were characterized by saccate-lobed-chained, thick-walled dense cytoplasmic and non-vacuolate features, width of 179–270 μm and various lengths up to 18 mm. Gemmae not only developed new thalli on PYGS agar or in PYGS broth, but also in sterilized synthetic seawater.

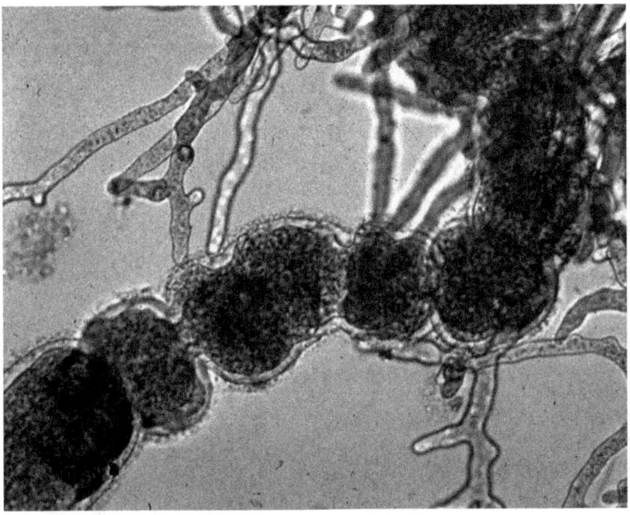

Fig. 2.10 Gemmae (*arrow*) spontaneously occurred in *Halocrusticida panulirata* culture in PYGS broth at 25°C

Halocrusticida okinawaensis was originally reported as *Atkinsiella okinawaensis* (Nakamura and Hatai 1995a). The new fungus was isolated from infected eggs and zoeae of the marine crab, *Portunus pelagicus*. Hyphae were stout, non-septate at first, irregularly branched with numerous shiny rod granules, 10–38 μm width. In seawater, hyphae were divided into subthalli with septa. Gemmae were present with thick walls, 22–190 μm in diameter. Zoosporangia were the same size and shapes as subthalli and gemmae. Discharge tubes were produced laterally or terminally from the sporangia, usually coiled or wavy. Each sporangium extended one to several discharge tubes. In the discharge tubes, zoospores were produced in more than two rows. The discharge tubes were 6–10 μm diameter and 40–510 μm length. Zoospores were laterally biflagellate, diplanetic, 4.7 × 6.3 μm on average. Germination was observed about 3 h after spores had encysted, with a hair-like filament measuring 5–190 μm length (Fig. 2.11).

2.2.4 Halioticida *Infection*

Halioticida infection was reported from abalone, *Haliotis* spp. in Japan (Muraosa et al. 2009). The genus was classified in Peronosporomycetes (formerly Oomycetes) as a new genus. The class Peronosporomycetes contains species that are pathogens of many commercially important plants, fish, and crustaceans (Kamoun 2003). Among the marine invertebrates, infections resulting from some members of the Peronosporomycetes cause problematic diseases, especially in the seed production of marine crustaceans such as shrimp and crabs. On the other hand,

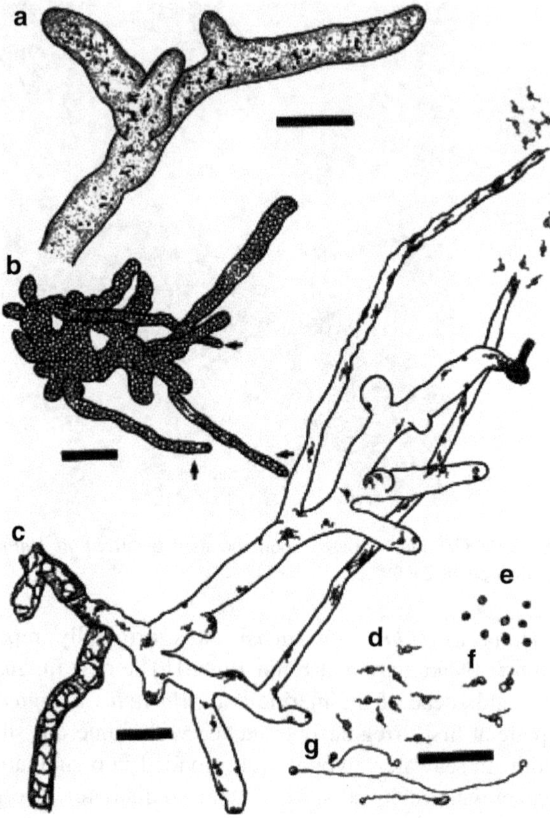

Fig. 2.11 Morphological characteristics of *Halocrusticida okinawaensis* isolated from a zoea of *P. pelagicus*. Scale = 50 μm. (**a**) Hyphae in PYGS broth; (**b**) A zoosporangium with three discharge tubes (*arrows*); (**c**) Zoospores released from the orifices of two discharge tubes. Another zoosporangium with one discharge tube is on the right; (**d**) Zoospores; (**e**) Encysted zoospores; (**f**) Secondary zoospores released from cysts; (**g**) Germination (Nakamura and Hatai 1995a)

Haliphthoros milfordens (Hatai 1982), *Halocrusticida awabi* (Kitancharoen et al. 1994), and *Atkinsiella dubia* (Nakamura and Hatai 1995b) have been reported as causative agents of such diseases in abalone, *Haliotis sieboldii*. Recently, a new fungus belonging to the Peronosporomycetes was isolated from white nodules found on the mantle of three species of abalone, *Haliotis midae* imported from the Republic of South Africa, *Haliotis rufescens* imported from the Republic of Chile and the United Mexican State, and *Haliotis sieboldii* collected in Japan. They were stocked for sale in the same tank and died from the infection. The daily mortality of stocked abalone was about 1%. The clinical sign of a moribund abalone was the presence of white nodules on the mantle (Fig. 2.12). In the lesions of the white nodules, thick and aseptate hyphae were present. The fungus was isolated from moribund abalones using PYGS agar. The manner of zoospore formation in the fungus isolated from the lesion was totally different from that of the genera *Halocrusticida* and *Atkinsiella*, but similar to that of the genus *Haliphthoros*. However, the isolate differed from the genus *Haliphthoros* as follows. In artificial seawater, the fragments were formed by constricting protoplasm in the hyphae such as in the genus *Haliphthoros*, but the protoplasm constriction was weaker, and fragments were longer, with smaller space between them, than those of

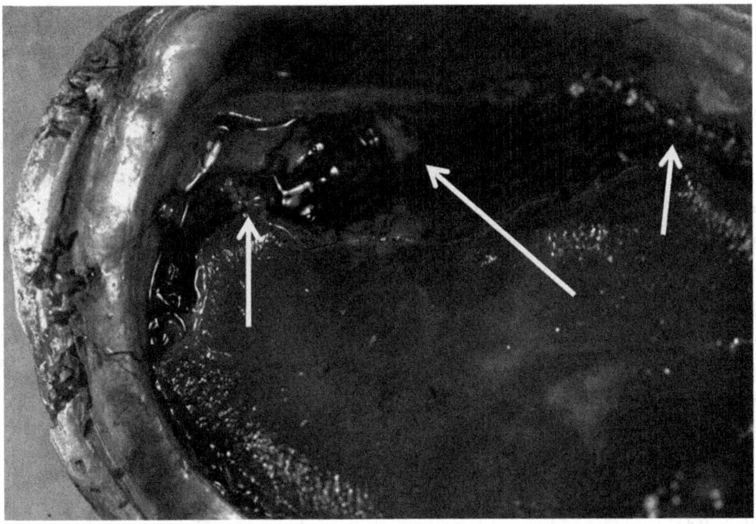

Fig. 2.12 Clinical sign of a moribund abalone. Note the presence of white nodules on the mantle (*arrows*)

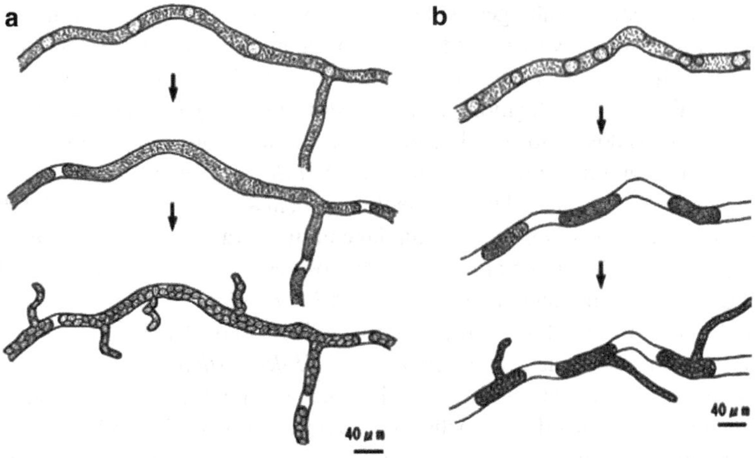

Fig. 2.13 Differences in zoospore formation between *Halioticida noduliformans* (**a**) and *Haliphthoros milfordensis* (**b**). (**a**) Fragments are longer, and spaces between adjacent fragments are smaller than those of *H. milfordensis*. One to several discharge tubes are formed. (**b**) Fragments with only one tube are shorter, and space between fragments are larger than those of *H. noduliformans*

Haliphthoros (Fig. 2.13). The species under the genus *Haliphthoros* form only one discharge tube from a zoosporangium (Vishniac 1958; Hatai et al. 1980, 1992, 2000; Nakamura and Hatai 1995a; Chukanhom et al. 2003), but the fungus from abalone has one or more discharge tubes formed from each zoosporangium.

The size of zoospores was 7.0–8.5 × 9.5–12.5 μm (width × length). From the results mentioned above, the isolate was recognized to have unique morphological characteristics in the family Haliphthoraceae.

Four isolates from white nodules and nine peronosporomycete species isolated from various marine invertebrate animals were used for analysis on the D1/D2 region of LSU rDNA. In the phylogenic tree based on LSU rDNA, the isolate was not classified into the subclass Peronosporomycetidae, Saprolegniomycetidae, or Rhipidiomycetidae, but as a new clade with the genera *Haliphthoros* and *Halocrusticida* (Fig. 2.14). Within this new clade, the four isolates from abalone, *Haliphthoros* spp. and *Halocrusticida* spp. were grouped in the respective independent subclade. *Atkinsiella dubia* and *Lagenidium* spp. were included in Saprolegniomycetidae and Peronosporomycetidae, respectively. The phylogenetic analysis also supported that the four isolates were classified into a new genus and species belonging to the family Haliphthoraceae based on their morphological characteristics. As a result, it named *Halioticida noduliformans* as new genus and species (Muraosa et al. 2009).

Dick (2001) proposed a new taxonomic system for Peronosporomycetes, in which Peronosporomycetes were subdivided into three subclasses: Peronosporomycetidae, Rhipidomycetidae, and Saprolegniomycetidae. Under this taxonomic system, the genera *Haliphthoros*, *Halocrusticida*, and *Atkinsiella* were classified in Haliphthoraceae – Salilagenidiales – Saprolegniomycetidae, and the genus *Salilagenidium*, which was named as a new genus by Dick (2001) for marine species of the genus *Lagenidium*, was classified in Salilagenidiaceae – Salilagenidiales – Saprolegniomycetidae. Molecular phylogenetic analysis by Muraosa et al. (2009) showed that only *Atkinsiella dubia* was included in the subclass Saprolegniomycetidae, but the genera *Haliphthoros*, *Halocrusticida*, and *Halioticida* were not included within the three subclasses proposed by Dick (2001). Furthermore, the genus *Lagenidium* (Salilagenidium) was included in the subclass Peronosporomycetidae in the analysis of Muraosa et al. (2009). Cook et al. (2001) also suggested that the genera *Atkinsiella* and *Lagenidium* (*Salilagenidium*) were classified into the subclass Saprolegniomycetidae and Peronosporomycetidae, respectively, and the genera *Haliphthoros* and *Halocrusticida* were not included in the three subclasses, according to their molecular phylogenetic analysis using the mitochondrially encoded cytochrome *c* oxidase subunit 2 (*cox*2) gene. Thus, the taxonomic position of genera *Haliphthoros*, *Halocrusticida*, *Atkinsiella*, and *Lagenidium* is still confusing.

In December 2006, a *Halioticida* infection was found in wild mantis shrimp, *Oratosquilla oratoria* in Tokyo Bay, Japan (Atami et al. 2009). Fungi were found in the gills of mantis shrimp (Fig. 2.15), isolated from lesions using PYGS agar, and identified by morphological observation and molecular analysis. The fungus formed fragments in the hyphae and several discharge tubes developed from each fragment. Zoospores were formed within the fragments and released into the seawater through the tops of discharge tubes. Based on the characteristics of zoospore production mode, the fungus was classified into the genus *Halioticida*. It was compared by molecular analysis of the D1/D2 region of the large subunit

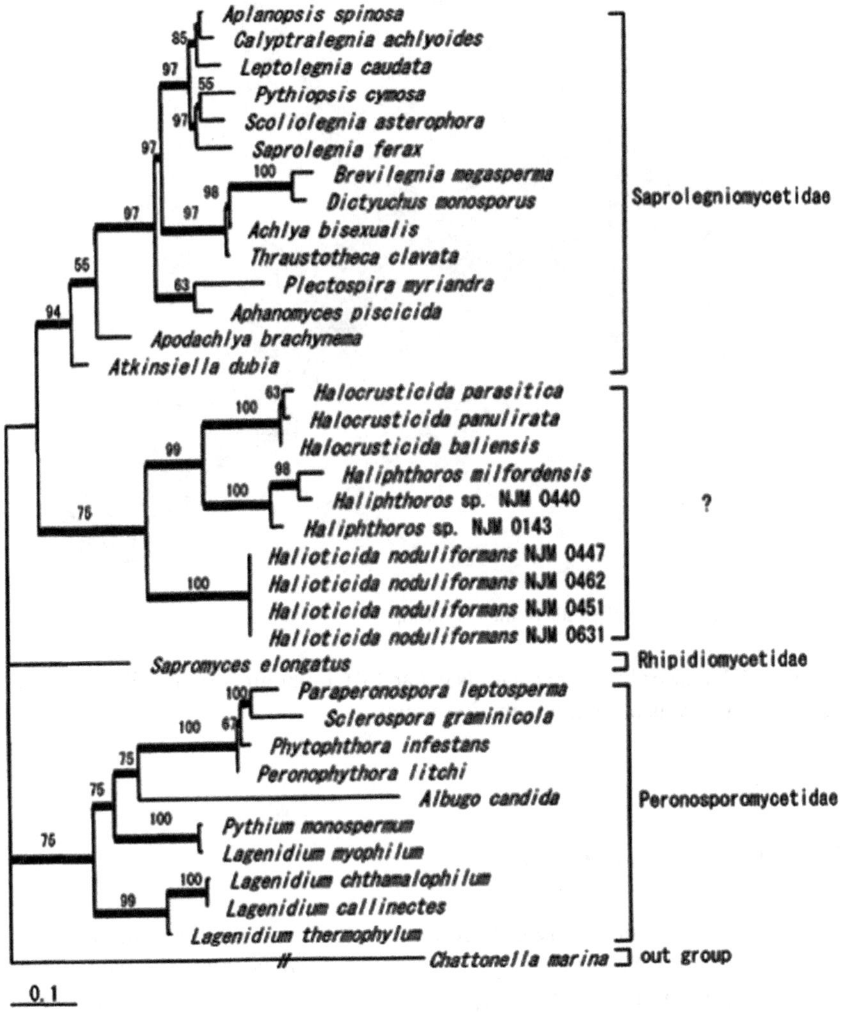

Fig. 2.14 Maximum-likelihood tree based on the D1/D2 region of LSU rDNA. Numbers on branches show bootstrap values (1,000 replicates, above 50% are indicated)

ribosomal RNA gene with *Halioticida noduliformans* isolated from abalone, *Haliotis* spp. (Muraosa et al. 2009). As a result, the sequences of the fungus isolated from mantis shrimp showed 99–100% homology at the D1/D2 domain of the large subunit ribosomal RNA gene sequence with *Halioticida noduliformans.* Histopathological observation indicated that the fungus grew in an aerobic environment, because the hyphae were found mainly in the gills and base of gills. The fungus grew well at 15–25°C, with optimal temperature of 20°C, which corresponds with *H. noduliformans* (Muraosa et al. 2009). The fungus could not

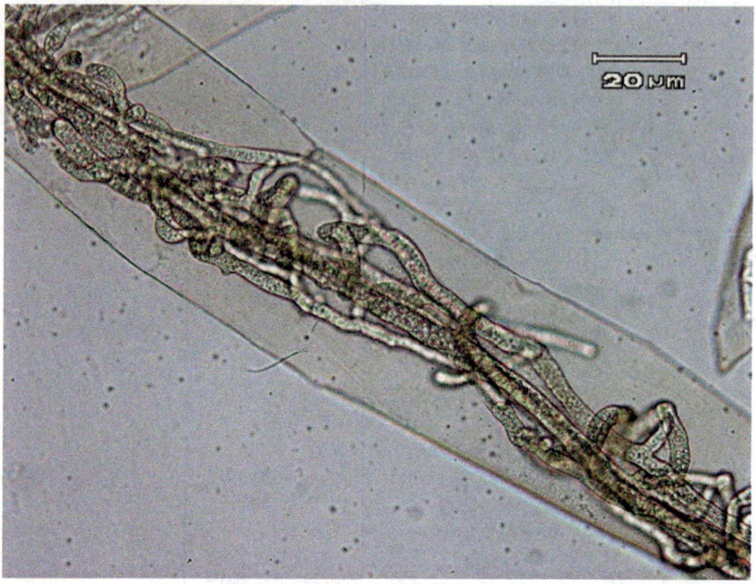

Fig. 2.15 Fungus found in the gills of mantis shrimp

grow on PYG agar or PYG agar with NaCl and KCl, but grew on PYGS agar. This suggested that it was an obligate marine fungus.

2.2.5 Atkinsiella *Infection*

Atkinsiella dubia was originally isolated from eggs of pea crab, *Pinnotheres pisum* in England (Atkins 1954), and assigned to the genus *Plectospira*. Atkins observed the same species on the eggs of *Gonoplax rhomboids* and succeeded in experimentally infecting the eggs of some species of crustaceans. The morphology of the fungal parasite on the eggs of crab was studied at that time. Later, Vishniac (1958) established a new family, Haliphthoraceae (Saprolegniales), for holocarpic biflagellate filamentous fungi, including *Haliphthoros milfordensis* and Atkins' fungus, which was renamed *A. dubia*, although she did not actually observe *A. dubia*. Its morphology and development in pure culture were followed by Fuller et al. (1964) and Sparrow (1973b) from marine algae and the eggs of various crabs, respectively.

Dick (2001) classified *Atkinsiella dubia* and *Haliphthoros* spp. into Saprolegniomycetidae, but at present the genus *Haliphthoros* is classified into different clade, *Haliphthoros*/*Halocrusticida* clade (Sekimoto et al. 2007), or unknown group (Muraosa et al. 2009), because they constructed different clades from phylogenetic analysis.

During the survey of the fungi belonging to Lagenidiales on marine animals without clinical signs, an interesting fungus was isolated from the mantle of abalone, *Haliotis sieboldii*, in Chiba Prefecture, Japan. The same fungus was also obtained from the gills of swimming crab, *Portunus trituberculatus* in Chiba Prefecture. The fungus was characterized by crystalline, tuberculate and moist colonies, dimorphic and diplanetic zoospores, and zoospores which remained in the zoosporangia during the first motile stage (Fig. 2.16), and identified as *Atkinsiella dubia*, new to Japan (Nakamura and Hatai 1995b).

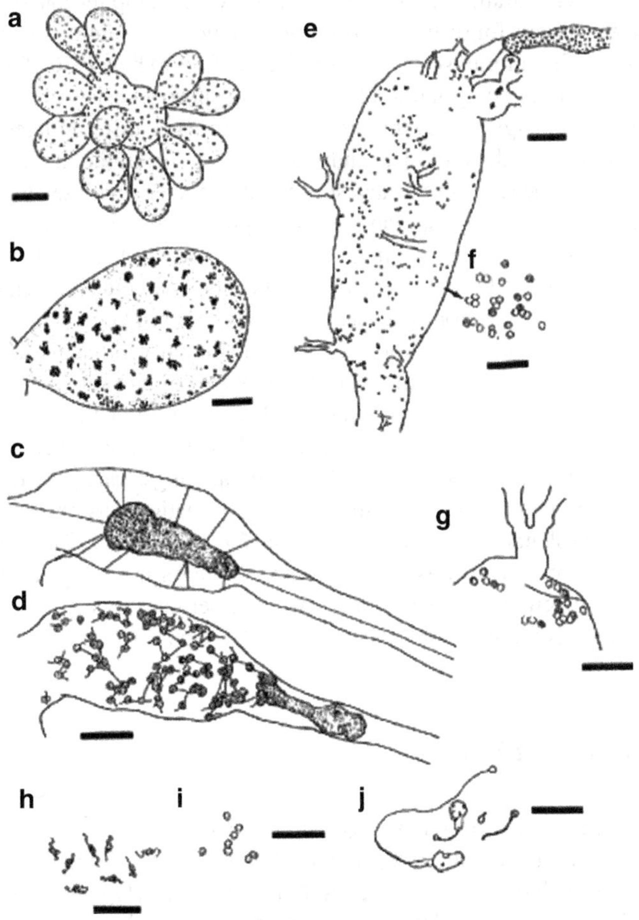

Fig. 2.16 *Atkinsiella dubia.* (**a**, **b**) Mycelium with granular clusters. (**c**) A protoplasmic mass supported by several protoplasmic thread. (**d**) Loose net-works of zoospores. These differentiated into free individual zoospores in the first motile stage. (**e**) A zoosporangium with branched discharge tubes. (**f**) Empty encysted zoospores, and encysted zoospores with protoplasm from which zoospores in the second motile stage will emerge. (**g**) A branched discharge tube with flared openings. (**h**) Zoospores in the second motile stage. (**i**) Encysted zoospores after the second motile stage. (**j**) Germination. *Scales*: (**a**, **e**) 150 μm; (**b**) 70 μm; (**c**, **d**, **g–j**) 50 μm; (**f**) = 40 μm

Roza and Hatai (1999) reported that heavy mortalities reaching 100% among larvae of the Japanese mitten crab, *Eriocheir japonicus*, occurred in Yamaguchi Prefecture, Japan. Under the microscope, infected zoeal larvae were filled with numerous aseptate hyphae. The infected fungus was inoculated on PYGS agar and incubated at 25°C for 7–10 days. Colonies on PYGS agar were attaining a diameter of about 25 mm in 15 days, crystalline, tuberculate, and moist; moderately heaped at the center. Mycelia in the broth were aseptate, radially branched, stout, swollen up to 150 μm in diameter, with clusters of shiny spherical granules, without oil droplets and vacuoles. Granular clusters were evenly distributed inside mycelia, generally consisting of several granules. Mycelia in seawater developing narrow branches (discharge tubes) were followed by zoospore production. Gemmae were present. Zoospores in the first motile stage were produced after 30 h at 25°C. Protoplasmic masses due to gathering of granular clusters on zoosporogenesis were supported at the center of zoosporangia by several protoplasmic threads; differentiated into loose networks of zoospores, then into free individual zoospores in the first motile stage. Zoosporangia were the same in size and shape as the mycelia, with several discharge tubes extending from each zoosporangium. Zoospores in the first motile stage were swimming dully and encysting within zoosporangia and discharge tubes, and biflagellate, subglobose to globose, 3–6 μm in size. Zoospores in the second motile stage were released one by one from encysted zoospores within zoosporangia and discharge tubes, swimming freely for a long time; laterally biflagellate, pyriform, slipper-shaped, isokont, 2–7 μm. Zoospores were dimorphic and diplanetic. Encysted spores were globose to subglobose, 3–7 μm in the first motile stage and 3.5–6 μm in the second motile stage. Discharge tubes were unbranched or occasionally branched, straight or tapering with flared openings, rarely with a central swelling, 4–9 μm in width, 5–16 μm in length. Germination was observed at 6–8 h after spores with slender germ tube were transferred to broth. This fungus was identified as *A. dubia*. This was the first report of mass mortality in crustaceans due to *A. dubia* infection. The optimum growth temperature was at 25°C, and grew only on PYG agar containing 2.5% NaCl and PYGS agar.

2.2.6 **Pythium** ***Infection***

This infection was first reported as *Lagenidium myophilum* infection from marine shrimp (Hatai and Lawhavinit 1988). Later, Muraosa et al. (2009) made clear that the fungus was included into the genus *Pythium* by phylogenic tree. *Pythium myophilum* (*Lagenidium myophilum*) infection occurred in the abdominal muscles and swimmerets of adult northern shrimp, *Pandalus borealis*, cultured at the Japan Seafarming Association (JASFA). Pure cultures of *P. myophilum* were consistently isolated from the partly blackened abdominal muscle (Fig. 2.17) and the inside of the swimmerets of the adult northern shrimps. Growth of the fungus on PYGS agar was observed at 2 days after incubation. Microscopical observation of the blackened areas of the lesions showed them to be filled with hyphae and the pathogenic

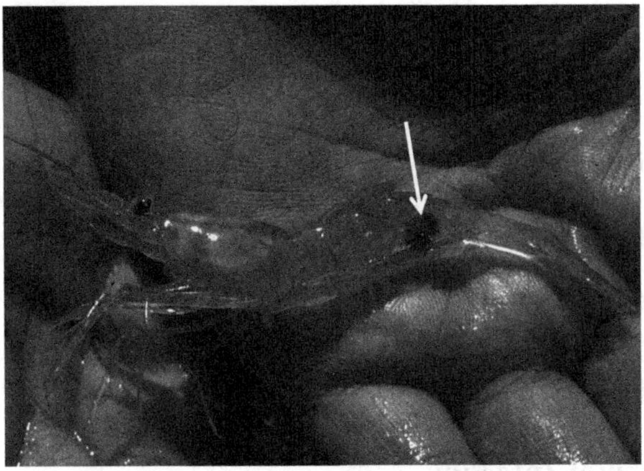

Fig. 2.17 *Pythium myophilum* isolated from the partly *blackened* abdominal muscle (*arrow*)

Fig. 2.18 A juvenile coonstripe shrimp infected with *Pythium myophilum*. The lesions look *whitish* (*arrows*)

fungus to grow only in the tissue of shrimp. The optimum temperature for growth of this fungus was 25°C , but it also grew at the low temperature of 5°C. It would thus be able to infect northern shrimps living in cold seawater; the temperature of the Japan Sea was approximately at 5°C. In pure culture, the hyphae were somewhat uniform with a diameter of 7–10 μm and generally vacuolated. Vesicle formed at the end of discharge tube were measuring 86–240 × 7–10 μm in diameter. Zoospores were 12.9 × 9.6 μm, globose, reniform, pyriform or elongate, monoplanetic and laterally biflagellate. Encysted zoospores were spherical, 5.5–12.0 μm in diameter. Sexual reproduction was not observed.

In 1991, a fungal infection occurred in the larvae of coonstripe shrims, *Pandalus hypsinotus*, artificially produced at Hokkaido in Japan. Mortality was 100%. In 1993, the infection also occurred in juvenile coonstripe shrimps (Fig. 2.18), which had been reared in tanks after seed production. Mortality was about 70%

(Nakamura et al. 1994a, b). The pathogenic fungi isolated from the lesions were same as those caused by *Pythium myophilum* reported by Hatai and Lawhavinit (1988). *P. myophilum* is pathogenic toward adult northern shrimp, larval and juvenile coonstripe shrimps and Hokkai shrimp, *Pandalus kessleri* (Hatai, unpublished). *P. myophilum* infections have only been in Japan, and these shrimps of the genus *Pandalus* are known to live only in the deep areas of the sea off the coast of Japan. It was interesting that these hosts seemed to be highly sensitive to *P. myophilum*.

2.3 Diseases of Fish and Shellfish Caused by Mitosporic Fungi

2.3.1 Fusarium *Infection*

Some species in the genus *Fusarium* such as *Fusarium solani*, *F. moniliforme*, and *F. oxysporum* have been isolated from kuruma prawn, *Penaeus japonicus*, with black gill in Japan. Among these species, *F. solani* subsequently was reported as an important pathogen.

F. solani was originally proposed as a genus *Fusarium* (Wollenweber 1913). Later the section included five species, ten varieties, and four forms (Wollenweber and Reinking 1935). Snyder and Hansen (1941) combined three species (*F. solani*, *F. martti*, and *F. coeruleum*) into *F. solani*. This taxonomy, however, was not approved by Gerlach and Nirenberg (1982). Booth (1971) and Gerlach and Nirenberg (1982), included four and six species in the section *Martiella*, respectively, from conidiogenesis and shapes of conidia which are major criterions for the classification. The main species, *F. solani*, in the genus *Fusarium* has been later reported in the literatures as *formae specials* (f. sp.), mating population (MP 1–MP VII), or anamorph of *Nectria haematococca* due to its polytypic appearances (Matuo and Snyder 1973). Because of its pathogenic importance, studies on the biological specification of *F. solani* species has also been developed (O'Donnell 2000). Previous studies on the taxonomy of this complex fungal species have contributed valuable information on the limits of the specification and evolutionary relationships within species, *F. solani* (Matuo and Snyder 1973; Hawthorne et al. 1992; Suga et al. 2000; O'Donnell and Gray 1995; O'Donnell 2000). However, *F. solani* isolated from aquatic animals including marine crustaceans and fishes have never been studied in detail in previous reports.

Black gill disease of pond-cultured kuruma prawn, *Penaeus japonicus*, was first reported in Japan (Egusa and Ueda 1972). They demonstrated that a fungus belonged to the genus *Fusarium* was the causative agent and gave the fungus a temporary designation, BG-Fusarium. Since their report, the disease has often broken out among pond cultured kuruma prawn in various districts. Hatai et al. (1978) investigated a taxonomical position of the BG-Fusarium isolated from gill lesions of kuruma prawn with black gill disease (Fig. 2.19). The fungus produced micro- and macro-conidia on conidiophores and chlamydospores. As a result, the

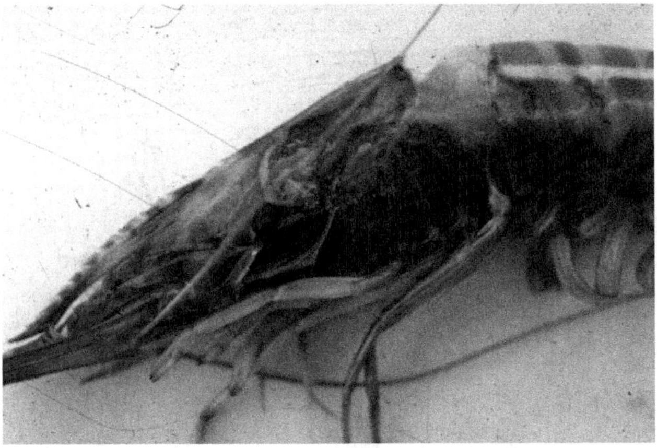

Fig. 2.19 Kuruma prawn infected with *Fusarium solani*. Note gills showing *black color*

BG-Fusarium was identified as *Fusarium solani* according to Booth (1971, 1977) from the characteristics of the fungus on Potato Sucrose Agar. They demonstrated that the pathogenic fungus could be isolated from wet sand in ponds with fungal infection, but not from it without fungal infection, and was capable of surviving for long time in wet sand. Khoa et al. (2005) also reported *Fusarium solani* infection of kuruma prawn (Fig. 2.20). They demonstrated that the fungus showed pathogenicity to kuruma prawn by intramuscular injection. Phylogenetic analyses based on the sequences of its internal transcribed spacer region, including 5.8 S ribosomal DNA and a partial 28 S ribosomal DNA region, showed that all strains tested were monophyletic. And the strains isolated from the diseased kuruma prawn and the phytopathogenic *Fusarium solani* were clearly distinguished by the morphological and phylogenetical characteristics (Khoa et al. 2005).

Fusarium moniliforme was also isolated from gill lesions of kuruma prawn with black gill disease at a private farm in Japan (Rhoobunjongde et al. 1991). The colonies of the fungus cultured on upper surface of potato dextrose agar (PDA) were floccose, creamy white, undersurface a lavender to violet, but did not grow on mycobiotic agar containing cycloheximide. Fungal hyphae were hyaline and 2.5–6.0 μm in diameter. Conidiogenous cells with long monophialides were abundantly formed laterally on aerial mycelium or on sympodially branched conidiophores, and were hyaline, subulate 2.0–4.0 μm in diameter. Macroconidia were present, but only rarely and their appearance varied from slightly sickle- to cigar-shaped, three to four septa, rarely five septa and 26.0–50.0 μm in length. Microconidia were abundant and variable on shape and size from ovoid to elliptical, zero to one septa, rarely two septa, 6.0–20.0 μm in length, and were produced in chains mostly from a simple conidiophores (Fig. 2.21) and false heads on PDA and KCl medium but especially with longer chains on the KCl medium. Chlamydospore was absent. This was the first case of *F. moliniforme* infection in crustacean.

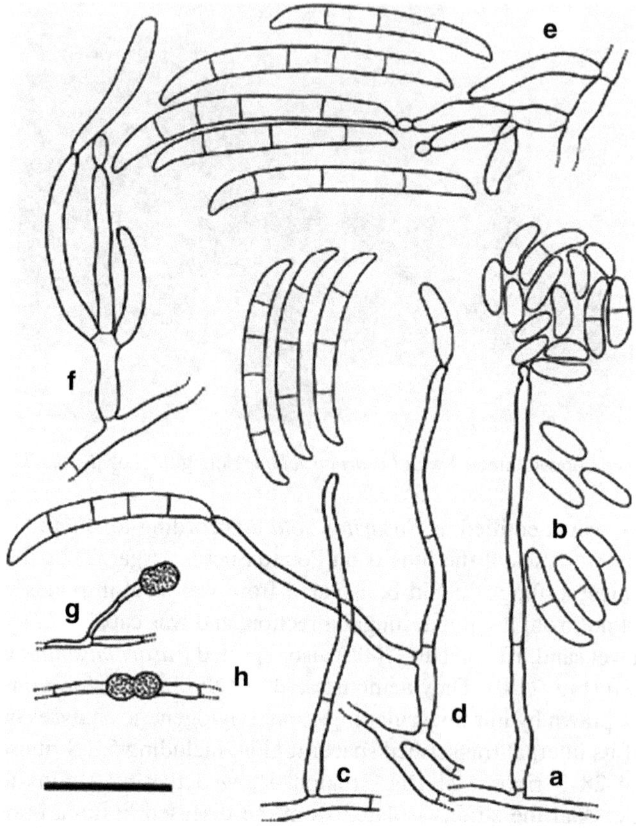

Fig. 2.20 Microscopic morphology of *Fusarium solani* isolated from an infected *Penaeus japonicus*. Scale bar = 25 μm. (**a**) Aerial conidiophores is long and unbranched, slightly narrow toward the apex, monophialidic, producing abundant zero to one-septate conidia that cohere in a false head. (**b**) Oval or ellipsoid one-cell conidia and subcylindric or slightly curved two-cells conodia. (**c**) Unbranched aerial conidiophores bear three to four-septate conidia and are slightly curved with a short and blunt apical cell and slightly notched basal cell. (**d**) A lateral branched aerial conidiophores producing one to four-septate conidia. (**e**) Irregularly and verticillately branched sporodochial conidiophores, bearing monophialides and producing three and four-septate conidia. Conidia extend from the basal part and curve to the apex. The dorsal side is more curved than ventral side, and there is a blunt apical cell and slightly notched basal cell. (**f**) A sporodochial conidiophores verticillately forming monophialides in the early stage of sporulation. (**g**) Terminal chlamydospore from conidiophores is smooth-walled, globose. (**h**) Intercalary chlamydospores in the hyphae are smooth-walled, globose, and in a pair

Khoa and Hatai (2005) reported *Fusarium oxysporum* infection in cultured kuruma prawn *Penaeus japonicus* in Japan. The infection was the first case in kuruma prawn. The infected prawn showed black gills, but the other apparently looked healthy. Fungal hyphae with septa and canoe-shaped conidia were clearly observed in wet-mount preparations of the prawns with black gills. Colony of the *F. oxysporum* grew well on PDA at 25°C. Mycelia were delicate, felt-like, and

Fig. 2.21 Microconidia are produced in chains mostly from simple conidiophores

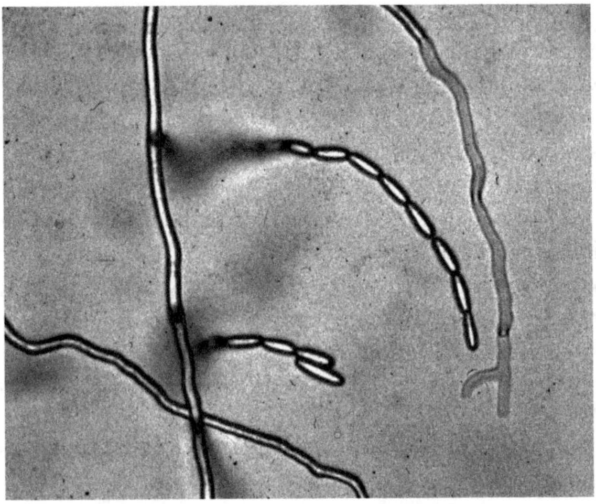

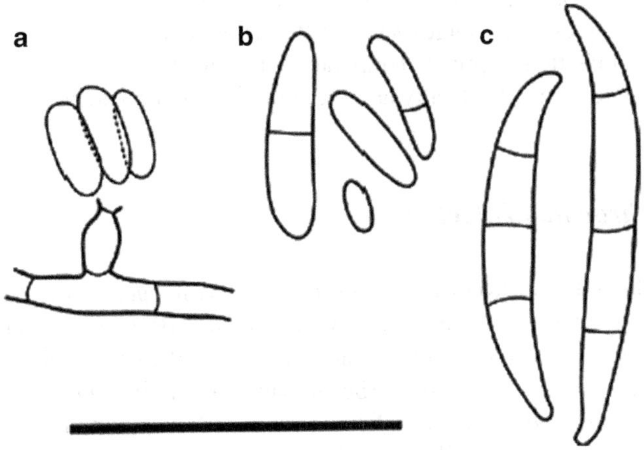

Fig. 2.22 Morphology of *Fusarium oxysporum* . Scale bar = 30 μm. (**a**) Short aerial conidiophore, unbranched, monophialidic conidiophore, producing one-cell conidia in false head. (**b**) Aerial conidia with one to two cells. (**c**) Three-septate sporodochial conidia, tapering toward both ends with a pointed apical cell and a slightly hooked basal cell

funiculous flat appressed. Pigment on PDA was white at first, gradually changed to pale beige in the center of the agar plate, and purple or dark violet in the aged cultures. Chlamydospores were not observed. Arial conidiophores were usually single, unbranched and mostly short, 5–15 μm in length, and produced one cell or two cells conidia in false head at 4-day culture (Fig. 2.22). Aerial conidia were usually oval, cylindrical or ellipsoid, straight or slightly curved. One-cell conidia were 5 ± 2.5 × 2.5 ± 0.5 μm, and two cells conidia were 8.5 ± 3.5 × 0.7 μm. Sporodochial conidiophores were occasionally observed on SNA cultures (Nirenberg 1990), and monophialidic, irregularly or verticillately branched, and produced three to five septate conidia, predominantly three septate conidia. Sporodochial conidia were usually curved, equally tapering toward both ends with a pointed apical cell and a slightly hooked basal cell. Three septate conidia were 27 ± 3.7 × 2.7 ± 0.3 μm. The prawns artificially injected with *F. oxysporum* showed typical black gills, and the clinical sign was similar to that of prawn naturally infected with fungus.

F. oxysporum infection was also found in cultured red sea bream, *Pagrus major*, in Japan (Hatai et al. 1986). In almost all cases, no external signs were observed, but kidneys of the fish were remarkably swollen and discolored. The other organs, however, appeared to be normal. The fungus was isolated by inoculating a piece of kidney on Sabouraud dextrose agar (SA agar) at 25°C, and a pure culture was obtained. The fungus was identified as *Fusarium oxysporum* as described by Booth (1971).

In Vietnam a new *Fusarium* infection occurred in black tiger shrimp, *Penaeus monodon* (Khoa et al. 2004). Infected shrimps showed typical signs of black gill disease and mortalities about a month prior to harvest. The isolated fungus was identified as *F. incarnatum* from the detailed morphological and molecular phylogenic analyses. The fungus showed the pathogenicity to kuruma prawn by experimental infection. Optimal temperature for the fungus ranged from 20 to 30°C. The fungus grew drastically at 35°C, but did not at 5 and 40°C.

2.3.2 *Ochroconis* Infection

The fungal infection in fishes caused by *Ochroconis humicola* was first reported from the kidney of coho salmon, *Oncorhynchus kisutch* (Ross and Yasutake 1973). Later, the infection was reported from rainbow trout, *Salmo gairdneri* (Ajello et al. 1977), Atlantic salmon, *Salmo salar* (Schaumann and Priebe 1994).

In Japan, *Ochroconis humicola* infection has been found in marine cultured fish. First description was from devil stinger, *Inimicus japonicas* (Wada et al. 1995). The diseased fish were about 1.4 g in body weight, and had some ulcers on the body surface. The fish examined showed little appetite, but no mortality was recorded. The center of the lesion was necrotic and sloughed, leaving trunk muscles exposed in a crater-shaped cavity surrounded by an erosion periphery. Direct microscopical examination of the exposed trunk muscles revealed numerous septate fungal hyphae. Fungal colonies were slow growing, slightly domed, velvety to floccose, and pale brown in color. Hyphae were septate, pale brown in color, and 1–2 μm in width. Conidia were usually sparse, 1.8–2.2 × 7.0–10.0 μm, two-celled, pale brown in color and cylindrical with rounded ends. The reproductive mode of the conidia was sympodial. The fungus was identified as *Ochroconis humicola* according to de Hoog and von Arx (1973) and Howard (1983). Later, *O. humicola* infection was found in marine-cultured fish, red sea bream, *Pagrus major*, and marbled rockfish, *Sebasticus marmoratus* (Wada et al. 2005). The average body weight of the fish examined was 1.2 g for red sea bream and 1.0 g for marble

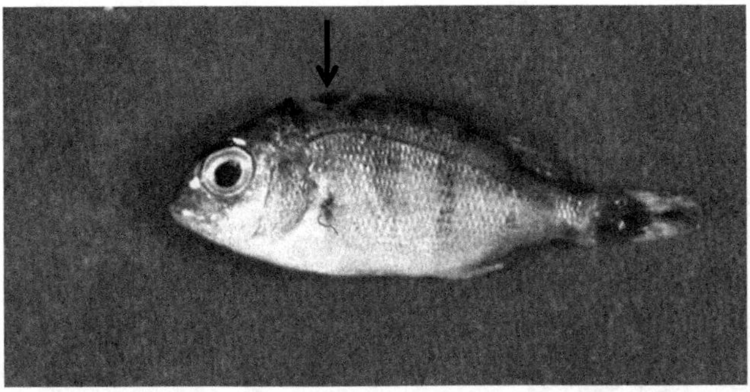

Fig. 2.23 Ochroconis infection in red sea bream. Severe ulceration (*arrow*) formed around the base of the dorsal fins

rockfish. Both cases showed apparent lesions on the body surfaces. In the red sea bream, severe ulceration was found around the base of the dorsal fin (Fig. 2.23), while erosive and/or ulcerative lesions mainly appeared at the mouth regions in the marbled rockfish.

In April 2004, a fungal infection occurred in cultured young striped jack, *Pseudocaranx dentex* at a fish farm in Japan (Munchan et al. 2006). The water temperature in this month was 17–18.5°C. The examined 0-year-old fish were 6–10 cm in body length and 5–10 g in body weight. Moribund fish with fungal infection showed disease sign such as distended abdomen kidney. Numerous brownish hyphae were found in squash preparation of the kidney under microscopy. The cumulative mortality of the disease reached 25% (62,000 out of 250,000 fish) for 1 month after the disease was recognized. Histopathology showed that fungal hyphae were found in the musculature, spleen (Fig. 2.24) and kidney. The granulomas consisted of massive fungal elements and outer layers surrounded by epitheloid cells. No bacteria or parasites were found in the examined tissues. Munchan et al. (2009a) compared the histopathology of young striped jack experimentally infected with dematiaceous fungus *O. humicola* with that of spontaneously infected fish. Moribund and freshly dead fish from both groups showed similar histopathology, and appeared to have been killed due to hyphae penetrating the visceral organs. Fish that survived the infection appeared to be able to suppress the fungal growth by well-established inflammatory reaction involving mycotic granulomas and granulation tissues. The results suggested that two types of *O. humicola* infection occur in young striped jack: an acute type infection, which is characterized by penetrating hyphae that cause direct tissue destruction and a chronic type infection, which is characterized by severe inflammatory reaction that causes function disorders of the affected organs. All fungi isolated from diseased fish were identified as the same fungus. Colony of the isolate showed dark brown to black color when observed from the reverse side of the plate and no visible exudates diffused into the medium, and it was flat, very slow-growing on PDA plate. Colony radii on PDA incubated at 25°C reached 30.1 mm after 4 weeks. Hyphae were septate, 2–3 μm in diameter, pale brown in color, and aerial hyphae

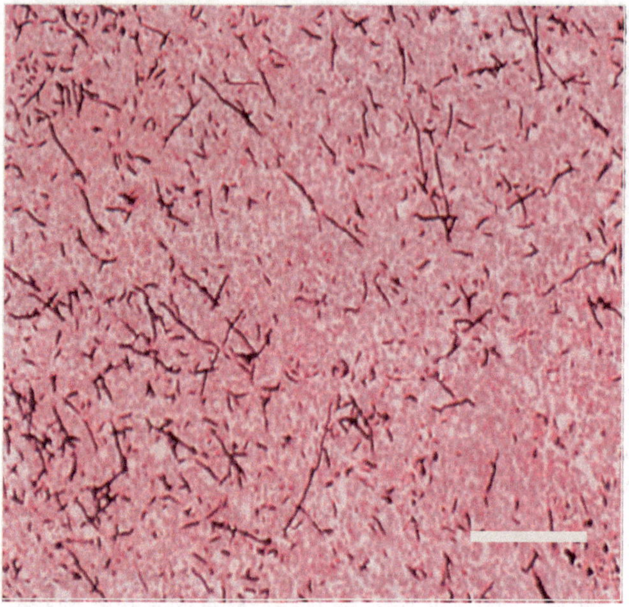

Fig. 2.24 Histopathological finding of spleen in diseased fish. Note many fungal hyphae in the spleen. Grocott-HE stain, Bar = 100 μm

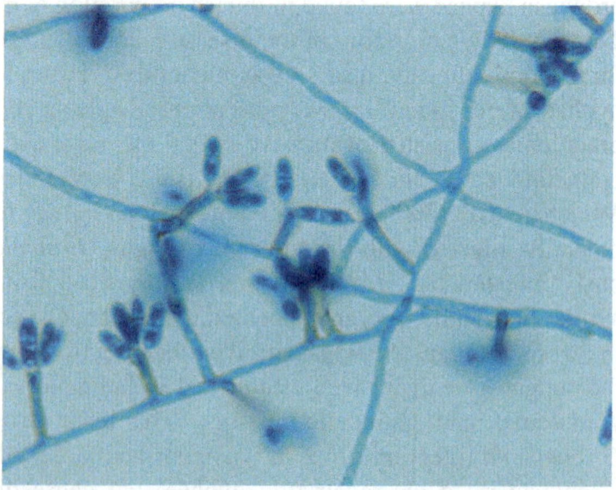

Fig. 2.25 Conidia of *Ochroconis humicola*: two-celled, cylindrical to oblong, constricted at the septum

were sparse. Conidiophores were predominantly cylindrical, average 2.5 × 7.5 μm had denticles at each tip. Conidia were holoblastic, two-celled, cylindrical to oblong with rounded ends, average 2.5–4.5 × 5.5–12.5 μm, smooth-walled, and pale brown in color (Fig. 2.25). The isolate was as *Ochroconis humicola* from these characteristics. The isolate grew at 10–30°C, but not at 35°C. The isolate could grow up to 9% NaCl indicating that *O. humicola* could grow in an environment with a wide range of salinity. Itraconazole (for oral administration), with an MIC (MFC) range of 0.06–0.13 (0.0625–0.125) μg/mL was chosen for in vivo treatment. In vivo treatment with itraconazole of striped jack experimentally infected with *O. humicola* was conducted for 50 days. No fish died, but gray to white nodules were found in the visceral membrane, kidney, liver, and spleen in the fish. Granulomatous inflammatory reactions were histopathologically found in all fish injected with conidia of *O. humicola*. Clinical signs and histopathological findings indicated that itraconazole showed no efficacy for curing the fish infected with *O. humicola* (Munchan et al. 2009b).

2.3.3 **Exophiala** ***Infection***

Fungal infection caused by the genus *Exophiala*, known as black yeast, has been reported in several species of fish. The first report was by Carmichael (1966) who described a systemic infection of cutthroat trout, *Salmo clarki*, and lake salmon, *Salvelinus namaycush*. The causative agent was initially named a *Phialophora*-like fungus but later classified as *Exophiala salmonis*. Fijan (1969) reported a systemic mycosis in channel catfish, *Ictalurus punctatus*, later the fungus was identified as *E. pisciphilus* (McGinnis and Ajello 1974). Later, *E. salmonis* infection was reported from Atlantic salmon, *Salmo salar* (Richard et al. 1978; Otis and Wolke 1985). On the other hand, *E. pisciphila* infection was reported from smooth dogfish, *Mustelus canis* (Gaskins and Cheung 1986), Atlantic salmon (Langdon and McDonald 1987). *E. psychrophila* infection was also reported from Atlantic salmon (Pedersen and Langvad 1989).

In Japan, *Exophiala* infection occurred in cultured striped jack, *Pseudocaranx dentex*, in 2005 (Munchan et al. 2009c). One hundred out of 35,000 fish died per day and mortalities continued for 1 month. Diseased fish showed swelling of the abdomen and kidney distension. Microscopic examination of the kidney of diseased fish revealed numerous septate hyphae, pale brown in color, in squash preparations. Histology revealed abundant fungal hyphae and conidia in gill, heart, and kidney. Fungal hyphae were accompanied by cell necrosis and in influx of inflammatory, mainly mononuclear cells. The fungus isolated from the diseased fish had septate hyphae, pale brown in color and 1.8–3.0 μm in diameter. The colony morphology of the fungus after 1 week of incubation on PDA at 25°C was initially a black yeast form. It then became woolly and velvety and olive brown in color but black on the reverse side after 4-week incubation. Conidiogenous cells were conspicuous annellides (Fig. 2.26), short or cylindrical or fusiform in shape. Conidia were one-celled, ellipsoidal with smooth walls, accumulated in balls at the apices of annellides that tended to slide down, 1.5–2.0 μm in width and 3.0–5.0 μm in length. The fungus was classified into the genus *Exophila* based on its morphology and as *Exophiala xenobiotica* based on sequences of the ITS1-5.8S-ITS2 regions of rDNA. This is the first record of this fungus in a marine fish.

Fig. 2.26 Conidiogenous cells: conspicuous annellated zones, short or cylindrical

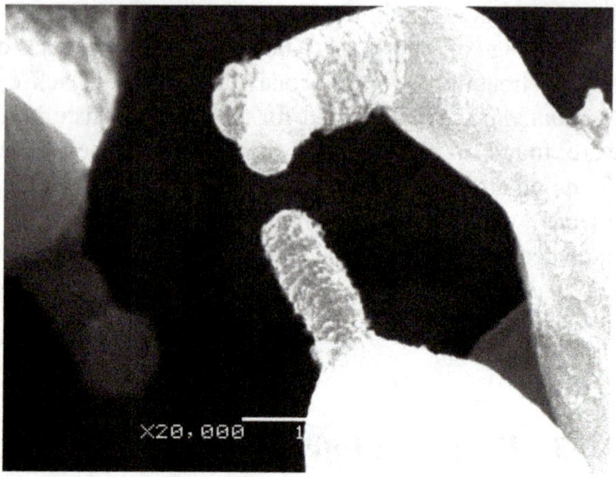

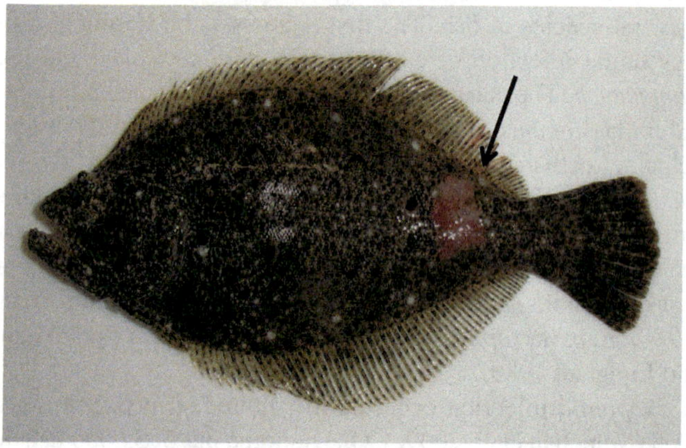

Fig. 2.27 The lesion was only limited in the skin, which is involving ulcerative skin lesion (*arrow*) in the fish

On the other hand, different *Exophiala* infection occurred in Japanese flounder, *Paralichthys olivaceus* in Japan (Kurata et al. 2008). The lesion was only limited in the skin, which is involving ulcerative skin lesions in the fish (Fig. 2.27). The water temperature during the period was approximately 17–21°C. A dematiaceous fungus was only isolated from the fish skin with ulcerative and erosion. The fungal colonies were dark brown to olive black in color. The fungus produced conidia (2.0–3.0 × 2.7–5.0 μm) of an elliptic or obovoid shape and with no or one septum. Conidia were formed as a cluster on the tip of conidiogenous cells. Annellations on the tip of conidiogenous cells were observed under scanning electron microscopy, but were inconspicuous under light microscopy. The fungus grew well at 25°C, but no growth was observed at 37°C. The fungus was identified as an *Exophiala* species, with different morphological, biological and molecular characteristics from three previously described pathogenic *Exophiala* species. The fungus had a high similarity of 99.6% with *Capronia coronate* from the phylogenetic tree of *Exophiala* spp. based on the sequence of the D1D2 domain of large subunit ribosomal DNA (LSU rDNA). Histology showed that fungal hyphae extended laterally in the dermis, and were absent from the epidermis and musculature of the skin lesions and kidneys of the diseased fish. An inflammatory response with granuloma occurred in the dermis involving accumulations of epitheloid cells around the hyphae. The granulomas were surrounded by lymphocyte-like cells. Epidermal degeneration was observed above the inflamed dermis, suggesting that the inflammatory response caused epidermal damage. Experimental infection reproduced hyphal extension and infiltration of inflammatory cells in the dermis of the flounder, confirming the pathogenicity of the fungus.

2.3.4 Scytalidium *Infection*

Iwatsu et al. (1990) reported first *Scytalidium* infection in striped jack, *Pseudocaranx dentex* with systemic mycosis in Japan. The external clinical signs were blackish patches and ulcers formed on the surface, especially at the basement of dorsal fin, at the tip of snout, and the anal area (Fig. 2.28). No apparent clinical signs were found in the internal organs. Numerous pale brown, septate hyphae, and arthroconidia were found in the lesions of the surface and various internal organs by direct microscopical examination. The fish was reared in sea water with a temperature of about 18°C. The mortality was about 6% of the original population. A fungus was isolated from the lesions of the surface and the internal organs. Experimental infection using striped jack showed that the fungus was a causal agent of the mycosis. The fungus was isolated on PYGS agar. The colonies were dark green and conidia showing dark green were abundantly produced. Mycelium immersed or superficial, composed of straight or sinuous, sometimes curled, smooth, cylindrical, hyaline to mid-brown, branched, rather thick-walled, septate. Stromata were absent. Conidiophores were micronematous, mononematous, straight or flexuous, hyaline to pale brown and branched or unbranched, smooth-walled. Conidiogenous cells were undifferentiated, fragmenting and forming

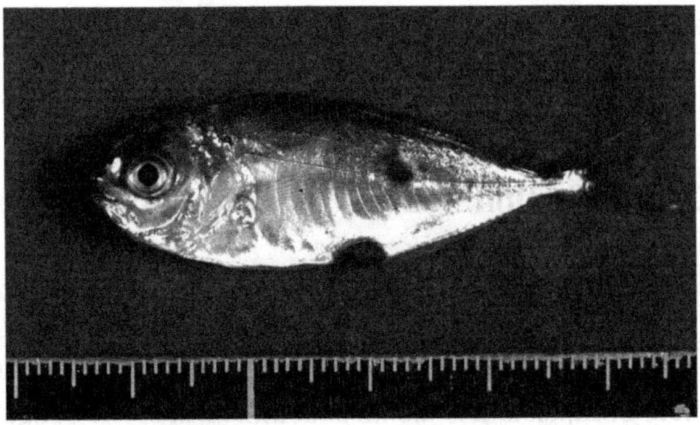

Fig. 2.28 *Scytalidium* infection in striped jack, *Pseudocaranx dentex* with systemic mycosis. The external clinical signs were *blackish patches* and ulcers formed on the surface, especially at the basement of dorsal fin, at the tip of snout, and the anal area

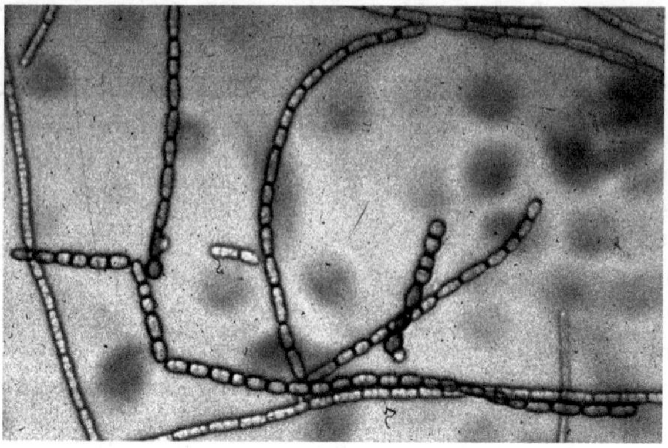

Fig. 2.29 Arthroconidia of *Scytalidium infestans* formed in extended chains

arthroconidia. Arthroconidia of one-type, formed in extended chains (Fig. 2.29), hyaline to mid-brown, dry, simple, rather thick-walled, smooth or verrucose, oblong, dolioform or broadly ellipsoidal, truncate at both ends, 0-1(-3) septate, not easily detached. Chlamydospores were absent. It did not grow at 37°C. As a result, the fungus was a new species of the genus, and named *S. infestans*.

The genus *Scytalidium* was originally erected by Pesante (1957), based on *S. lignicola*. The fungus was characterized by possession of arthroconidia of two types: hyaline, thin-walled, cylindrical ones formed by fragmentation of undifferentiated hyphae, and brown, thick-walled, broadly ellipsoid ones borne in an intercalary fashion. The generic concept was expanded when Sigler and Carmichael (1974) described *S. acidophilum*, a species possessing only dematiaceous arthroconidia. They considered that the genus was characterized by dematiaceous intercalary or terminal arthroconidia formed by fragmentation of undifferentiated hyphae and that the presence of hyaline arthroconidia was not essential for the genus delimitation.

Futhermore, *Scytalidium infestans* infection was first found in red sea bream, *Pagrus major* (Hanjavanit et al. 2004) in Japan. Ulcerative lesions were observed from head to dorsal part of the body surface of red sea bream. Histopathologically, numerous, frequently septate fungal hyphae were observed in the lesions. A fungus was isolated in pure culture from each lesion using PYGS agar. Colonies were dark green in color, and arthroconidia formed in extended chains. It was identified as *S. infestans* according to Iwatsu et al. (1990).

2.4 Infection in Mantis Shrimp by Mitosporic Fungi

The mantis shrimp, *Oratosquilla oratoria*, is an economically important and delicious culinary crustacean species. One of the famous Japanese, sushi dishes, is made from the meat of mantis shrimp. This shrimp is living in mud in the coastal areas of Japan and is the most dominant species. Fungal infection of mantis shrimp has never been reported in Japan, but it has been known that many mantis shrimp died and the production decreased from 1991. Moribund mantis shrimp were sampled and examined. As a result, it was made clear that the mortality was caused by fungal infection (Duc et al. 2009). They had fungal infection in the gills. Gills of almost mantis shrimp with naturally fungal infection showed brown discoloration (Fig. 2.30). Some gills disappeared due to the fungal infection. Numerous conidia and hyphae inside the gill lamella were observed under microscope (Fig. 2.31). The results of histological examination showed that fungal elements were present in the gills. Fungal hyphae were encapsulated in base of gills.

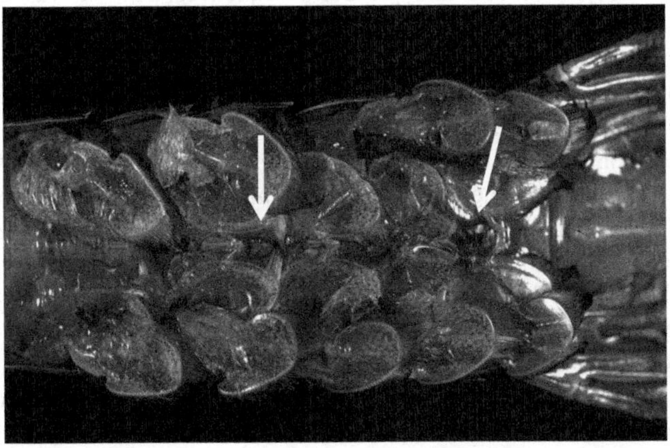

Fig. 2.30 Gills of most mantis shrimp with naturally fungal infection showed *brown discoloration* (*arrows*)

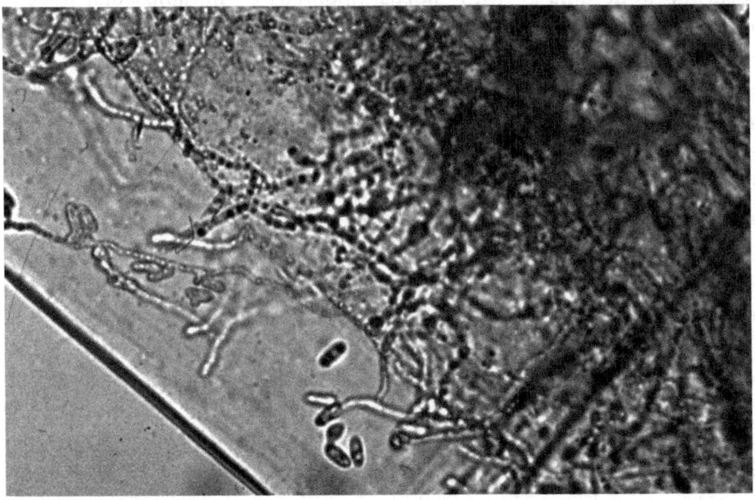

Fig. 2.31 Numerous conidia and hyphae inside the gill lamella observed under microscope

Infected gills were washed three times in sterile physiological saline (0.85% NaCl) and inoculated on PYGS agar. Ampicillin and streptomycin sulfate were added to the medium to inhibit bacterial growth. Plates were incubated at 25°C for 2–4 days. The single spore culture method was applied to obtain pure culture (Ho and Ko 1997). As a result, two kinds of fungi were isolated from the lesions. They were easy to recognize in culture by their growth. The morphological and physiological characteristics, and DNA analysis and sequencing were examined. One of them showed slow growth, and was identified as a new species, *Plectosporium oratosquillae*. The other one exhibited fast growth, and was identified to *Acremonium* sp. (a member of the *Emericellopsis* marine clade) from phylogenetic analysis of ITS and β-tubulin sequences. And it was also a new species, but teleomorph development in culture failed. It was the first report on fungal infection in mantis shrimp (Duc et al. 2009).

Duc and Hatai (2009) carried out experiments to determine pathogenicity of anamorphic fungi *Plectosporium oratosquillae* and *Acremonium* sp., which were isolated from gills of marine shrimp *Oratosquilla oratoria* caught in Japan. Cumulative mortality of the mantis shrimp injected with a high dose (5.0×10^6 conidia/mL) and a low dose (5.0×10^4 conidia/mL) of *P. oratosquillae* reached 100% and 60% at day 25, respectively. Cumulative mortality of the shrimp injected with the high dose and the low dose of the *Acremonium* sp. reached 100% and 80% at day 25, respectively. The gill lesions in the shrimp experimentally infected with the fungi showed many brown spots in the gill filaments, which were similar to the clinical sign of mantis shrimp naturally infected with the fungi. Histopathologically, the hyphae and conidia were found in the gill filaments and heart, and the hyphae were encapsulated by hemocytes in the gill filaments and the base of gills. The result confirmed that these two anamorphic fungi were pathogenic to mantis shrimp.

Duc et al. (2010a) demonstrated the pathogenicity of both the fungi isolated from mantis shrimp to kuruma prawn *Penaeus japonicus* by intramuscular injection of conidial suspensions. These fungi caused mortality in the injected kuruma prawn. Especially cumulative mortality in kuruma prawn injected with 0.1 mL of a conidial suspension with 5×10^6 conidial/mL of *Acremonium* sp. reached 100%. The results indicated that the both fungi were also pathogenic to kuruma prawn. The prawn is important cultured crustacean in Japan, and lives in the same environmental conditions.

Acremonium sp. isolated from diseased mantis shrimp was susceptible *in vitro* to three kinds of antifungal agents: voriconazole, amphotericin B, and terinafine hydrochloride (Duc et al. 2010b). They selected voriconazole to treat kuruma prawn, which had been intramuscularly injected with 0.1 mL of 5.0×10^4 conidia/mL of *Acremonium* sp. Voriconazole was administered orally at doses of six and 2 mg/kg body weight per 7 consecutive days, or intramuscularly injected at doses of 4 and 2 mg/kg body weight per day for 3 consecutive days. Both treatments were started 6 h after injection of the conidial suspension. They demonstrated that voriconazole was an efficient antifungal agent against *Acremonium* sp. from the gross features, mortality, and histopathological observations.

2.5 Future Research

The culture of fish and shellfish and their seed production are important industries in Japan. Diseases caused by fungi result in significant economic losses. Thorough descriptions of many important diseases of fish and shellfish and procedures for identification have been presented in this chapter. It is hoped that these data will be helpful for future research in Japan and in other parts of the world desiring to increase production of commercial fisheries. Some of these obligate host–parasite associations offer excellent tools for research on disease development at cellular and molecular levels. The defense reactions of animals that escape infection will be worth investigating.

References

Ajello L, McGinnis MR, Camper J (1977) An outbreak of phaeohyphomycosis in rainbow trout caused by *Scolecobasidium humicola*. Mycopathologia 62:15–22

Armstrong DA, Buchanan AV, Caldwell RS (1976) A mycosis caused by *Lagenidium* sp. in laboratory-reared larvae of the Dungeness crab, *Cancer magister*, and possible chemical treatment. J Invertebr Pathol 28:329–336

Atami H, Muraosa Y, Hatai K (2009) *Halioticida* infection found in wild mantis shrimp *Oratoaquilla oratoria* in Japan. Fish Pathol 44:145–150

Atkins D (1954) A marine fungus *Plectospira dubia* n. sp. (Saprolegniaceae), infecting crustacean eggs and small Crustacea. J Mar Biol Assoc UK 33:721–732

Bian BZ, Egusa S (1980) *Atkinsiella hamanaensis* sp. nov. isolated from cultivated ova of the mangrove crab, *Scylla serrata* (Forsskål). J Fish Dis 3:373–385

Bian BZ, Hatai K, Po GL, Egusa S (1979) Studies on the fungal diseases in Crustaceans. I. *Lagenidium scyllae* sp. nov. isolated from cultivated ova and larvae of the mangrove crab (*Scylla serrata*). Trans Mycol Soc Jpn 20:115–124

Bland CE, Amerson HV (1973) Observation on *Lagenidium callinectes*: isolation and sporangial development. Mycologia 65:310–320

Booth C (1971) The genus *Fusarium*. Commonwealth Mycological Institute, Kew, Surrey

Booth C (1977) *Fusarium*. Commonwealth Mycological Institute, Kew, Surrey

Carmichael JW (1966) Cerebral mycetoma of trout due to a *Phialophora*-like fungus. Sabouraudia 6:120–123

Chukanhom KP, Borisutpeth P, Khoa L, Hatai K (2003) *Haliphthoros milfordensis* isolated from black tiger prawn larvae (*Penaeus monodon*) in Vietnam. Mycoscience 44:123–127

Cook KL, Deborah DSS, Hudspeth MES (2001) A *cox*2 phylogeny of terrestrial marine Peronosporomycetes (Oomycetes). Nova Hedwigia 122:231–243

Couch JN (1935) A new saprophytic species of *Lagenidium*, with notes on other forms. Mycologia 27:376–387

Couch JN (1942) A new fungus on crab eggs. J Elisha Mitchell Sci Soc 58:158–162

de Hoog GS, von Arx JA (1973) Revision of *Scolecobasidium* and *Pleurophragminium*. Kavaka 1:55–60

Dick MW (2001) Straminipilous fungi: systematics of the Peronosporomycetes, including accounts of the marine straminipilous protists, the Plasmodiophorids and similar organisms. Kluwer Academic Publishers, Dordrecht, The Netherland

Duc PM, Hatai K (2009) Pathogenicity of anamorphic fungi *Plectosporium oratosquillae* and *Acremonium* sp. to mantis shrimp *Oratosquilla oratoria*. Fish Pathol 44:81–85

Duc PM, Hatai K, Kurata O, Tensha K, Yoshitaka U, Yaguchi T, Udagawa S-I (2009) Fungal infection of mantis shrimp (*Oratosquilla oratoria*) caused by two anamorphic fungi found in Japan. Mycopathologia 167:229–247

Duc PM, Wada S, Kurata O, Hatai K (2010a) Pathogenicity of *Plectosporium oratosquillae* and *Acremonium* sp. isolated from mantis shrimp *Oratosquilla oratoria* against kuruma prawn *Penaeus japonicus*. Fish Pathol 45:133–136

Duc PM, Wada S, Kurata O, Hatai K (2010b) *In vitro* and *in vivo* efficacy of antifungal agents against *Acremonium* sp. Fish Pathol 45:109–114

Egusa S, Ueda T (1972) A *Fusarium* sp. associated with black gill disease of the kuruma prawn, *Penaeus japonicus* Bate. Bull Jap Soc Sci Fish 38:1253–1260

Fijan N (1969) Systemic mycosis in channel catfish. Bull Wildl Dis Associat 5:109–110

Fisher WS, Nilson EH, Shleser RS (1975) Effect of fungus *Haliphthoros milfordensis* on the juvenile stages of the American lobster, *Homarus americanus*. J Invertebr Pathol 26:41–45

Fuller MS, Fowles BE, McLaughlin DJ (1964) Isolation and pure culture study of marine Phycomycetes. Mycologia 56:745–756

Gaskins JE, Cheung PJ (1986) *Exophiala pisciphila*, a study of its development. Mycopathologia 93:173–184

Gerlach W, Nirenberg HI (1982) The genus *Fusarium*: a pictorial atlas. Mitt Biol Bundesanst Land u Forstwirtsch 209:155–161

Hanjavanit C, Wada S, Kurata O, Hatai K (2004) Mycotic disease of juvenile red sea bream, *Pagrus major*, caused by *Scytalidium infestans*. Suisanzoshoku 52:421–422

Hatai K (1982) On the fungus *Haliphthoros milfordensis* isolated from temporarily held abalone (*Haliotis sieboldii*). Fish Pathol 17:199–204

Hatai K, Lawhavinit O-R (1988) *Lagenidium myophilum* sp. nov., a new parasite on adult northern shrimp (*Pandalus borealis* Kroyer). Trans Mycol Soc Jpn 29:175–184

Hatai K, Furuya K, Egusa S (1978) Studies on the pathogenic fungus associated with black gill disease of kuruma prawn, *Penaeus japonicus* – I. Isolation and identification of the BG-Fusarium. Fish Pathol 12:219–224

Hatai K, Bian BZ, Baticados MCL, Egusa S (1980) Studies on the fungal diseases in Crustaceans. II. *Haliphthoros philippinensis* sp. nov. isolated from cultivated larvae of the jumbo tiger prawn (*Penaeus monodon*). Trans Mycol Soc Jpn 21:47–55

Hatai K, Kubota SS, Kida N, Udagawa S-I (1986) *Fusarium oxysporum* in red sea bream (*Pagrus* sp.). J Wildl Dis 22:570–571

Hatai K, Rhoobunjongde W, Wada S (1992) *Haliphthoros milfordensis* isolated from gills of juvenile kuruma prawn (*Penaeus japonicus*). Trans Mycol Soc Jpn 33:185–192

Hatai K, Roza D, Nakamura K (2000) Identification of lower fungi isolated from larvae of mangrove crab, *Scylla serrata*, in Indonesia. Mycoscience 41:565–572

Hawthorne BT, Rees-George J, Boradhurst PG (1992) Mating behavior and pathogenicity of New Zealand isolates of *Nectria haematococca* (*Fusarium solani*). NZ J Crop Hort Sci 20:51–57

Ho WC, Ko WH (1997) A simple method for obtaining single spore isolates of fungi. Bot Bull Acad Sin 38:41–44

Howard DH (1983) Fungi pathogenic for human and animals, Part A. Biology. Marcel Dekker, New York

Iwatsu T, Udagawa S-I, Hatai K (1990) *Scytalidium infestans* sp. nov., isolated from striped jack (*Pseudocaranx dentex*) as a causal agent of systemic mycosis. Trans Mycol Soc Jpn 31:389–397

Johnson TW Jr (1958) A fungus parasite in ova of the barnacle *Chthamalus fragilis denticulate*. Biol Bull (Woods Hole) 114:205–214

Johnson TW Jr, Bonner RR (1960) *Lagenidium callinesctes* Couch in barnacle ova. J Elisha Mitchell Sci Soc 76:147–149

Kamoun S (2003) Molecular genetics of pathogenic oomycetes. Eukaryotic Cell 2:191–199

Khoa LV, Hatai K (2005) First case of *Fusarium oxysporum* infection in cultured kuruma prawn *Penaeus japonicus* in Japan. Fish Pathol 40:195–196

Khoa L, Hatai K, Aoki T (2004) *Fusarium incarnatum* isolated from black tiger shrimp, *Penaeus monodon* Fabricius, with black gill disease cultured in Vietnam. J Fish Dis 27:507–515

Khoa LV, Hatai K, Yuasa A, Sawada K (2005) Morphology and molecular phylogeny of *Fusarium solani* isolated from kuruma prawn *Penaeus japonicus* with black gills. Fish Pathol 40:103–109

Kitancharoen N, Hatai K (1995) A marine oomycete *Atkinsiella panulirata* sp. nov. from philozoma of spiny lobster, *Panulirus japonicas*. Mycoscience 36:97–104

Kitancharoen N, Nakamura K, Wada S, Hatai K (1994) *Atkinsiella awabi* sp. nov. isolated from stocked abalone, *Haliotis sieboldii*. Mycoscience 35:265–270

Kurata O, Munchan C, Wada S, Hatai K, Miyoshi Y, Fukuda Y (2008) Novel *Exophila* infection involving ulcerative skin lesions in Japanese flounder *Paralichthys olivaceus*. Fish Pathol 43:35–44

Langdon JS, McDonald WL (1987) Cranial *Exophiala pisciphila* infection in *Salmo salar* in Australia. Bull Eur Assoc Fish Pathol 7:35–37

Lightner DV, Fontaine CT (1973) A new fungus disease of the white shrimp *Penaeus setiferus*. J Invertebr Pathol 22:94–99

Martin WW (1977) The development and possible relationships of a new *Atkinsiella* parasitic in insect eggs. Am J Bot 64:760–769

Matuo T, Snyder WC (1973) Use of morphology and mating populations in the identification of *formae specials* in *Fusarium solani*. Phytopathology 63:562–565

McGinnis MR, Ajello L (1974) A new species of *Exophiala* isolated from channel catfish. Mycologia 66:518–520

Munchan C, Kurata O, Hatai K, Hashiba N, Nakaoka N, Kawakami H (2006) Mass mortality of young striped jack *Pseudocaranx dentex* caused by a fungus *Ochroconis humicola*. Fish Pathol 41:179–182

Munchan C, Kurata O, Wada S, Hatai K, Nakaoka N, Kawakami H (2009a) Histopathology of striped jack *Pseudocaranx dentex* experimentally infected with *Ochroconis humicola*. Fish Pathol 44:128–132

Munchan C, Hatai K, Takagi S, Yamashita A (2009b) *In vitro* and *in vivo* effectiveness of itraconazole against *Ochroconis humicola* isolated from fish. Aquacul Sci 57:399–404

Munchan C, Kurata O, Wada S, Hatai K, Sano A, Kamei K, Nakaoka N (2009c) *Exophiala xenobiotica* infection in cultured striped jack. *Pseudocaranx dentex* (Bloch & Schneider), in Japan. J Fish Dis 32:893–900

Muraosa Y, Lawhavinit O-R, Hatai K (2006) *Lagenidium thermophilum* isolated from eggs and larvae of black tiger shrimp *Penaeus monodon* in Thailand. Flsh Pathol 41:35–40

Muraosa Y, Morimoto K, Sano A, Nishimura K, Hatai K (2009) A new peronosporomycete, *Halioticida noduliformans* gen. et sp. nov., isolated from white nodules in the abalone *Haliotis* spp. from Japan. Mycoscience 50:106–115

Nakamura K, Hatai K (1994) *Atkinsiella parasitica* sp. nov. isolated from a rotifer, *Brachionus plicatilis*. Mycoscience 35:383–389

Nakamura K, Hatai K (1995a) Three species of Lagenidiales isolated from the eggs and zoeae of the marine crab, *Portunus pelagicus*. Mycoscience 36:87–95

Nakamura K, Hatai K (1995b) *Atkinsiella dubia* and its related species. Mycoscience 36:431–438

Nakamura K, Nakamura M, Hatai K (1994a) *Atkinsiella* infection in the rotifer, *Brachionus plicatilis*. Mycoscience 35:291–294

Nakamura K, Wada S, Hatai K, Sugimoto T (1994b) *Lagenidium myophilum* infection in the coonstripe shrimp, *Pandalus hypsinotus*. Mycoscience 35:99–104

Nakamura K, Nakamura M, Hatai K, Zafran (1995) *Lagenidium* infection in eggs and larvae of mangrove crab (*Scylla serrata*) produced in Indonesia. Mycoscience 36:399–404

Nilson EH, Fisher WS, Shleser RA (1976) A new mycosis of larval lobster (*Homarus americanus*). J Invertebr Pathol 27:177–183

Nirenberg HI (1990) Recent advances in the taxonomy of *Fusarium*. Stud Mycol 32:91–101

O'Donnell K (2000) Molecular phylogeny of the *Nectria haematococca – Fusarium solani* species complex. Mycologia 95:919–938
O'Donnell K, Gray LE (1995) Phylogenetic relationships of the soybean sudden death syndrome pathogen *Fusarium solani* f. sp. *phaseoli* from rDNA sequence data and PCR primer for its identification. Mycologia 90:465–493
Otis EJ, Wolke RE (1985) Infection of *Exophiala salmonis* in Atlantic salmon (*Salmo salar* L.). J Wildl Dis 21:61–64
Pedersen OA, Langvad F (1989) *Exophiala psychrophila* sp. nov., a pathogenic species of the black yeast isolated from farmed Atlantic salmon. Mycol Res 92:153–156
Pesante A (1957) Osservazioni su una carie del platano. Ann Sper Agric n s 11:264–265
Rhoobunjongde W, Hatai K, Wada S, Kubota SS (1991) *Fusarium moniliforme* (Sheldon) isolated from gills of kuruma prawn *Penaeus japonicus* (Bate) with black gill disease. Nippon Suisan Gakkaishi 57:629–635
Richard RH, Holliman A, Helgason S (1978) *Exophiala salmonis* infection in Atlantic salmon *Salmo salar* L. J Fish Dis 1:357–368
Rogers-Talbert R (1948) The fungus *Lagenidium callinectes* (1942) on eggs of the blue crab in Cheaspeake Bay. Biol Bull (Woods Hole) 95:214–228
Ross AJ, Yasutake WT (1973) *Scolecobasidium humicola*, a fungal pathogen of fish. J Fish Res Board Can 30:994–995
Roza D, Hatai K (1999) *Atkinsiella dubia* infection in the larvae of Japanese mitten crab, *Eriocheir japonicus*. Mycoscience 40:235–240
Schaumann K, Priebe K (1994) *Ochroconis humicola* causing muscular black spot disease of Atlantic salmon (*Salmo salar*). Can J Bot 72:1629–1634
Sekimoto S, Hatai K, Honda D (2007) Molecular phylogeny of an unidentified *Haliphthoros*-like marine oomycete and *Haliphthoros milfordensis* inferred from nuclear-encoded small- and large-subunit rRNA genes and mitochondrial-encoded cox2 gene. Mycoscience 48:212–221
Sigler L, Carmichael JW (1974) A new acidophilic *Scytalidium*. Can J Microbiol 20:267–268
Snyder WC, Hansen HN (1941) The species concept in *Fusarium* with reference to section *Martiella*. Am J Bot 28:738–742
Sparrow FK (1960) Aquatic phycomycetes, 2nd edn. University of Michigan Press, Ann Arbor, MI
Sparrow FK (1973a) Lagenidiales. In: Ainsworth GC, Sparrow FK, Sussman AS (eds) The fungi, IV B. Academic Press, London, pp 159–163
Sparrow FK (1973b) The peculiar marine phycomycete *Atkinsiella dubia* from crab eggs. Arch Mikrobiol 93:137–144
Suga H, Hasegawa T, Mitsui H, Kageyama K, Hyakumachi M (2000) Phylogenetic analysis of the phytopathogenic fungus *Fusarium solani* based on the rDNA-ITS region. Mycol Res 104:1175–1183
Tharp TP, Bland CE (1977) Biology and host range of *Haliphthoros milfordensis*. Can J Bot 55:2936–2944
Vishniac HS (1958) A new marine Phycomycete. Mycologia 50:66–79
Wada S, Nakamura K, Hatai K (1995) First case of *Ochroconis humicola* infection in marine cultured fish in Japan. Fish Pathol 30:125–126
Wada S, Hanjavanit C, Kurata O, Hatai K (2005) *Ochroconis humicola* infection in red sea bream *Pagrus major* and marble rockfish *Sebastiscus marmoratus* cultured in Japan. Fish Sci 71:682–684
Wollenweber HW (1913) Studies on the *Fusarium* problem. Phytopathology 3:24–50
Wollenweber HW, Reinking OA (1935) Die Fusarien, ihre Beschreibung, Schadwirkung und Bekämpfung. Paul Pare, Berlin

Chapter 3
Fungal Endosymbionts of Seaweeds

T.S. Suryanarayanan

Contents

Abstract Seaweeds are being studied for their role in supporting coastal marine life and nutrient cycling and for their bioactive metabolites. For a more complete understanding of seaweed communities, it is essential to obtain information about their interactions with various other components of their ecosystem. While interactions of seaweeds with herbivores such as fish and mesograzers and surface colonizers such as bacteria and microalgae are known, their interactions with marine and marine-derived fungi are little understood. This chapter highlights the need for investigations on the little-known ecological group of fungi, viz. the fungal endosymbionts, that have intimate associations with seaweeds.

3.1 Introduction

Seaweeds are constituents of highly productive ecosystems and play a major role in nutrient cycling and stabilizing the habitat (Mann 1973). While a lot of information has accumulated on the roles played by fungi in structuring and maintaining

T.S. Suryanarayanan (✉)
Vivekananda Institute of Tropical Mycology (VINSTROM), Ramakrishna Mission Vidyapith, Chennai 600004, India
e-mail: t_sury2002@yahoo.com

C. Raghukumar (ed.), *Biology of Marine Fungi*,
Progress in Molecular and Subcellular Biology 53,
DOI 10.1007/978-3-642-23342-5_3,

terrestrial ecosystems like forests (Rayner 1998), their role in orchestrating an important and integral part of the marine ecosystem consisting of seaweeds is not clear. In this context, there is a need to study the biology of fungal endosymbionts of seaweeds with the ultimate aim of aiding in the maintenance and utilization of seaweeds.

3.2 Endomycobiota of Seaweeds

Fungi form parasitic, saprobic or asymptomatic associations with seaweeds. Most of the studies on fungal associations with seaweeds are concerned with the first two ecological groups of fungi (Kohlmeyer 1968; Raghukumar 2006, 2008; Kohlmeyer and Volkmann-Kohlmeyer 2003; Ramaiah 2006; Solis et al. 2010). There are only a few studies on the asymptomatic fungal endosymbionts of seaweeds (Raghukumar et al. 1992; Jones et al. 2008; Zuccaro et al. 2008; Loque et al. 2010; Suryanarayanan et al. 2010b); these studies are based on cultural and molecular methods and have shown that Ascomycetes and anamorphic fungi are the predominant endosymbionts of seaweeds (Fries 1979; Zuccaro et al. 2003, 2008; Tsuda et al. 2004; Schulz et al. 2008; Harvey and Goff 2010). Our study on the endosymbionts of several brown, red and green algae of southern India also showed that anamorphic fungi are dominant in all these algae (Suryanarayanan et al. 2010b). Molecular analysis using LSU rRNA PCR-DGGE showed that the brown alga *Fucus serratus* harbours fungi belonging to the Halosphaeriaceae, Lulworthiaceae, Hypocreales and Dothidiomycetes (Zuccaro et al. 2008). It appears that certain anamorphs of Ascomycetes such as *Sigmoidea marina* could even grow systemically inside algal thalli (Zuccaro et al. 2008). With reference to the species diversity of the endomycobiota harboured by seaweeds, the red and brown algae support a higher species diversity than the green algae (Suryanarayanan et al. 2010b, Table 3.1). The short life cycle of some of the green algal species and the characteristically slow growth of the endosymbionts could together be responsible for the green algae harbouring a low diversity of these fungi (Zuccaro and Mitchell 2005). Although the common endophyte genera such as *Phyllosticta*, *Colletotrichum*, *Phoma* and *Pestalotiopsis* observed among terrestrial plants are absent from the seaweeds, the pattern of distribution of endosymbionts is similar to that of the endophyte assemblages in terrestrial plants with a few dominant fungal species of low host specificity able to colonize taxonomically disparate seaweeds (Suryanarayanan et al. 2010b). Marine-derived fungi are more commonly isolated as endosymbionts of seaweeds than true marine fungi; it is not clear what factors determine the diversity and distribution of this ecological group of organisms. Harvey and Goff (2010) state that differences among algal host species are less important than geographic isolation in determining genetic covariation of their fungal endosymbionts. Environmental factors such as salinity gradient may select fungal species that colonize seaweeds (Mohamed and Martiny 2010). Another factor that can determine the composition of the endosymbiont assemblage is the antifungal metabolites produced by seaweeds to deter fungal colonization

Table 3.1 Fungal symbionts of some seaweeds

S. No	Fungus	Host
1.	*Acremoniella* sp.	Brown alga: *Padina tetrastromatica*[a]
2.	*Acremonium* sp., *Acroconidiella* sp., *Asteromyces cruciatus, Botrytis cinerea, Dendryphiella salina, Emericellopsis minima, Epicoccum* sp., *Geomyces* sp., *Gliocladium* sp., *Humicola fuscoatra, Lindra obtusa, Microascus* sp., *Mycosphaerella* sp., *Nodulisporium* sp., *Periconia* sp., *Scopulariopsis* sp., *Sigmodocea marina, Verticillium cinnabarinum*	Brown alga: *Fucus serratus*[b]
3.	*Alternaria* spp.	Green alga: Ulva linza[c]
		Brown algae: *Stoechospermum marginatum*[a], *Sargassum* sp.[a], *Turbinaria conoides*[a], *Fucus serratus*[b]
		Red alga: *Gelidiella acerosa*[a]
4.	*Aphanocladium* sp.	Red alga: *Gracilaria* sp.[a]
5.	*Arthrinium* sp.	Brown algae: *Fucus* sp.[c], *Fucus serratus*[b]
6.	*Aspergillus* spp.	Green algae: *Caulerpa racemosa*[a], *C. scalpelliformis*[a], *C. sertularioides*[a], *Halimeda macroloba*[a], *Ulva fasciata*[a], *U. lactuca*[a]
		Brown algae: *Dictyota dichotoma*[a], *Lobophora variegata*[a], *Padina tetrastromatica*[a], *P. gymnospora*[a], *Stoechospermum marginatum*[a], *Sargassum ilicifolium*[a], *Sargassum* sp.[a], *Sargassum wightii*[a], *Turbinaria* sp.1[a], *T. conoides*[a], *T. decurrens*[a], *Fucus serratus*[b]
		Red algae: *Portieria hornemanii*[a], *Gelidiella acerosa*[a], *Gracilaria crassa*[a], *G. edulis*[a], *Grateloupia lithophila*[a], *Halymenia* spp.[a]
7.	*Aureobasidium pullulans*	Green alga: *Caulerpa racemosa*[a]
8.	*Chaetomium* spp.	Green algae: *Caulerpa racemosa*[a], *C. sertularioides*[a], *H. macroloba*[a], *Ulva fasciata*[a], *U. lactuca*[a]
		Brown algae: *Lobophora variegata*[a], *Padina tetrastromatica*[a], *P. gymnospora*[a], *Stoechospermum marginatum*[a], *Turbinaria conoides*[a], *T. decurrens*[a], *Ceramium* sp.[c], *Turbinaria* sp.1[a], *Fucus serratus*[b]
9.	*Cladosporium* spp.	Green algae: *Caulerpa racemosa*[a], *C. sertularioides*[a], *U. lactuca*[a]

(continued)

Table 3.1 (continued)

S. No	Fungus	Host
		Brown algae: *Padina gymnospora*[a], *Sargassum wightii* [a], *Turbinaria* sp.1[a], *T. conoides*[a], *Fucus serratus*[b]
		Red algae: *Portieria hornemanii*[a], *Grateloupia lithophila*[a], *Halymenia* spp.[a]
10.	*Colletotrichum* sp.	Brown algae: *Sargassum* sp.[a], *Turbinaria* sp.1[a]
11.	*Coniothyrium* sp.	Brown alga: *Fucus serratus*[b]
		Red alga: *Corallina elongata*[c]
12.	*Corollospora* sp.	Brown algae: *Sargassum* sp.[c], *Fucus serratus*[b]
13.	*Curvularia* spp.	Green algae: *Caulerpa scalpelliformis*[a], *Ulva fasciata*[a]
		Brown algae: *Turbinaria* sp.1[a], *T. decurrens*[a]
14.	*Drechslera* spp.	Brown algae: *Sargassum* sp.[a], *Turbinaria* sp.1[a], *T. decurrens*[a]
		Red alga: *Halymenia* spp.[a]
15.	*Emericella nidulans*	Brown algae: *Lobophora variegata*[a], *Stoechospermum marginatum*[a], *Turbinaria* sp.1[a], *T. conoides*[a], *T. decurrens*[a], *Sargassum wightii*[a]
		Red alga: *Halymenia* sp.2[a]
16.	*Fusarium* spp.	Green algae: *Caulerpa racemosa*[a], *C. sertularioides*[a]
		Brown algae: *Turbinaria* sp.1[a], *Fucus serratus*[b]
17.	*Geniculosporium* sp.	Red alga: Polysiphonia sp.[c]
18.	*Haloguignardia* sp.	Brown algae: *Cystoseira* sp.[d], *Halidrys* sp.[d]
19.	*Helicosporium* sp.	Brown alga: *Sargassum* sp.[a]
20.	*Lindra thalassiae*	Brown alga: *Sargassum cinereum*[e]
21.	*Memnoniella* sp.	Red alga: *Portieria hornemanii*[a]
22.	*Monilia* sp.	Red alga: *Gracilaria* sp.[a]
23.	*Monodictys* sp.	Brown algae: *Padina tetrastromatica*[a], *Turbinaria* sp.1[a]
24.	*Mucor* sp.	Green alga: *Caulerpa racemosa*[a]
		Brown alga: *Fucus serratus*[b]
25.	*Myrothecium* sp.	Green alga: *Caulerpa sertularioides*[a]
26.	*Nigrospora* sp.	Green alga: *Ulva lactuca*[a]
		Brown algae: *Lobophora variegata*[a], *Stoechospermum marginatum*[a], *Sargassum* sp.[a], *S. wightii*[a], *Turbinaria decurrens*[a]

(continued)

Table 3.1 (continued)

S. No	Fungus	Host
27.	*Oidiodendron* sp.	Brown algae: *Sargassum wightii*[a], *Fucus serratus*[b]
28.	*Paecilomyces* spp.	Green algae: *Caulerpa scalpelliformis*[a], *Halimeda macroloba*[a], *Ulva fasciata*[a]
		Brown algae: *Padina tetrastromatica*[a], *Sargassum wightii*[a], *Turbinaria* sp.1[a], *Fucus serratus*[b]
		Red algae: *Gracilaria edulis*[a], *Gracilaria* sp.[a], *Halymenia* sp.1[a]
29.	*Penicillium* spp.	Green algae: *Caulerpa racemosa*[a], *C. scalpelliformis*[a], *C. sertularioides*[a], *H. macroloba*[a], *U. lactuca*[a]
		Brown algae: *Padina tetrastromatica*[a], *P. gymnospora*[a], *Sargassum* sp.[a], *S. wightii*[a], *Fucus serratus*[b], *F. vesiculosus*[c]
		Red algae: *Portieria hornemanii*[a], *Gelidiella acerosa*[a], *Halymenia* sp.2[a]
30.	*Phaeotrichoconis* sp.	Brown alga: *Turbinaria* sp.1[a]
31.	*Phialophora* sp.	Green alga: *Caulerpa sertularioides*[a]
		Brown algae: *Turbinaria conoides*[a], *T. decurrens*[a]
32.	*Phoma* sp.	Green alga: *Caulerpa racemosa*[a]
		Brown algae: *Padina gymnospora*[a], *Turbinaria conoides*[a], *Fucus* sp.[c], *F. serratus*[b], *F. vesiculosus*[c]
		Red algae: *Gelidiella acerosa*[a], *Halymenia* sp.1[a]
33.	*Phomopsis* sp.	Brown alga: *Fucus serratus*[b]
		Red alga: *Portieria hornemanii*[a]
34.	*Pyrenochaeta* sp.	Green alga: *Caulerpa racemosa*[a]
35.	*Taeniolella* sp.	Brown alga: *Sargassum* sp.[a]
36.	*Tolypocladium inflatum*	Brown alga: *Fucus vesiculosus*[c]
37.	*Trichoderma* sp.	Green alga: *Halimeda macroloba*[a]
		Brown algae: *Dictyota dichotoma*[a], *Padina gymnospora*[a], *Fucus serratus*[b]
		Red alga: *Gracilaria crassa*[a]
38.	*Trimmatostroma* sp.	Red alga: *Portieria hornemanii*[a]
39.	*Varicosporium* sp.	Brown algae: *Sargassum* sp.[a], *Turbinaria* sp.1[a]

[a]Suryanarayanan et al. (2010b)
[b]Zuccaro et al. (2008)
[c]Schulz et al. (2008)
[d]Harvey and Goff (2010)
[e]Raghukumar et al. (1992)

(Kubanek et al. 2003; König et al. 2006; Lam et al. 2008). Adaptations by a few fungal species to tolerate or detoxify these compounds along with the constant availability of host tissue for colonization in seaweed beds can lead to the evolution of generalist endosymbionts capable of colonizing taxonomically unrelated seaweeds. Such adaptations can possibly explain the constant presence of certain marine-derived fungal genera such as *Aspergillus*, *Cladosporium* and *Penicillium* as endosymbionts in different seaweeds (König et al. 2006; Zuccaro et al. 2008). For instance, Suryanarayanan et al. (2010b) isolated three species of *Cladosporium* from *Caulerpa racemosa*, *C. sertularioides*, *Ulva lactuca*, *Padina gymnospora*, *Sargassum wightii*, *Turbinaria conides*, *Turbinaria* sp.1, *Portieria hornemanii*, *Grateloupia lithophila* and *Halymenia* spp. Zuccaro et al. (2008) isolated several *Cladosporium* sp. from *Fucus serratus*. *Aspergillus terreus* was most frequently isolated as an endosymbiont from *Caulerpa scalpelliformis*, *Helimeda macroloba*, *Ulva lactuca*, *Ulva fasciata*, *Lobophora variegata*, *Padina gymnospora*, *Stoechospermum marginatum*, *Sargassum ilicifolium*, *Portieria hornemanni* and *Gracilaria edulis* (Suryanarayanan et al. 2010b). Marine-derived fungi such as *Aspergillus* spp., apart from dominating the endosymbiont assemblage of seaweeds (Suryanarayanan et al. 2010b), also dominate the fungal assemblages of marine invertebrates of different geographical locations such as the North Sea, the Mediterranean, the Caribbean and the great Barrier Reef (Höller et al. 2000) attesting to their adaptation to occupy such a niche as the inner tissues of seaweeds or marine animals. Such a widespread occurrence of marine-derived fungi may be indicative of their passive migration from terrestrial habitats (Alva et al. 2002). However, since these fungi are better adapted to marine environments than their terrestrial conspecifics (Zuccaro et al. 2004; König et al. 2006) and survive in seaweeds which produce antifungal metabolites (Kubanek et al. 2003; Lam et al. 2008), it is likely that they are not casual residents of the seas but have coevolved with the seaweeds (Zuccaro et al. 2004; Suryanarayanan et al. 2010b). Seaweeds are known to produce secondary metabolites that deter microbial colonization; these metabolites determine the abundance and composition of the microbial community associated with the seaweeds as they select those microbes that can tolerate or detoxify the chemicals (Nylund et al. 2010). The observation that the spores of *Acremonium fuci* associated with *F. serratus* can germinate only in the presence of tissues of the alga and not in seawater alone points to host specificity among seaweed endosymbionts and adds credence to the view that marine-derived fungi are not mere migrants from land habitats (Zuccaro et al. 2004). Furthermore, the widespread occurrence of marine-derived fungi as endosymbionts as opposed to the few obligate marine fungi such as *Ascochyta salicorniae* (König et al. 2006), *Lindra* sp., *Lulworthia* sp. and *Corollospora* sp. (Zuccaro et al. 2008) suggests that the endosymbiotic marine-derived fungi constitute a special community. It is pertinent to mention here that genetic variations exist in a single morphological species of seaweed fungal endosymbiont (Harvey 2002). Molecular studies to discern variations in genes and gene expressions between fungal conspecifics of marine and terrestrial habitats are needed to substantiate this hypothesis.

3.3 Fungal Endosymbionts of Seaweeds as Drug Source

Although a simple procedure of inoculating surface-sterilized segments of algal thalli on suitable growth media yields several endosymbiotic fungi, the nature of association of these fungi with their algal hosts is not clear. As has been reported for endophytes of terrestrial plants (Ganley et al. 2004; Suryanarayanan and Murali 2006), the endosymbiont assemblage of an alga can consist of different ecological groups of fungi such as latent pathogens, true endosymbionts or casual residents. Irrespective of the nature of association that these fungi have with their seaweeds, their capacity to produce secondary metabolites is considerable (Bugni and Ireland 2004; Ebel 2010). They produce novel metabolites showing various bioactivities which are of pharmaceutical importance (Saleem et al. 2007). The extraordinary synthetic ability of these fungi and the fact that natural products are superior to combinatorial synthesis of drugs due to greater number of chiral centres, enhanced steric complexity and capacity to bind to targets (Feher and Schmidt 2003; Koehn and Carter 2005) have renewed the search for novel drugs from marine fungi (Jones et al. 2008; Suryanarayanan et al. 2010a).

Endosymbionts of seaweeds produce, among other bioactive metabolites, strong antioxidants (Suryanarayanan et al. 2010b). An *Epicoccum* sp. associated with *Fucus vesiculosus* produces epicoccone, a novel antioxidant (Abdel-Lateff et al. 2003a). *Wardomyces anomalus*, an endosymbiont of green algae produces a novel xanthone with antioxidant and radical scavenging properties (Abdel-Lateff et al. 2003b). Synthesis of antioxidants may be a strategy among endosymbionts to counter the defence reactions of their algal hosts (Dring 2005). Many fungi isolated from green, red and brown algae also produce antialgal, antifungal and antiinsect metabolites (Suryanarayanan et al. 2010b) which may help in deterring colonization of algal thalli by other microbes thereby reducing competition and in warding off herbivores.

Although marine-derived fungi are marine isolates of terrestrial species, they are prolific producers of novel metabolites of unprecedented structures (Bugni and Ireland 2004; Oh et al. 2006; Meyer et al. 2010; Kjer et al. 2010; Gao et al. 2011) which are not produced by their terrestrial conspecifics (Osterhage et al. 2000). These metabolites exhibit anticancer, antibacterial, antiplasmodial, anti-inflammatory, nematicidal, antiviral and antiangiogenic activities (Bhadury et al. 2006; Newman and Hill 2006; Bhatnagar and Kim 2010; Ohkawa et al. 2010). Marine-derived fungi associated with marine algae have been reported to produce novel metabolites including aromatic polyketides (Pontius et al. 2008), alkaloids (Tsuda et al. 2004), sesquiterpenes (Bugni and Ireland 2004) and terpenes (Klemke et al. 2004; König et al. 2006; Lösgen et al. 2007). These fungi are thus an important source of pharmacologically active metabolites or potential lead compounds for drug synthesis (Jensen and Fenical 2002; Blunt et al. 2008, 2010) (Table 3.2). Production of bioactive metabolites by fungi depends largely on culture conditions. Manipulating physical factors such as pH, salinity and aeration (Bode et al. 2002) and altering growth medium composition or adding certain inhibitors and precursors to the medium (Llorens et al. 2004) increase or induce the synthesis of bioactive

Table 3.2 Some bioactive metabolites isolated from fungal endosymbionts of seaweeds

Fungus	Host	Secondary metabolite(s)	Activity	References
Aspergillus versicolor	*Penicillus capitatus*	Sesquiterpenoid nitrobenzoyl esters	Antitumour	Belofsky et al. (1998)
Unidentified fungus	*Ceralodictyon spongiosum*	Anthracycline-derived pentacyclic metabolite	–	Shigemori et al. (1999)
Acremonium sp.	*Cladostephus spongiosus*	Hydroquinone	Antioxidant	Abdel-Lateff et al. (2002)
Drechslera dematioidea	*Liagora viscida*	10 new sesquiterpenoids	Antimalarial	Osterhage et al. (2002)
Aspergillus terreus	*Marine alga*	Chiral dipyrrolobenzoquinone derivatives	UV-A protection	Lee et al. (2003)
Penicillium citrinum	*Actinotrichia fragilis*	Alkaloid	Anticancer	Tsuda et al. (2004)
Apiospora montagnei	*Polosiphonia violacea*	Diterpene	Anticancer	Klemke et al. (2004)
Penicillium sp.	*Dictyosphaeria versluyii*	Decalactones	–	Bugni et al. (2004)
Unidentified fungus	*Gracillaria verrucosa*	Cyclopentenone derivatives	–	Li et al. (2004)
Myrothecium sp.	*Enteromorpha compressa*	Cyclopentenone derivatives	Tyrosinase inhibition	Li et al. (2005)
Geniculosporium sp	*Polysiphonia sp.*	11 new tricyclic sesquiterpenes	Antimicrobial	Krohn et al. (2005)
Fusarium sp.	*Codium fragile*	Cyclic tetrapeptide	Anticancer	Ebel (2006)
Emericella nidulans var. *acristata*	Green alga	Prenylated Polyketides	–	Kralj et al. (2006)
Chaetomium globosum	*Polysiphonia urceolata*	Benzaldehyde derivative	Antitumour and antioxidant	Wang et al. (2006)
Botrytis sp.	*Enteromorpha compressa*	Cyclopentenone derivatives	–	Li et al. (2007)
Aspergillus niger	*Colpomenia sinuosa*	Diels–Alder adduct of a steroid and maleimide	–	Zhang et al. (2007a)
Aspergillus niger	*Colpomenia sinuosa*	Naphthoquinoneimine	Antifungal	Zhang et al. (2007b)
Aspergillus niger	*Colpomenia sinuosa*	Novel sphingolipids	Antifungal	Zhang et al. (2007c)
Aspergillus flavus	*Enteromorpha tubulosa*	2-Pyrone derivatives	Induction of cAMP production	Lin et al. (2008)

Beauveria felina	*Caulerpa* sp.	Cyclodepsipeptides	Antimycobacterial, Antibacterial, Antifungal and Cytotoxic	de Vita-Marques et al. (2008)
Pseudoanguillospora sp.	*Polyides rotundus*	New isochromans	Antimicrobial	Kock et al. (2009)
Penicillium chrysogenum	*Laurencia* sp.	Polyketide and glycerol derivatives	Antifungal	Gao et al. (2011)
Cadophora malorum.	*Enteromorpha* sp.	New hydroxylated sclerosporin derivatives	Inhibition of fat deposition in cells	Almeida et al. (2010)
Curvularia sp.	*Gracilaria folifera*	New macrolides	Antimicrobial	Dai et al. (2010)
Chaetomium globosum	*Ulva pertusa*	7 new cytochalasan derivatives	–	Cui et al. (2010)
Aspergillus oryzae	*Heterosiphonia japonica*	2 new indoloditerpenes derivatives	Antibiotic	Qiao et al. (2010)
Exophiala oligosperma	*Laurencia similis*	Phenoxynaphthalene	–	Li et al. (2011)
Coniothyrium cereale	*Enteromorpha* sp.	New phenalenone derivatives	Antibiotic and inhibitor of human leukocyte elastase	Elsebai et al. (2011)

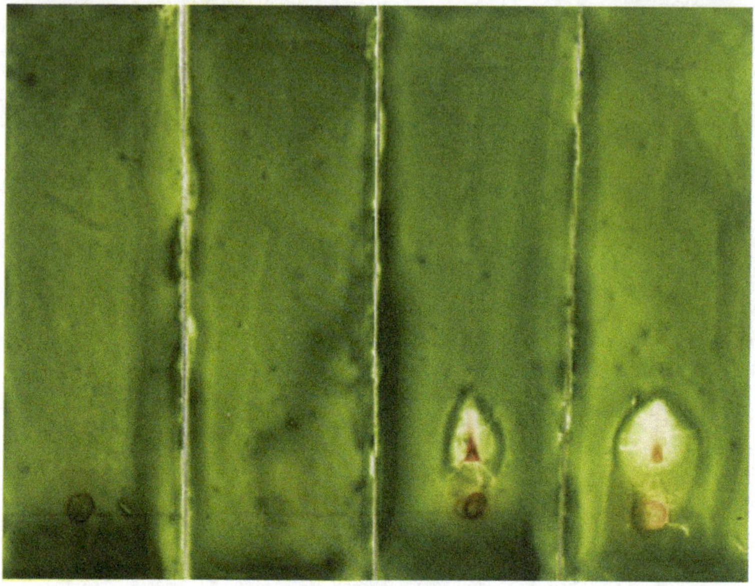

Fig. 3.1 Autobiogram showing the production of antialgal substance(s) by *Aspergillus terreus* endosymbiont of *Ulva lactuca* in NaCl-amended medium

metabolites by the fungi. For example, Miao et al. (2006) reported that the antibiotic activity of a marine-derived fungus increased with salinity of the growth medium. Extending the argument that marine-derived fungi are not recent chance immigrants from terrestrial habitats, we studied the secondary metabolite production in *Aspergillus terreus* isolated from the alga *Ulva lactuca* as influenced by sodium chloride in the growth medium. The fungus produces antialgal substance(s) only when grown in a medium with 3.5 or 7% NaCl and not in the absence of the salt (Fig. 3.1) (Suryanarayanan and Venkatachalam, unpublished). This suggested that cultivation of marine fungi in NaCl medium can result in the production of metabolites not synthesized or only synthesized in undetectable levels in normal growth medium (Bugni and Ireland 2004). Similarly, co-cultivation of marine-derived fungi with microbes of marine habitats activates the silent gene clusters of the fungi enabling them to synthesize novel metabolites (Brakhage and Schroeckh 2011). A marine-derived *Emericella* sp. synthesizes emericellamides when co-cultured with the marine actinomycete *Salinispora arenicola* due to enhanced expression of emericellamide gene cluster (Oh et al. 2007). *Pestalotia* sp., another marine-derived fungus, synthesizes pesatlone, an antibiotic effective against methicillin-resistant *Staphylococcus aureus* when cultured with a marine bacterium (Cueto et al. 2001). The synthetic ability of the fungi can also be enhanced by the presence of small molecule elicitors in the growth medium as they specifically effect the transcription of secondary metabolite gene clusters (Pettit 2010). For instance, Christian et al. (2005) reported that the marine-derived fungus *Phomopsis asperagi* produces novel chaetoglobosins when the growth medium is amended with the sponge metabolite jasplakinolide. The addition of small molecule elicitors to the fermentation medium specifically influences the synthesis of bioactive molecules by fungi (Pettit 2010). Thus, the synthetic ability of seaweed endosymbionts can be enhanced by altering the culture conditions. In this context, Kjer et al. (2010) have recently described methods for isolating marine-derived endophytic fungi and extracting bioactive metabolites from them.

3.4 Lessons from Fungal Endophytes

We have very limited information on the diversity and nature of association of fungal endosymbionts of seaweeds. Culture methods should be complemented by molecular methods to obtain a more complete picture of the diversity of the endosymbionts as has been done for endophyte of terrestrial plants (Arnold and Lutzoni 2007; Unterseher and Schnittler 2010). Although direct comparisons of seaweeds with terrestrial plants may not be ideal due to differences in their tissue construction, environment and propagule density and survival (Schiel and Foster 2006), the more intense studies conducted on endophytes of terrestrial plants provide us with blueprints for future lines of investigations on seaweed endosymbionts. For instance, endophyte infections render plants more tolerant to abiotic stress (Rodriguez et al. 2008). Endophytes elaborate metabolites that make their hosts more resistant to biotic stress such as infection by pathogens (Arnold et al. 2003) or damage by herbivores (Wagner and Lewis 2000). The strong antioxidants produced by endophytes also play a role in protecting the plant hosts from stresses (White and Torres 2010). Since many endosymbionts of seaweeds produce antimicrobial metabolites and strong antioxidants and since some marine algae such as *Ulva* lack structural and chemical defences (Moksnes et al. 2008), it would be worthwhile determining the role of endosymbionts in stress tolerance and survival of seaweeds. Endophyte infections can also negatively affect a terrestrial plant host by increasing its susceptibility to drought (Arnold and Engelbrecht 2007) and reducing its photosynthetic efficiency (Pinto et al. 2000). There is no information regarding the cost and benefit of endosymbiont association with seaweeds. Some fungal pathogens survive as symptomless endophytes in tissues of terrestrial plants and induce disease under suitable environmental conditions or when the host's defence is weakened (Suryanarayanan and Murali 2006; A'lvarez-Loayza et al. 2011). If pathogenic fungi can exist as latent pathogens in seaweeds, they can be harnessed to control undesirable and invasive seaweeds which are detrimental to marine ecosystems (Anderson 2007). Plant endophytes are known to produce numerous bioactive compounds including those elaborated by their hosts (Stierle et al. 1993; Weber 2009). Though many seaweeds produce antibacterial, antiviral and cytotoxic compounds (Kamenarska et al. 2009), there is no information on how their synthetic ability is affected by the presence of fungal endosymbionts. If a fungal endosymbiont produces the same bioactive metabolite of industrial importance as its algal host, extraction and purification of the chemical will be easier since marine-derived fungi can be cultured easily. Genotypically similar endophytes of land plants vary in their secondary metabolite profile, indicating that screening of a fungal species once will not capture the entire metabolic spectrum of the endophyte (Seymour et al. 2004). This may be true of the fungal endosymbionts of seaweeds as well since we observed variation in the biological activities of the metabolites of *Aspergillus terreus* isolated from different seaweeds (Suryanarayanan et al. 2010b). Another important aspect of seaweed–endosymbiont association is the diversity and distribution of these fungi. Information on land plant–endophyte diversity provide us with a background with which this question can be tackled. Firstly, both molecular and culture methods should be employed to assess the endosymbiont diversity since they complement each other to provide a more complete data as has been demonstrated for terrestrial plants and endophyte association (Arnold et al. 2007). Furthermore, different organs or tissues of seaweeds with complex structures will have to be screened since different tissues of a single plant species can harbour different endophyte species (Kumaresan et al. 2002). It is also necessary to screen seaweeds during different seasons as the endophyte assemblage of plants is altered by seasons (Suryanarayanan and Thennarasan 2004). Foliar endophytes persist in fallen leaves and initiate tissue degradation by functioning as pioneer leaf litter fungi (Sun et al. 2011). In this context, the role of fungal endosymbionts in the degradation of old and dead seaweed tissues has to be addressed.

3.5 Future Directions

Considering the economic importance of seaweeds and the significant roles they play in maintaining the coastal marine ecosystem, their biology, including their interactions with other organisms need to be investigated in detail. Among the various types of organisms that are known to be associated with seaweeds, the endosymbiotic fungi warrant special attention since their biosynthetic potential is unique and many of them synthesize novel metabolites of unprecedented molecular architecture. The bioactivities of such metabolites indicate that they have pharmaceutical and agricultural applications. Studying the diversity and distribution of seaweed endosymbionts and using modern methods of genome-guided search for discovering their bioactive metabolites will help in harnessing the biotechnological potential of these fungi.

References

A'lvarez-Loayza P, Jr White JF, Torres MS et al (2011) Light converts endosymbiotic fungus to pathogen, influencing seedling survival and niche-space filling of a common tropical tree, *Iriartea deltoidea*. PLoS One 6(1):e16386

Abdel-Lateff A, Koenig GM, Fisch KM et al (2002) New antioxidant hydroquinone derivatives from the algicolous marine fungus *Acremonium* sp. J Nat Prod 65:1605–1611
Abdel-Lateff A, Fisch KM, Wright AD et al (2003a) A new antioxidant isobenzofuranone derivative from the algicolous marine fungus *Epicoccum* sp. Planta Med 69:831–834
Abdel-Lateff A, Klemke C, König GM et al (2003b) Two new xanthone derivatives from the algicolous marine fungus *Wardomyces anomalus*. J Nat Prod 66:706–708
Almeida C, Eguereva E, Kehraus S et al (2010) Hydroxylated sclerosporin derivatives from the marine-derived fungus *Cadophora malorum*. J Nat Prod 73:476–478
Alva P, Mckenzie EHC, Pointing SB et al (2002) Do seagrasses harbour endophytes? In: Hyde KD (ed) Fungi in marine environments, Fungal Diversity Research Series. Hong Kong University Press, Hong Kong, UK
Anderson LWJ (2007) Control of invasive seaweeds. Bot Mar 50:418–437
Arnold AE, Engelbrecht BMJ (2007) Fungal endophytes nearly double minimum leaf conductance in seedlings of a neotropical tree species. J Trop Ecol 23:369–372
Arnold AE, Lutzoni F (2007) Diversity and host range of foliar fungal endophytes: are tropical leaves biodiversity hotspots? Ecology 88:541–549
Arnold AE, Mejia LC, Kyllo D et al (2003) Fungal endophytes limit pathogen damage in a tropical tree. Proc Natl Acad Sci U S A 100:15649–15654
Arnold AE, Henk DA, Eells RL et al (2007) Diversity and phylogenetic affinities of foliar fungal endophytes in loblolly pine inferred by culturing and environmental PCR. Mycologia 99:185–206
Belofsky GN, Jensen PR, Renner MK et al (1998) New cytotoxic sesquiterpenoid nitrobenzoyl esters from a marine isolate of the fungus *Aspergillus versicolor*. Tetrahedron 54:1715–1724
Bhadury P, Mohammad BT, Wright PC et al (2006) The current status of natural products from marine fungi and their potential as anti-infective agents. J Ind Microbiol Biotechnol 33:325–337
Bhatnagar I, Kim SK (2010) Immense essence of excellence: marine microbial bioactive compounds. Mar Drugs 8:2673–2701
Blunt JW, Copp BR, Hu WP et al (2008) Marine natural products. Nat Prod Rep 25:35–94
Blunt JW, Copp BR, Munro MH et al (2010) Marine natural products. Nat Prod Rep 27:165–237
Bode HB, Bethe B, Höfs R et al (2002) Big effects from small changes: possible ways to explore nature's chemical diversity. Chem Bio Chem 3:619–627
Brakhage AA, Schroeckh V (2011) Fungal secondary metabolites – strategies to activate silent gene clusters. Fungal Genet Biol 48:15–22
Bugni TS, Ireland CM (2004) Marine-derived fungi: a chemically and biologically diverse group of microorganisms. Nat Prod Rep 21:143–163
Bugni TS, Janso JE, Williamson RT et al (2004) Dictyosphaeric acids A and B: new decalactones from an undescribed *Penicillium* sp. obtained from the alga *Dictyosphaeria versluyii*. J Nat Prod 67:1396–1399
Christian OE, Compton J, Christian KR et al (2005) Using Jasplakinolide to turn on pathways that enable the isolation of new Chaetoglobosins from *Phomospis asparagi*. J Nat Prod 68:1592–1597
Cueto M, Jensen PR, Kauffman C et al (2001) Pestalone, a new antibiotic produced by a marine fungus in response to bacterial challenge. J Nat Prod 64:1444–1446
Cui CM, Li XM, Li CS et al (2010) Cytoglobosins A-G, Cytochalasans from a marine-derived endophytic fungus, *Chaetomium globosum* QEN-14. J Nat Prod 73:729–733
Dai J, Krohn K, Flörke U et al (2010) Curvularin-type metabolites from the fungus *Curvularia* sp. isolated from a marine alga. Eur J Org Chem 2010:6928–6937
Dring MJ (2005) Stress resistance and disease resistance in seaweeds: the role of reactive oxygen metabolism. Adv Bot Res 43:175–207
Ebel R (2006) Secondary metabolites from marine-derived fungi. In: Proksch P, Müller WEG (eds) Frontiers in marine biotechnology. Horizon Bioscience, Norwich, UK
Ebel R (2010) Terpenes from marine-derived fungi. Mar Drugs 8:2340–2368

Elsebai MF, Kehraus S, Lindequist U et al (2011) Antimicrobial phenalenone derivatives from the marine-derived fungus *Coniothyrium cereale*. Org Biomol Chem 9:802–808
Feher M, Schmidt JM (2003) Property distributions: differences between drugs, natural products, and molecules from combinatorial chemistry. J Chem Infect Comp Sci 43:218–227
Fries N (1979) Physiological characteristics of *Mycosphaerella ascophylli*, a fungal endophyte of the marine brown alga *Ascophyllum nodosum*. Physiol Plantarum 45:117–121
Ganley RJ, Brunsfeld SJ, Newcombe G (2004) A community of unknown, endophytic fungi in western white pine. Proc Natl Acad Sci U S A 101:10107–10112
Gao SS, Li XM, Du FY et al (2011) Secondary metabolites from a marine-derived endophytic fungus *Penicillium chrysogenum* QEN-24 S. Mar Drugs 9:59–70
Harvey JBJ (2002) Intraspecific variation in *H. irritans*, a fungal endosymbiont of marine brown algae of the North American Pacific. J Phycol 38:16
Harvey JBJ, Goff LJ (2010) Genetic covariation of the marine fungal symbiont *Haloguignardia irritans* (Ascomycota, Pezizomycotina) with its algal hosts *Cystoseira* and *Halidrys* (Phaeophyceae, Fucales) along the west coast of North America. Fungal Biol 114:82–95
Höller U, Wrigh AD, Matthee GF et al (2000) Fungi from marine sponges: diversity, biological activity and secondary metabolites. Mycol Res 104:1354–1365
Jensen PR, Fenical W (2002) Secondary metabolites from marine fungi. In: Hyde KD (ed) Fungi in marine environments. Fungal Diversity Research Series, Hong Kong, UK
Jones EBG, Stanley SJ, Pinruan U (2008) Marine endophyte sources of new chemical natural products: a review. Bot Mar 51:163–170
Kamenarska Z, Serkedjieva J, Najdenski H et al (2009) Antibacterial, antiviral, and cytotoxic activities of some red and brown seaweeds from the Black sea. Bot Mar 52:80–86
Kjer J, Debbab A, Aly AH et al (2010) Methods for isolation of marine-derived endophytic fungi and their bioactive secondary products. Nat Protoc 5:479–490
Klemke C, Kehraus S, Wright AD et al (2004) New secondary metabolites from the endophytic fungus *Apiospora montagnei*. J Nat Prod 67:1058–1063
Kock I, Draeger S, Schulz B et al (2009) Pseudoanguillosporin A and B: two new isochromans isolated from the endophytic fungus *Pseudoanguillospora* sp. Eur J Org Chem 2009:1427–1434
Koehn FE, Carter GT (2005) The evolving role of natural products in drug discovery. Nat Rev Drug Discov 4:206–220
Kohlmeyer J (1968) Revisions and descriptions of algicolous marine fungi. J Phytopathol 63:342–363
Kohlmeyer J, Volkmann-Kohlmeyer B (2003) Marine ascomycetes from algae and animal hosts. Bot Mar 34:1–35
König GM, Kehraus S, Seibert SF et al (2006) Natural products from marine organisms and their associated microbes. Chembiochem 7:229–238
Kralj A, Kehraus S, Krick A et al (2006) Arugosins G and H: prenylated polyketides from the marine-derived fungus *Emericella nidulans* var. *acristata*. J Nat Prod 69:995–1000
Krohn K, Dai J, Flörke U et al (2005) Botryane metabolites from the fungus *Geniculosporium* sp. isolated from the marine red alga *Polysiphonia*. J Nat Prod 68:400–405
Kubanek J, Jensen PR, Keifer PA et al (2003) Seaweed resistance to microbial attack: a targeted chemical defense against marine fungi. Proc Natl Acad Sci U S A 100:6916–6921
Kumaresan V, Suryanarayanan TS, Johnson JA (2002) Ecology of mangrove endophytes. In: Hyde KD (ed) Fungi in marine environments. Fungal Diversity Research Series, Hong Kong, UK
Lam C, Stang A, Harder T (2008) Planktonic bacteria and fungi are selectively eliminated by exposure to marine macroalgae in close proximity. FEMS Microbiol Ecol 63:283–291
Lee SM, Li XF, Jiang H et al (2003) Terreusinone, a novel UV-A protecting dipyrroloquinone from the marine algicolous fungus *Aspergillus terreus*. Tetrahedron Lett 44:7707–7710
Li XF, Kim SK, King JS et al (2004) Polyketide and sesquiterpenediol metabolites from a marine-derived fungus. Bull Korean Chem Soc 25:607–608
Li XF, Kim MK, Lee U et al (2005) Myrothenones A and B, cyclopentenone derivatives with tyrosinase inhibitory activity from the marine-derived fungus *Myrothecium* sp. Chem Pharm Bull 53:453–455

Li XF, Zhang DH, Lee U et al (2007) Bromomyrothenone B and botrytinone, cyclopentenone derivatives from a marine isolate of the fungus *Botrytis*. J Nat Prod 70:307–309
Li F, Li K, Li X et al (2011) Chemical constituents of marine algal-derived endophytic fungus *Exophiala oligosperma* EN-21. Chin J Ocean Limn 29:63–67
Lin A, Lu X, Fang Y et al (2008) Two new 5-Hydroxy-2-pyrone derivatives isolated from a marine-derived fungus *Aspergillus flavus*. J Antibiot 61:245–249
Llorens A, Matco R, Hinojo MJ et al (2004) Influence of the interactions among ecological variables in the characterization of Zearalenone producing isolates of *Fusarium* spp. Syst Appl Microbiol 27:253–260
Loque CP, Medeiros AO, Pellizzari FM (2010) Fungal community associated with marine macroalgae from Antarctica. Polar Biol 33:641–648
Lösgen S, Schlörke O, Meindl K et al (2007) Structure and biosynthesis of chatocyclinones, new polyketides produced by and endosymbiotic fungus. Eur J Org Chem 13:2191–2196
Mann KH (1973) Seaweeds: their productivity and strategy for growth. Science 182:975–981
Meyer SW, Mordhorst TF, Choonghwan Lee C et al (2010) Penilumamide, a novel lumazine peptide isolated from the marine-derived fungus, *Penicillium* sp. CNL-338. Org Biomol Chem 8:2158–2163
Miao L, Theresa FN, Qian KPY (2006) Effect of culture conditions on mycelial growth, antibacterial activity, and metabolite profiles of the marine-derived fungus *Arthrinium c.f. saccharicola*. Appl Microbiol Biotechnol 72:1063–1073
Mohamed DJ, Martiny JBH (2010) Patterns of fungal diversity and composition along a salinity gradient. ISME J 5(3):379–388
Moksnes PO, Gullström M, Tryman K et al (2008) Trophic cascades in a temperate seagrass community. Oikos 117:763–777
Newman DJ, Hill RT (2006) New drugs from marine microbes: the tide is turning. J Ind Microbiol Biotechnol 33:539–544
Nylund GM, Persson F, Lindegarth M et al (2010) The red alga *Bonnemaisonia asparagoides* regulates epiphytic bacterial abundance and community composition by chemical defence. FEMS Microbiol Ecol 71:84–93
Oh DC, Jensen PR, Fenical W (2006) Zygosporamide, a cytotoxic cyclic depsipeptide from the marine-derived fungus *Zygosporium masonii*. Tetrahedron Lett 47:8625–8628
Oh DC, Kauffman CA, Jensen PR et al (2007) Induced production of Emericellamides A and B from the marine-derived fungus *Emericella* sp. in competing co-culture. J Nat Prod 70:515–520
Ohkawa Y, Miki K, Suzuki T et al (2010) Antiangiogenic metabolites from a marine-derived fungus, *Hypocrea vinosa*. J Nat Prod 73:579–582
Osterhage C, Schwibbe M, König GM et al (2000) Differences between marine and terrestrial *Phoma* species as determined by HPLC-DAD and HPLC-MS. Phytochem Anal 11:1–7
Osterhage C, König GM, Höller U et al (2002) Rare sesquiterpenes from the algicolous fungus *Drechslera dematioidea*. J Nat Prod 65:306–313
Pettit RK (2010) Small-molecule elicitation of microbial secondary metabolites. Microb Biotechnol 4(4):471–478
Pinto LSRC, Azevedo J, Periera O et al (2000) Symptomless infection of banana and maize by endophytic fungi impairs photosynthetic efficiency. New Phytol 147:609–615
Pontius A, Krick A, Mesry R et al (2008) Monodictyochromes A and B, dimeric xanthone derivatives from the marine algicolous fungus *Monodictys putredinis*. J Nat Prod 71:1793–1799
Qiao MF, Ji NY, Liu XH et al (2010) Indoloditerpenes from an algicolous isolate of *Aspergillus oryzae*. Bioorg Med Chem Lett 20:5677–5680
Raghukumar C (2006) Algal-fungal interactions in the marine ecosystem: symbiosis to parasitism. In: Tewari A (ed) Recent advances on applied aspects of Indian marine algae with reference to global scenario. Central Salt and Marine Chemicals Research Institute, Gujarat, India
Raghukumar C (2008) Marine fungal biotechnology: an ecological perspective. Fungal Divers 31:19–35

Raghukumar C, Nagarkar S, Raghukumar (1992) Association of thraustochytrids and fungi with living marine algae. Mycol Res 96:542–546

Ramaiah N (2006) A review on fungal diseases of algae, marine fishes, shrimps and corals. Ind J Mar Sci 35:380–387

Rayner ADM (1998) Fountains of the forest – the interconnectedness between trees and fungi. Mycol Res 102:1441–1449

Rodriguez RJ, Henson J, Volkenburgh EV et al (2008) Stress tolerance in plants via habitat-adapted symbiosis. ISME J 2:404–416

Saleem M, Ali MS, Hussain S et al (2007) Marine natural products of fungal origin. Nat Prod Rep 24:1142–1152

Schiel DR, Foster MS (2006) The population biology of large brown seaweeds: ecological consequences of multiphase life histories in dynamic coastal environments. Annu Rev Ecol Evol Syst 37:343–372

Schulz B, Draeger S, Cruz TE, Rheinheimer J, Siems S, Loesgen K, Bitzer J, Schloerke O, Zeeck A, Kock I, Hussain H, Dai J, Krohn K (2008) Screening strategies for obtaining novel, biologically active, fungal secondary metabolites from marine habitats. Bot Mar 51:219–234

Seymour FA, Cresswell JE, Fisher PJ et al (2004) The influence of genotypic variation on metabolite diversity in populations of two endophytic fungal species. Fungal Genet Biol 41:721–734

Shigemori H, Komatsu K, Mikamiaa Y et al (1999) Seragakinone A, a new pentacyclic metabolite from a marine-derived fungus. Tetrahedron 55:14925–14930

Solis MJL, Draeger S, dela Cruz TEE (2010) Marine-derived fungi from *Kappaphycus alvarezii* and *K. striatum* as potential causative agents of ice-ice disease in farmed seaweeds. Bot Mar 53:587–594

Stierle A, Strobel G, Stierle D (1993) Taxol and taxane production by *Taxomyces andreanae*, an endophytic fungus of Pacific Yew. Science 260:214–216

Sun X, Guo LD, Hyde KD (2011) Community composition of endophytic fungi in *Acer truncatum* and their role in decomposition. Fungal Divers 47(1):85–95

Suryanarayanan TS, Murali TS (2006) Incidence of *Leptosphaerulina crassiasca* in symptomless leaves of peanut in southern India. J Basic Microbiol 46:305–309

Suryanarayanan TS, Thennarasan S (2004) Temporal variation in endophyte assemblages of *Plumeria rubra* leaves. Fungal Divers 15:195–202

Suryanarayanan TS, Ravishankar JP, Muruganandam V (2010a) Drug discovery: going with the tide. Curr Sci 99:1308

Suryanarayanan TS, Venkatachalam A, Thirunavukkarasu N et al (2010b) Internal mycobiota of marine macroalgae from the Tamilnadu coast: distribution, diversity and biotechnological potential. Bot Mar 53:457–468

Tsuda M, Kasai Y, Komatsu K et al (2004) Citrinadin A, a novel pentacyclic alkaloid from amrine-derived fungus *Penicillium citrinum*. Org Lett 6:3087–3089

Unterseher M, Schnittler M (2010) Species richness analysis and ITS rDNA phylogeny revealed the majority of cultivable foliar endophytes from beech (*Fagus sylvatica*). Fungal Ecol 4:366–378

Vita-Marques AM, Lira SP, Berlinck RGS et al (2008) A multi-screening approach for marine-derived fungal metabolites and the isolation of cyclodepsipeptides from *Beauveria feline*. Quím Nova 31:1099–1103

Wagner BL, Lewis LC (2000) Colonization corn, *Zea mays*, by the endopathogenic fungus *Beauveria bassiana*. Appl Environ Microbiol 66:3468–3473

Wang S, Li X, Teuscher F et al (2006) Chaetopyranin, a benzaldehyde derivative, and other related metabolites from *Chaetomium globosum*, an endophytic fungus derived from the marine red alga *Polysiphonia urceolata*. J Nat Prod 69:1622–1625

Weber D (2009) Endophytic fungi, occurrence and metabolites. In: Anke T, Weber D (eds) The mycota XV. Springer, Berlin, Heidelberg

White JF Jr, Torres MS (2010) Is plant endophyte-mediated defensive mutualism the result of oxidative stress protection? Physiol Plant 138:440–446

Zhang Y, Li XM, Proksch P (2007a) Ergosterimide, a new natural Diels–Alder adduct of a steroid and maleimide in the fungus *Aspergillus niger*. Steroids 72:723–727

Zhang Y, Li XM, Wang CY et al (2007b) A new naphthoquinoneimine derivative from the marine algal-derived endophytic fungus *Aspergillus niger* EN-13. Chin Chem Lett 18:951–953

Zhang Y, Wang S, Li XM et al (2007c) New sphingolipids with a previously unreported 9-methyl-C20-sphingosine moiety from a marine algous endophytic fungus *Aspergillus niger* EN-13. Lipids 42:759–764

Zuccaro A, Mitchell JI (2005) Fungal communities of seaweeds. In: Dighton J, White JF, Oudeman P (eds) The fungal community: its organization and role in the ecosystem, 3rd edn. CRC, Boca Raton

Zuccaro A, Schulz B, Mitchell JI (2003) Molecular detection of ascomycetes associated with *Fucus serratus*. Mycol Res 107:1451–1466

Zuccaro A, Summerbell RC, Gams W et al (2004) A new *Acremonium* species associated with *Fucus* spp., and its affinity with a phylogenetically distinct marine *Emericellopsis* clade. Stud Mycol 50:283–297

Zuccaro A, Schoch CL, Spatafora JW, Kohlmeyer J, Draeger S, Mitchell JI (2008) Detection and identification of fungi intimately associated with the brown seaweed *Fucus serratus*. Appl Environ Microbiol 74:931–941

Chapter 4
Diversity and Biogeochemical Function of Planktonic Fungi in the Ocean

Guangyi Wang, Xin Wang, Xianhua Liu, and Qian Li

Contents

Abstract Microbial communities play critical biogeochemical roles in the functioning of marine ecosystems. Recent advances in molecular methods and environmental genomics have greatly advanced our understanding of microbial prokaryotes and their diversity and functional ecology in the world's oceans. Large populations of heterotrophic eukaryotes are well documented in the oceans and yet, their diversity and function remain relatively unknown. Particularly, large populations of planktonic fungi have long been known to exist in coastal and oceanic waters but the diversity and ecology of planktonic fungi remain one of the most under-studied microbial topics. Recent studies have revealed novel diversity and interesting ecological functions of planktonic fungi and suggest that they are a potentially important component in marine microbial food web. This chapter

G. Wang (✉)
School of Environment and Energy, Peking University Shenzhen Graduate School, Shenzhen 518055, China

Department of Oceanography, University of Hawaii at Manoa, Honolulu, HI 96822, USA
e-mail: guangyi@hawaii.edu; gywang@pkusz.edu.cn

C. Raghukumar (ed.), *Biology of Marine Fungi*,
Progress in Molecular and Subcellular Biology 53,
DOI 10.1007/978-3-642-23342-5_4,

will review the diversity and ecology of planktonic fungi in the world's oceans and discuss their significance in ocean carbon and nutrient cycling.

4.1 Introduction

Fungi have long been known to exist in marine environments. Obligate marine fungi are distinct from terrestrial and freshwater fungi in their taxonomy and morphology as well as their adaptations to an aquatic habitat (Kohlmeyer and Kohlmeyer 1979; Moss 1986; Jones and Mitchell 1996; Meyers 1996; Jones 2000). They are more precisely defined as fungi that can complete their entire life cycles within the sea, i.e., those that can grow and sporulate exclusively in a marine or estuarine habitat (Kohlmeyer and Volkmann-Kohlmeyer 2003). Facultative marine fungi refer those originating from freshwater or terrestrial environment that are also capable of growth and sporulation in the sea (Kohlmeyer and Kohlmeyer 1979). In addition, fungi from marine environments are often referred to as "marine-derived fungi," which include obligate and facultative marine fungi. In this chapter, the term "marine fungi" mostly refers to both obligate and facultative marine fungi unless otherwise specified. Marine fungi primarily include the *Ascomycetes*, the *Basidiomycetes*, and the *Chytridiomycetes* (Kohlmeyer and Kohlmeyer 1979; Kohlmeyer and Volkmann-Kohlmeyer 1991; Fell and Newell 1998; Hyde and Sarma 2000; Kohlmeyer et al. 2004). The *Oomycetes* and *Labyrinthulomycetes*, formerly known as pseudofungi or fungal-like organisms, were originally classified as fungi and have now been reclassified under the new Kingdom of Straminipila (Wang and Johnson 2009). Nevertheless, they are still commonly studied by marine mycologists and mycoplankton ecologists (Fell and Newell 1998; Raghukumar 2002; Kis-Papo 2005; Hibbett et al. 2007). Some marine fungi may occur in freshwater or terrestrial habitats. For example, both *Savoryella lignicola* and *Lignincola leavis* have been found in both marine and fresh habitats (Jones 2000). So far, there are about 467 described obligate or facultative higher marine fungi from 244 genera largely derived from marine substrata using cultivation-based methods (Kohlmeyer and Volkmann-Kohlmeyer 1991, 2003; Hyde and Sarma 2000; Rand 2000; Vrijmoed 2000; Kohlmeyer et al. 2004; Kis-Papo 2005; Pang and Mitchell 2005; Wang et al. 2008a, b). They are divided into a majority of *Ascomycota* (97%), a few *Basidiomycota* (~2%), and anomorphic fungi (<1%). The largest order of marine ascomycetes is the *Halosphaeriales*, with some 45 genera and 119 species. It is estimated that approximately 1.5 million fungal species exist, but only 70,000 species (less than 5%) have been described (Hawksworth 1991). Of these 70,000 species, 468 are marine fungi (0.67%). This suggests that, based on this ratio, there are a total of 10,029 species of marine fungi (Kis-Papo 2005). Hence, the vast majority of marine fungi remain to be discovered and described, particularly planktonic forms in the water column or symbiotic ones in marine animals. This chapter provides a review of the diversity and functional ecology of fungi in marine environments with a focus on water column.

4.2 Marine Fungal Taxonomy

Fungi have historically been identified and classified using morphological characteristics, such as sexual structure. However, fungi, especially microscopic forms, often have few useful morphological features and show pronounced morphological variability (Brasier 1997; Petersen and Hughes 1999; Höller et al. 2000; Burnett 2003). Many fungi are anamorphic without sexual production, yet possess a surprisingly high level of genetic variation (Harrington and Rizzo 1999; Talhinhas et al. 2002). In addition, an increasing number of morphologically indistinguishable (cryptic) species have recently been described by Grunig et al. (2007). Accordingly, the use of morphology for fungal identification and classification purposes can be severely biased. In the past decade, the ability to identify species at the molecular level has changed our understanding of the species concept for different groups of fungi (Hibbett et al. 2007). Particularly, molecular phylogenetic approaches avoid subjectivity in determining the limits of a species by relying on the concordance of more than one gene genealogy and have been approved as well suited for teleomorphic and anamorphic fungi (Taylor et al. 2000).

Ribosomal DNA (rDNA) sequence has been proven to be one of the most useful techniques in aiding fungal identification and resolving taxonomic positions at different levels (Hillis and Dixon 1991). The ribosomal gene cluster comprises three regions coding for the 5.8 S, 18 S, and 28 S rRNA. Some of the rDNA fragments are useful for delineation of higher taxonomic ranks, such as classes and phyla (5.8 rRNA gene, parts of the Large Subunit rRNA (LSU) 1 and 2 regions and Small Subunit rRNA (SSU) genes) (Bruns et al. 1991; Sugiyama 1994), and others are used for separation of genera and species (e.g., Internally Transcribed Spacer (ITS) 1 and 2 regions, Domain 2 of the LSU) (White et al. 1990; Bruns et al. 1991; Hillis and Dixon 1991; Sugiyama 1994). Some of these regions have successfully been used to study fungal communities in marine environments (Pang and Mitchell 2005; Gao et al. 2008, 2010; Li and Wang 2009).

In six decades of marine mycology research, identification of fungi derived from marine environments had largely used morphology-based approaches (Hyde 2002; Li and Wang 2009; Wang and Johnson 2009), which have significantly contributed to the development of marine mycology. However, the majority of fungi derived from marine environments (e.g., sponges) are mitosporic; about one-third of these fungi cannot be identified using the morphology-based approach (Morrison-Gardiner 2002). Furthermore, these approaches are well known to detect only a small fraction of marine fungal communities in natural environments. Hence, identification of marine fungi from natural environments is at best difficult, and in most cases not possible (Gao et al. 2008). Recently, a combination of the morphology-based approach with the molecular method has identified an unprecedented diversity of fungi from marine environments (e.g., sponges) (Li and Wang 2009; Wang and Johnson 2009). In particular, molecular methods have identified a novel diversity of unculturable fungal communities in marine habitats (Pang and Mitchell 2005; Gao et al. 2008, 2010).

4.3 Fungi in the Water Column

Planktonic fungi (also called mycoplankton) include filamentous free-living fungi and yeasts, and those associated with planktonic particles or phytoplankton (Wang and Johnson 2009; Gao et al. 2010). Fungi have long been known for their important role in processing detrital organic matter from terrestrial plants (Carlile et al. 2001). Although they occur in different marine ecosystems and have been postulated to play a major role in the detrital food web (Fell and Newell 1998; Raghukumar 2005; Gulis et al. 2006; Gessner et al. 2007), their role in decomposing marine organic matter and in marine biogeochemical cycles is still poorly understood in comparison with those in terrestrial environments (Fell and Newell 1998; Clipson et al. 2006; Gutierrez et al. 2010). Particularly, in coastal environments the productivity often exceeds the consumption of herbivores and a large fraction of primary organic matter become available to consumers as detritus (Newell 1982). In such ecosystems, productivity depends on the mineralization of wood, leaves, and other detrital plant materials. Fungi play an essential part in this process in terrestrial and freshwater systems (Hyde et al. 1998; Gulis et al. 2006). Hence, the ecological function of fungi in a water column can be significant, but has been largely ignored. To that end, this section is dedicated to the fungal communities (culturable and unculturable) and ecology of fungal populations in the water columns of the world's oceans.

4.3.1 Abundance and Existence of Living Fungi in Seawater

A typical milliliter of seawater contains over 10^3 fungal cells (or propagules) (Gao et al. 2008). Although marine yeasts are ubiquitous in seawater (Vogel et al. 2007; Kutty and Philp 2008; Chen et al. 2009), direct detection of living fungal mycelia using the Calcofluor White stain method has occurred only recently (Fig. 4.1). The recent discovery of the active forms of filamentous fungi in seawater provides an unequivocal piece of evidence for their potential significance in the marine detrital food web. In freshwater ecosystems, fungal biomass can account for as much as 18–23% of the total mass of detritus (Gessner and Chauvet 1994; Methvin and Suberkropp 2003; Gessner et al. 2007). Fungal biomass associated with leaf detritus reaches annual maximum values after leaf litter enters a stream and becomes colonized (Suberkropp 1991). Maximum fungal biomass associated with decaying leaves typically is one to two orders of magnitude greater than bacterial biomass found on the same leaves and, therefore, microbial biomass occurring in leaves is also dominated by fungi in freshwater ecosystems (Gulis et al. 2006). Recently, fungal mycelia, ranging from 1 to 3 μm in diameter and from 10 to >200 μm in length were observed as individual filaments or aggregates in the coastal upwelling ecosystem off central Chile (Gutierrez et al. 2010) and the Hawaiian coastal waters (Want et al. unpublished data) using the fluorescent

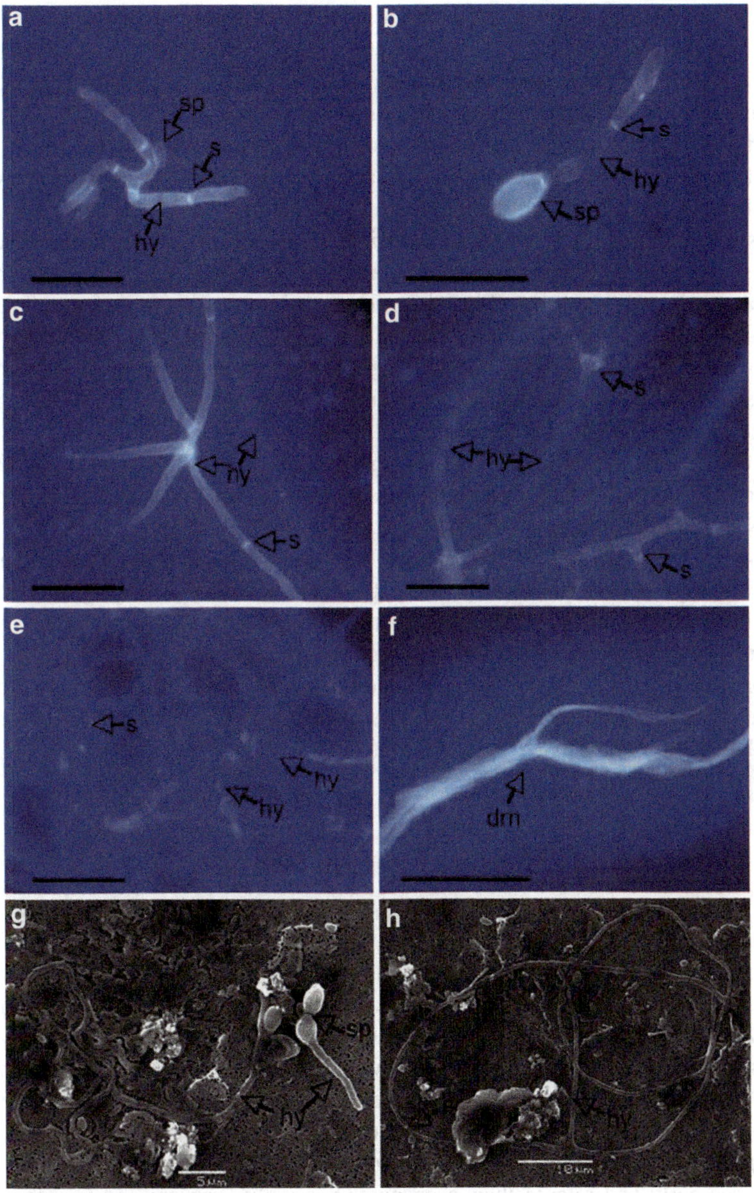

Fig. 4.1 Fungal mycelia detected by epifluorescence (**a–e**) and scanning electron microscopy (**g**, **h**) in the water column of the coastal upwelling ecosystem off central Chile during austral spring of 2008 (October) and summer of 2009 (early March). Fungal structures were clearly distinguished from detrital material (**f**). Scale bars correspond to 10 μm. *sp* spores; *hy* hyphae; *s* hyphal septa; *dm* detrital material (Gutierrez et al. 2011)

staining method. These filaments or aggregates can reach up to 20 μm in diameter and over 50 μm long, similar to fungal mycelia detected in deep-sea sediments in the Indian Basin (Damare and Raghukumar 2008) and water-stable aggregates associated with mycorrhizal fungi in soils (Tisdall and Oades 1982). Because organic aggregates in seawater represent a major growth habitat for planktonic microbial communities, the combined metabolic activities of fungi and prokaryotic microbes can promote a highly efficient mineralization of particulate organic matter to dissolved organic matter in seawater (Kiorboe and Jackson 2001; Gutierrez et al. 2010). Furthermore, fungal biomass displays spatial and temporal variations in the coastal marine ecosystems (Fig. 4.2). Fungal biomass ranged from 0.03 to 0.12 μg C/l of seawater in the top 15 m with a decreased trend toward bottom water to values between 0.006 and 0.01 μg C/l of seawater in the coastal upwelling ecosystem off central Chile during the winter (Gutierrez et al. 2010). During the summer, fungal biomass in the coastal water can reach ~6 μg C/l of seawater at the depth of 7 m and range from 0.19 to 0.44 μg C/l of seawater below 30 m (Fig. 4.2). In addition, the vertical profile of fungal biomass concurred with those of chlorophyll *a* and physical parameters of the water columns (e.g., temperature and oxygen) (Fig. 4.2c, d). Hence, like heterotrophic microplankton, fungi are active in the water column and respond to primary production activity and organic matter availability (Gutierrez et al. 2010). Further investigation of the biological activities of planktonic fungal communities could significantly improve our understanding of ecological function of marine microbial communities in the ocean carbon cycling and decipher the potentially important role of fungi in the current model of marine microbial food web.

4.4 Diversity of Culturable Fungi in Seawater

Because of the unique physio-chemical environments in seawater, planktonic fungi can be quite distinct from those occurring in the marine substrata. Several marine fungi, including *Antennospora quadricornuata*, *Arenariomyces* spp., *Corollospora* spp., and *Torpedospora radiate*, are commonly found in coastal and oceanic waters (Jones and Hyde 1990; Jones 2000). However, some fungi, such as *Lignincola laevis* and *Periconia prolifica*, occur in both seawater and substrata (i.e., mangroves). Because of their adaption to the oceanic environments, planktonic fungi (e.g., *Halosphaeriales*) can grow under submerged conditions and usually have asci that deliquesce early and release their ascospores passively (Fazzani and Gareth Jones 1977). They even produce ascospores with elaborate appendages, which enhance their floatation, impaction and attachment to planktonic particles (Jones 2000).

Compared with fungi in other marine habitats, the fungal populations in seawater are poorly investigated. As of now, over 186 fungal species have been isolated from seawater and are members of the *Asomycota*, *Basidiomycota*, and *Zygomycota* phyla (Kutty and Philp 2008; Roth et al. 1964; Steele 1967; Zhang et al. 1989). The first extensive investigation of fungal populations in seawater occurred in the

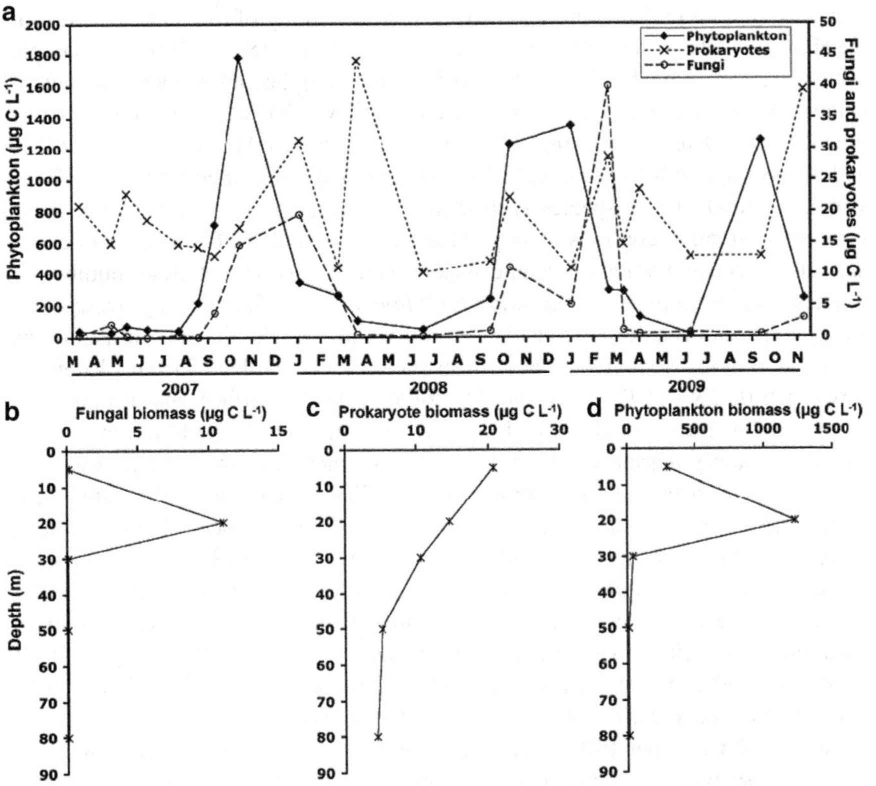

Fig. 4.2 Temporal variation of phytoplankton, fungal and bacterial biomass in surface waters (**a**) and vertical biomass profiles of fungi (**b**), prokaryote (**c**) and phytoplankton (**d**) during austral spring (October) of 2008 at station 18. The coefficient of variation of duplicate samples was less than 20% for fungal biomass (Gutierrez et al. 2011)

northwestern subtropical Atlantic Ocean using cultivation methods (Roth et al. 1964). All eulittoral water samples contained relatively high fungal populations with a minimum of ten different fungal species, ranging from 187 to 296 colonies per liter. Water samples obtained more distant from land and in deep waters displayed a more than tenfold decrease in average fungal population density and about a 50% reduction in the number of species. A similar trend was observed with progression into abyssal regions. Fungi were found in 80% of the 227 water samples collected from offshore stations (i.e., at least 3 miles from land), with most of these yielding less than five viable propagules per liter and giving rise to three or fewer fungal species. The deepest sample, which yielded *Deuteromycetes* (also called the Fungi Imperfecti because they have no sexual reproduction), was collected at the depth of 4,450 m. Three genera, *Cladosporium* spp., *Alternaria* spp., and *Nigrospora* spp., were the dominant fungi in eulittoral and oceanic waters. Both *Aureobasidium pullulans* and *Dendryphiella arenaria* were commonly present in seawater samples collected in this region. In another study of fungi in the waters of the Hawaiian Islands and elsewhere in the central Pacific (Steele 1967), all sample stations yielded fungal colonies. Of the 59 water samples, 44 of them yielded 50 colonies or less and the remainder gave count of over 50. The average number of fungal isolates from these samples ranged from 0.06 to 3.94 isolates per milliliter with the average of 0.14 per sample. The number of species ranged from 1 to 17 per sample. A total of 126 species representing 59 genera were isolated from these samples. Generally, coastal waters yielded a higher number of isolates than either the surf or oceanic waters; oceanic waters gave a very low average number of isolates. Furthermore, *Aureobasidium pullulans* and *Rhodotorula* spp. were commonly found in oceanic waters and also among those predominant in all isolations from water. In the South China Sea, water samples were collected from 24 stations between 6°02′03″–11°00′11″N and 112°24′55″E at the surface and at depths of 25 m and 100 m. Culturable fungi derived from the membrane filtration method contained mainly marine yeasts, but very few filamentous forms (Vrijmoed et al. 1993). The numbers of colony forming units (CFUs) varied among the stations for both filamentous and yeast forms. The fungal counts (CFUs per 100 ml seawater) decreased with increasing depth with average values of 196, 123, and 104 CFUs per 100 ml seawater in water samples collected at the surface and at the depths of 25 and 100 m, respectively. Among the cultural fungi, yeasts were the dominant populations at each individual depth and present at almost all three depths of all the stations and only water samples collected at two stations at 100 m did not yield yeast forms. The yeast counts decreased with increasing depth with the average counts of 196 CFUs per 100 ml seawater collected from the surface and 104 CFUs per 100 ml seawater collected at 100 m. On the other hand, the filamentous forms were much lower in numbers and were identified at both the surface and at 25 m collected from 11 stations (45.8%), and at 100 m from 9 stations (37.5%). Filamentous forms of these cultural fungi were dominated by *Aspergillus* spp. and *Cladosporium* spp. with many unidentified mycelial and non-sporulating state. Furthermore, fungi were also identified in the water samples collected from the ocean area of the Northwest Antarctic Peninsula (Zhang et al. 1989). Finally, a temporal study of fungal population dynamics in the Lagos lagoon of Nigeria (north of the Gulf of Guinea, Atlantic Ocean) revealed the seasonal variations of fungal populations (Akpata and Ekundayo 1983). The increased fungal populations were observed in the rainy season (March–October) when salinity was low and the concentration of organic matter was high. The low fungal populations in the lagoon occurred in the dry season (November–February) when salinity was generally high and the concentration of organic matter was low. All the isolated fungi had terrestrial origins with *Aspergillus* spp. and *Penicillium* spp. as the dominant fungal populations. Fungal populations in the lagoon were negatively correlated with water turbidity. Existing studies revealed higher yeast populations in the seawater samples than those of filamentous forms of fungi in the seawater samples (Vrijmoed et al. 1993; Kutty and Philp 2008). Generally, populations of planktonic fungi decrease with increased distance from land and increased depths, with certain species dominating in heavily contaminated seawaters. As such, some of the planktonic fungi could merely be contaminants from terrestrial sources, surviving passively in the water column. Nevertheless, several lines of evidence suggest the existence of indigenous marine fungi in the water columns of the world's oceans (Gao et al. 2008).

Fungal growth in seawater is influenced by various physio-chemical parameters. Temperature, organic matter content, and total ionic concentration influence the population of *Saprolegniacea* in tropical waters (Ritchie 1959). *Zygomycetes* (formerly called *Phycomycetes*) were only observed in estuarine waters of low salinity (<0.5%) (Johnson and Sparrow 1961). The diversity of culturable marine fungi is likely controlled by a wide range of biological, chemical, and physical factors (Jones 2000). Because the cultured fungi only represent a small fraction (less than 5%) of all fungal communities, the intrinsic relationship between physio-chemical environmental factors and planktonic fungi is largely unknown.

4.4.1 Diversity of Uncultured Fungi in Water Column

Molecular methods have been commonly used to reveal or survey uncultured microbial communities. These methods combine PCR-amplified ribosomal RNA gene (or spacer regions) sequences with screening strategies (e.g., DGGE) for quick assessment of diversity and identification of molecular phylotypes in environmental samples (Pang and Mitchell 2005). These methods were originally developed for the assessment of bacterial communities and then adapted for the analysis of fungal communities. Indeed, the recent development and application of molecular methods have significantly advanced our understanding of the diversity and ecology of fungal communities in terrestrial habitats (e.g., soils). Application of molecular approaches enables unculturable fungal communities to be directly detected from environmental samples. The most commonly used molecular methods include, but are not limited to, rRNA clone library construction, terminal restriction fragment length polymorphism (T-RFLP), denaturing gradient gel electrophoresis (DGGE), temperature gradient gel electrophoresis (TGGE), amplified rDNA restriction analysis (ARDRA), and amplified ribosomal intergenic spacer analysis (ARISA) (Gao et al. 2008). These methods have been mainly used to analyze fungal communities associated with marine substrata and yielded novel fungal diversity (Pang and Mitchell 2005). Application of these methods for the analysis of fungal communities in seawater is rare.

The first molecular analysis of fungal communities in water columns was carried out for water samples collected from the Hawaiian coast (Gao et al. 2010). In that analysis, DGGE, in conjunction with clone library construction, revealed interesting spatial diversity and novel phylotypes of planktonic fungi. Mycoplankton communities displayed distinct lateral and vertical variations in their diversity and composition. Nearshore waters had greater diversity and species richness than those of oceanic water samples (Fig. 4.3). However, no difference in the diversity of planktonic fungal communities was observed among

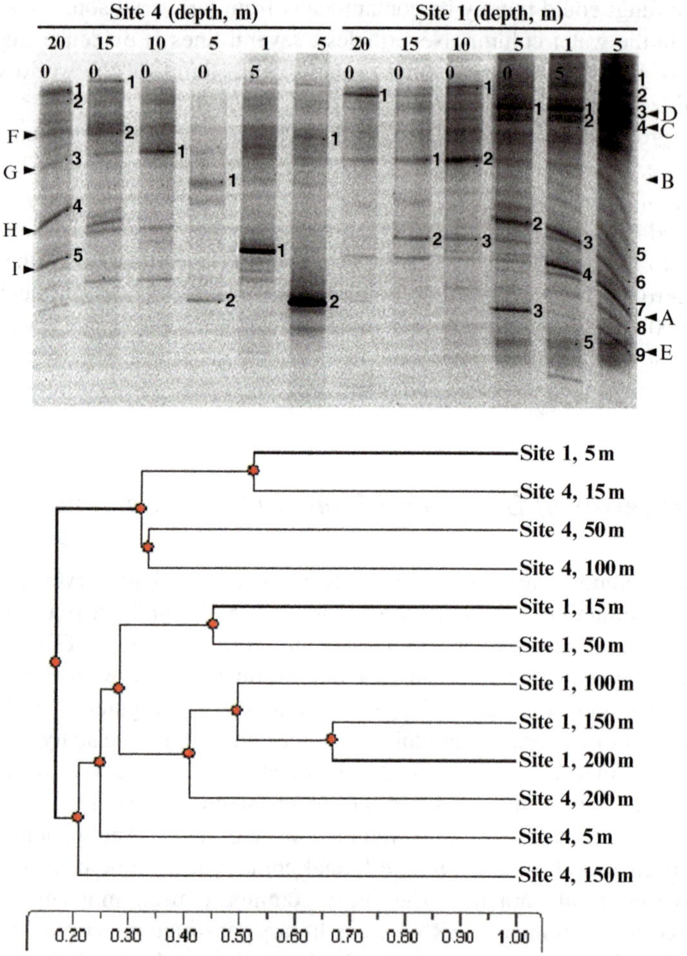

Fig. 4.3 DGGE analysis of planktonic fungi derived from the coastal waters of the island of Oahu (**a**). DGGE bands cloned are side-labeled with dot and number. UPMA tree based on matrix of distance calculated from the difference in DGGE band patterns among water samples from sites 1 (coastal) and 4 (open ocean) (**b**) (Gao et al. 2010)

nearshore and open ocean stations for water samples from the depths below 150 m. Interestingly, DGGE analysis of water samples from the coastal zone off central Chile revealed the similar trend of lateral variations with higher number of fungal phylotypes at surface of coastal stations than in offshore areas (Gutierrez et al. 2010) (Fig. 4.4). Of the 46 identified fungal phylotypes in the Hawaiian coastal waters, 42 are members of *Basidiomycota* and the rest belong to *Ascopmycota* (Fig. 4.5). All the ascomycete phylotypes are members of the class *Dothideomycetes*. In particular, one of the ascomycete phylotypes was

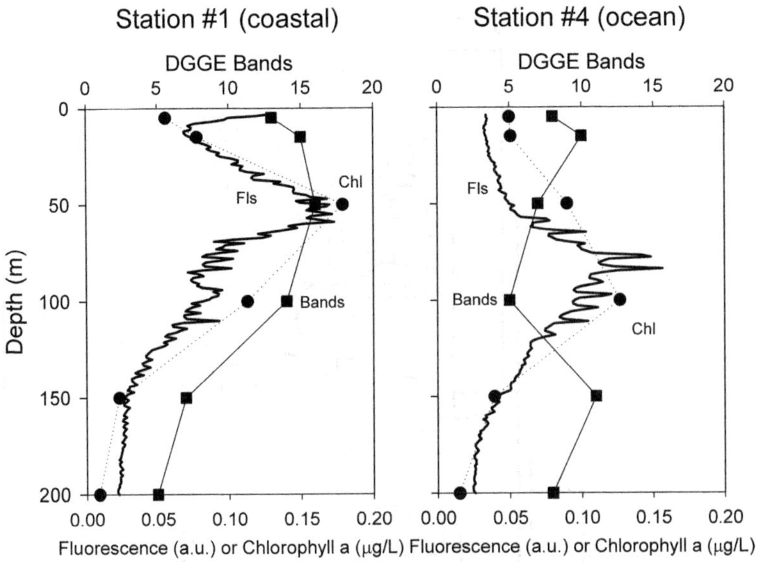

Fig. 4.4 Vertical profile of DGGE band numbers, fluorescence, and chlorophyll-*a* of Hawaiian waters (Gao et al. 2010)

closely matched to the marine yeast *Aureobasidium pullulans*, a complex commonly found in marine environments. The majority (93%) of basidiomycete phylotypes were not affiliated with any member of the described fungal taxa and are likely to be new fungal phylotypes (Fig. 4.5). Furthermore, the results of this study suggest that high primary production likely supports high population of fungal communities and plays a more dominant role in determining mycoplankton diversity in deep-water samples than in shallow areas (Figs. 4.2 and 4.4). Biological, geochemical, and chemical factors, which govern the diversity and composition of mycoplankton in the nearshore waters, may differ from those in the open ocean (Fig. 4.4). Because of the dynamic nature of ocean ecosystems, the relationship between planktonic fungi and their physio-chemical environments can be very complicated. Overall, the rich information on mycoplankton derived from molecular analysis confirms the significance and necessity for the further exploration of marine mycoplankton in the world's oceans. At the same time, these molecular studies clearly indicate that the development of fungal primers with a satisfactory degree of success in discriminating against other marine organisms while maintaining a broad range of compatibility with planktonic fungal communities can significantly benefit the study of unculturable fungi in water column. Novel fungal phylotypes have been detected from the deep-sea waters (3,000 and 4,000 m) with yeasts as the predominant forms (Bass et al. 2007). However, molecular surveys indicate that fungi are relatively rare in the deep-sea habitats (Bass et al. 2007; Raghukumar et al. 2010).

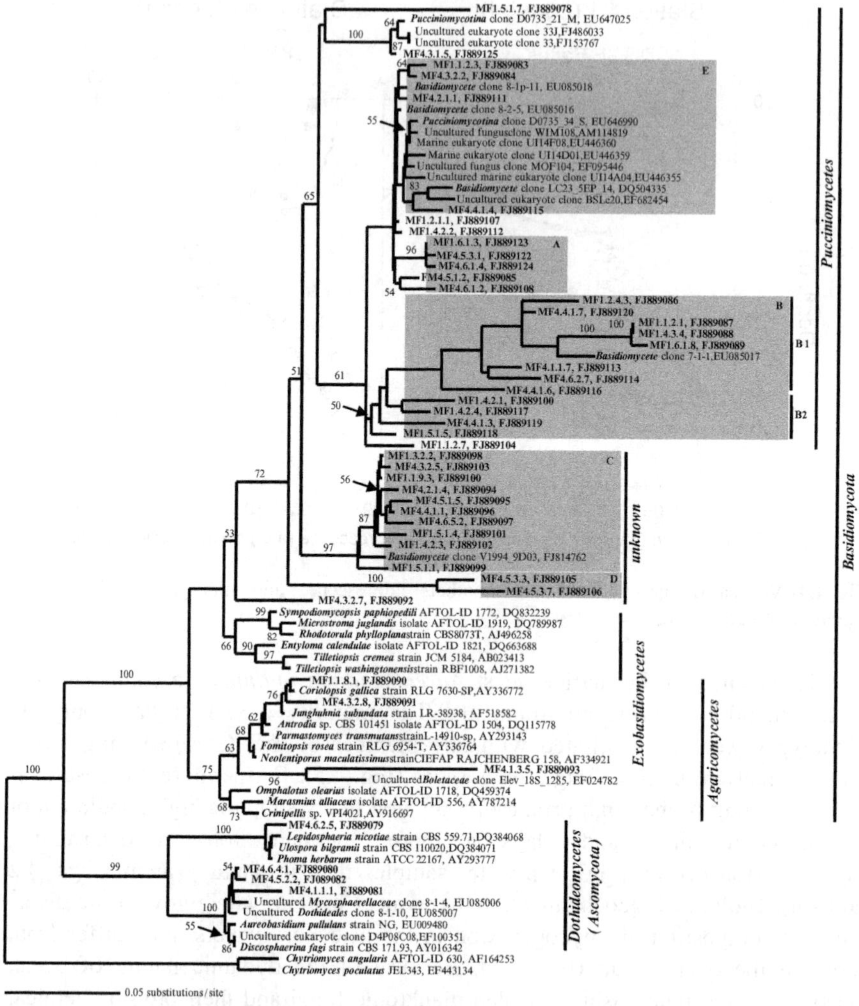

Fig. 4.5 Neighbor-joining tree based on fungal 18 S rRNA gene sequences derived from Hawaiian waters (Gao et al. 2010)

4.5 Biogeochemical Functions of Planktonic Fungi

The ecological importance of fungi in the marine ecosystems is often underestimated or ignored completely, and yet they represent a diverse range of saprobes, pathogens, and symbionts that form an integral part of coastal systems (Newell 1994; Hyde et al. 1998). Of their many ecological functions, marine fungi are best known to be major decomposers of woody and herbaceous substrates in coastal marine ecosystems (Hyde et al. 1998). However, these functions are largely derived from marine substratum studies or in vitro data and remain to be proven in the water column (Fell and Newell 1998; Raghukumar 2004).

4.5.1 Nutrient Cycling

Fungal distribution in marine ecosystems is largely dependent on the quality and availability of organic carbon (Clipson et al. 2006). Fungi are well known to secrete a battery of hydrolytic enzymes with particular capability to degrade and recycle organic matter, including recalcitrant materials that are less suitable for mineralization by bacteria and other saprotrophic groups. It should be recognized that bacteria and fungi can occupy different ecological niches in detritus (Newell 1994; Raghukumar 2005). Bacteria live on the surface of particulate detritus and have high substrate affinity with the capability to utilize even low concentrations of nutrients and to multiply quickly (Raghukumar 2005). Furthermore, small detrital particles of high surface to volume ratios tend to support more bacteria than fungi. Although filamentous fungi grow more slowly when compared to bacteria, they indeed have the ability to penetrate relatively persistent particulate detritus much more efficiently than the latter by virtue of their filamentous habit and apical growth (Raghukumar 2005). Fungi, including filamentous *Oomycetes*, are particularly adept at utilizing large lignocellulose-predominated substrates of high C:N ratio. The unicellular yeasts are considered to occupy a similar niche as bacteria and live on the surface of particles, or even in the presence of high concentrations of dissolved organic matter (DOM) (Raghukumar 2005). Fungal biomass was significantly greater than bacteria on leaves and wood, confirming the predominance of fungal biomass in larger sized organic matter in freshwater detrital ecosystems (Findlay et al. 2002).

Fungi have generally been considered unimportant in open water detrital processes. Several lines of evidence indicate that thraustochytrids fungi play a major role in detrital dynamics of water columns (Raghukumar 2002; also, refer the corresponding chapter on thraustochytrids). Open ocean waters are mostly low in available organic matter with low fungal biomass. Nevertheless, open ocean waters can have relatively high levels of organic matter when algal blooms occur. Little effort has been made to study fungal biomass of open ocean waters under the algal bloom conditions. On the other hand, coastal seawaters contain much higher biomass availability and complexity, particularly in the coastal areas with highly productive habitats such as salt marshes and mangroves. Application of the fungal sterol, ergosterol, in the quantification of fungal biomass associated with decaying plant liter and the technique for measuring in situ instantaneous growth rates of fungi from rates of [^{14}C]acetate incorporation into ergosterol have yielded rich information on the assessment of the magnitude of fungal contributions to the cycling of carbon and the flow energy in freshwater ecosystems (Gulis et al. 2006). These studies indicated that fungi are a significant decomposer of particulate detritus and play a key role in detrital food webs of freshwater ecosystems. It would not be surprising to find that fungi play a similar ecological role in the marine detrital food webs using these approaches. Indeed, direct detection in conjunction with microscopic counting indicates that fungal biomass in surface waters of the upwelling ecosystems off central-southern Chile is comparable to that of prokaryotes (Bacteria plus Archea) during active upwelling period (Gutierrez et al. 2011). In general, fungi can convert plant carbon into fungal biomass and also into CO_2 as a result of their respiratory activities (Gessner et al. 2007). Additionally, fungal decomposition of plant liter and feeding activities of detritivore consumers (invertebrates) can facilitate in the export of plant carbon as either fine particulate organic matter (FPOM) or dissolved organic matter (DOM) (Gulis et al. 2006). The role of fungi in the biogeochemical cycle of nitrogen has been grossly underestimated and their biomass can account for virtually all of the nitrogen present in decaying standing plant biomass in salt marshes (Newell 1996). Overall, studies of planktonic fungi in nutrient recycling are rare and would be an interesting research topic for future study.

4.5.2 Marine Microbial Food Webs

Large populations of fungal cells exist in coastal seawaters. Several lines of evidence support that planktonic fungi play an important role within the microbial food web in the coastal oceans (Gao et al. 2008). First, fungi have been reported as parasites of large marine phytoplankton (e.g., diatoms and dinoflagellates) (Johnson and Sparrow 1961; Drebes 1974; Raghukumar 1986; Elbrachter and Schnepf 1998; Ramaiah 2006; Wang and Johnson 2009). Some of these parasitic fungi can infect one and/or multiple marine phytoplankton species, up to 75% of the total natural marine phytoplankton populations (Tillmann et al. 1999). The fast-growing fungal parasites can consume the major part of the host phytoplankton within the water column, imposing a large impact on trophic web structure by directly contributing to phytoplankton lysis or death (Walsh 1983). Therefore, fungal parasites can be successful competitors against zooplankton in controlling energy flow and food web dynamics (Tillmann et al. 1999). Second, infection by planktonic fungal parasites can be important in regulating the dynamics of algal blooms. For example, parasitic *oomycetes* and/or chytrids have been observed to infect *Pseudo-nitzschia multiseries* and *P. pungens* cells in Cardigan Bay of Eastern Prince Edward Island (PEI), Canada (Bates et al. 1998; Elbrachter and Schnepf 1998). Lastly, zoospores of chytrid fungi and fungal-like *oomycetes* contain large amount of polyunsaturated fatty acids (PUFAS), docosahexaenoic acid (DHA), and eicosapentaenoic acid (EPA), with a high cholesterol concentration (Raghukumar et al. 1994; Raghukumar 2002; Kagami et al. 2007). Spores of these fungi (2–3 μm) are well within the preferred range of food particle size and can be grazed efficiently by zooplankton (Geller and Müller 1981; Kagami et al. 2007). Evidence accumulating from some investigations supports the idea that these spores are food of good quality for zooplankton (Müller-Navarra et al. 2000). During an epidemic, the density of these spores can be on the same order of magnitude or even more than the density of edible phytoplankton cells (i.e., several thousands per milliliter) in freshwater ecosystems (Kagami et al. 2007). Hence, planktonic fungi may enhance the growth and reproduction of some zooplankton, especially when large inedible phytoplankton species dominate the marine phytoplankton community (Kagami et al. 2007).

4.6 Conclusion

Evidence accumulating from culturable and molecular analyses clearly indicates that large populations of fungi with novel diversity exist in the water columns of the world's oceans. These planktonic fungi display interesting temporal and spatial (lateral and vertical) variations. Higher diversity and a greater abundance of fungi are found in the surface and coastal waters than in open-ocean and deep water samples. The diversity and abundance of planktonic fungi positively correlate with the primary production and availability of organic matter. The physio-chemical parameters of seawater (e.g., temperature and salinity) also impact the distribution of planktonic fungal communities. Several lines of evidence also support the idea that planktonic fungi may play significant role in microbial food webs through parasitism of marine phytoplankton. Given the significant contribution of fungi in nutrient (e.g., carbon and nitrogen) cycling of freshwater ecosystems, fungi may play an important role in ocean nutrient cycles, particularly in the detritus marine ecosystems (e.g., coastal waters). Overall, our understanding of the diversity and functional aspects of planktonic fungi is rather limited. Further investigation of this important, but largely understudied, eukaryotic microbial group will surely contribute to the global efforts to understand the role of microbes in ocean carbon cycling as well as global climate changes.

References

Akpata TVI, Ekundayo JA (1983) Occurrence and periodicity of some fungal populations in the Lagos lagoon. Trans Br Mycol Soc 80:347–352

Bass D, Howe A, Brown N, Barton H, Demidova M, Michelle H et al (2007) Yeast forms dominate fungal diversity in the deep oceans. Proc R Soc Lond B Biol Sci 274:3069–3077

Bates SS, Garrison DL, Horner RA (1998) Bloom dynamics and physiology of domoic-acid-producing *Pseudo-nitzschia* species. In: Anderson DM, Cembella AD, Hallegraeff GM (eds) Physiological ecology of harmful algal blooms. Springer, Heidelberg

Brasier CM (1997) Fungal species in practice: identifying species units in fungi. In: Claridge MF, Dawah HA, Wilson MR (eds) Species: the units of biodiversity, Systematics association special volume series 54. Chapman & Hall, London

Bruns TD, White TJ, Taylor JW (1991) Fungal molecular systematics. Annu Rev Ecol Syst 22:525–564

Burnett J (2003) Fungal populations and species. Oxford University Press, New York

Carlile MJ, Watkinson SC, Gooday GW (2001) The fungi. Academic Press, San Diego
Chen YS, Yanagida F, Chen LY (2009) Isolation of marine yeasts from coastal waters of northeastern Taiwan. Aquat Biol 8:55–60
Clipson N, Otte M, Landy E (2006) Biogeochemical roles of fungi in marine and estuarine habitats. In: Gadd GM (ed) Fungi in biogeochemical cycles. Cambridge University Press, New York
Damare S, Raghukumar C (2008) Fungi and macroaggregation in deep-sea sediments. Microb Ecol 56:168–177
Drebes G (1974) Marines phytoplankton. Georg Thieme-Verlag, Stuttgart
Elbrachter M, Schnepf E (1998) Parasites of harmful algae. In: Anderson DM, Cembella AD, Hallegraeff GM (eds) Physiological ecology of harmful algal blooms. Springer, Heidelberg
Fazzani K, Gareth Jones EB (1977) Spore release and dispersal in marine and brackish water fungi. Mater Org 12:235–248
Fell JW, Newell SY (1998) Biochemical and molecular methods for the study of marine fungi. In: Cooksey KE (ed) Molecular approaches to the study of the ocean. Chapman & Hall, London, pp 259–283
Findlay S, Tank J, Dye S, Valett HM, Mulholland PJ, McDowell WH et al (2002) A cross-system comparison of bacterial and fungal biomass in detritus pools of headwater streams. Microb Ecol 43:55–66
Gao Z, Li BL, Zheng CC, Wang G (2008) Molecular detection of fungal communities in the Hawaiian marine sponges *Suberites zeteki* and *Mycale armata*. Appl Environ Microbiol 74:6091–6101
Gao Z, Johnson ZI, Wang G (2010) Molecular characterization of the spatial diversity and novel lineages of mycoplankton in Hawaiian coastal waters. ISME J 4:111–120
Geller W, Müller H (1981) The filtration apparatus of cladocera – filter mesh-sizes and their implications on food selectivity. Oecologia 49:316–321
Gessner MO, Chauvet E (1994) Importance of stream microfungi in controlling breakdown rates of leaf litter. Ecology 75:1807–1817
Gessner MO, Gulis V, Kuehn K, Chauvet E, Suberkropp K (2007) Fungal decomposer of plant litter in aquatic ecosystems. In: Kubicek CP, Druzhinina IS (eds) The mycota. Springer, Berlin
Grunig CR, Patrick BC, Duo A, Sieber TN (2007) Suitability of methods for species recognition in the *Phialocephala fortinii-Acephala applanata* species complex using DNA analysis. Fungal Genet Biol 44:773–788
Gulis V, Kuehn K, Suberkropp K (2006) The role of fungi in carbon and nitrogen cycles in freshwater ecosystems. In: Gadd GM (ed) Fungi in biogeochemical cycles. Cambridge University Press, New York
Gutierrez MH, Pantoja S, Quinones RA, Gonzalez RR (2010) First record of filamentous fungi in the coastal upwelling ecosystem off central Chile. Gayana 74:66–73
Gutierrez MH, Pantoja S, Tejos E, Quinones RA (2011) The role of fungi in processing marine organic matter in the upwelling ecosystem off Chile. Mar Biol 158:205–219
Harrington TC, Rizzo DM (1999) Defining species in the fungi. In: Worrall JJ (ed) Structure and dynamics of fungal populations. Kluwer Press, Dordrecht, The Netherlands
Hawksworth DL (1991) The fungal dimension of biodiversity: magnitude, significance and conservation. Mycol Res 95:641–655
Hibbett DS, Binder M, Bischoff JF, Blackwell M, Cannon PF, Eriksson OE et al (2007) A higher-level phylogenetic classification of the fungi. Mycol Res 111:509–547
Hillis DM, Dixon MT (1991) Ribosomal DNA molecular evolution and phylogenetic inference. Q Rev Biol 66:411–454
Höller U, Wright AD, Matthee GF, König GM, Draeger S, Aust HJ, Schulz B (2000) Fungi from marine sponges: diversity, biological activity and secondary metabolites. Mycol Res 104:1354–1365
Hyde KD (2002) Fungi in marine environments. Fungal Diversity Press, Hong Kong
Hyde KD, Sarma VV (2000) Pictorial key to higher marine fungi. In: Hyde KD, Pointing SB (eds) Marine mycology: a practical approach. Fungal Diversity Press, Hong Kong

Hyde KD, Jones EBG, Leano E, Pointing SB, Poonyth AD, Vrijmoed LLP (1998) Role of fungi in marine ecosystems. Biodivers Conserv 7:1147–1161
Johnson TW, Sparrow FK (1961) Fungi in oceans and estuaries. J. Cramer Publisher, Weinheim
Jones EBG (2000) Marine fungi: some factors influencing biodiversity. Fungal Divers 4:53–73
Jones EBG, Hyde KD (1990) Observations on poorly known mangrove fungi and a nomenclatural correction. Mycotaxon 37:197–201
Jones EBG, Mitchell JI (1996) Biodiversity of marine fungi. In: Cimerman A, Gunde-Cimerman N (eds) Biodiversity: international biodiversity seminar. National Institute of Chemistry and Slovenia National Commission for UNESCO, Ljubljana
Kagami M, de Bruin A, Ibelings BW, Van Donk E (2007) Parasitic chytrids: their effects on phytoplankton communities and food-web dynamics. Hydrobiologia 578:113–129
Kiorboe T, Jackson GA (2001) Marine snow, organic solute plumes, and optimal chemosensory behavior of bacteria. Limnol Oceanogr 46:1309–1318
Kis-Papo T (2005) Marine fungal communities. In: Dighton J, White JF, Oudemans P (eds) The fungal community: its organization and role in the ecosystem. Taylor & Francis, Boca Raton, pp 61–92
Kohlmeyer J, Kohlmeyer E (1979) Marine mycology: the higher fungi. Academic Press, New York; London
Kohlmeyer J, Volkmann-Kohlmeyer B (1991) Illustrated key to the filamentous higher marine fungi. Bot Mar 34:1–35
Kohlmeyer J, Volkmann-Kohlmeyer B (2003) Mycological research news. Mycol Res 107:385–387
Kohlmeyer J, Volkmann-Kohlmeyer B, Newell SY (2004) Marine and estuaring mycelial Eumycota and Oomycota. In: Mueller GM, Bills GG, Foster MS (eds) Biodiversity of fungi: inventory and monitoring methods. Elsevier Academic Press, New York
Kutty SN, Philp R (2008) Marine yeasts – a review. Yeast 25:465–483
Li Q, Wang G (2009) Diversity of fungal isolates from three Hawaiian marine sponges. Microb Res 164:233–241
Methvin BR, Suberkropp K (2003) Annual production of leaf-decaying fungi in 2 streams. J N Am Benthol Soc 22:554–564
Meyers SP (1996) Fifty years of marine mycology: highlights of the past, projections for the coming century. SIMS News 46:119–127
Morrison-Gardiner S (2002) Dominant fungi from Australian coral reefs. Fungal Divers 9:105–121
Moss ST (1986) The biology of marine fungi. Cambridge University Press, Cambridge
Müller-Navarra DC, Brett MT, Liston AM, Goldman CR (2000) A highly unsaturated fatty acid predicts carbon transfer between primary producers and consumers. Nature 403:74–77
Newell RC (1982) The energetics of detritus utilization in coastal lagoons and nearshore waters. In: Laserre P, Postma H (eds) Proceedings of international symposium on coastal lagoons, Oceanologica Acta (special issue) Bordeaux, France, 8–14 September 1981
Newell SY (1994) Ecomethodology for organoosmotrophs: prokaryotic unicellular versus eukaryotic mycelia. Microb Ecol 28:151–157
Newell SY (1996) Established and potential impacts of eukaryotic mycelial decomposers in marine/terrestrial ecotones. J Exp Mar Biol Ecol 200:187–206
Pang KL, Mitchell JI (2005) Molecular approaches for assessing fungal diversity in marine substrata. Bot Mar 48:332–347
Petersen RH, Hughes KW (1999) Species and speciation in mushrooms: development of a species concept poses difficulties. Biosci 49:440–452
Raghukumar C (1986) Fungal parasites of the marine green-algae, *Cladophora* and *Rhizoclonium*. Bot Mar 29:289–297
Raghukumar S (2002) Ecology of the marine protists, the *Labyrinthulomycetes* (*Thraustochytrids* and *Labyrinthulids*). Eur J Protistol 38:127–145
Raghukumar S (2004) The role of fungi in marine detrital processes. In: Ramaiah N (ed) Marine microbiology: facets and opportunities. National Institute of Oceanography, Goa

Raghukumar S (2005) The role of fungi in marine detrital processes. In: Ramaiah N (ed) Marine microbiology: facts and opportunities. National Institute of Oceanography, Goa

Raghukumar S, Sharma S, Raghukumar C, Sathepathak V, Chandramohan D (1994) Thraustochytrid and fungal component of marine detritus IV. Laboratory studies on decomposition of leaves of the mangrove Rhizophora apiculata blume. J Exp Mar Biol Ecol 183:113–131

Raghukumar C, Damare SR, Singh P (2010) A review on deep-sea fungi: occurrence, diversity and adaptations. Bot Mar 53:479–492

Ramaiah N (2006) A review on fungal diseases of algae, marine fishes, shrimps and corals. Indian J Mar Sci 35:380–387

Rand TG (2000) Diseases of animals. In: Hyde KD, Pointing SB (eds) Marine mycology: a practical approach. Fungal Diversity Press, Hong Kong

Ritchie DON (1959) The effect of salinity and temperature on marine and other fungi from various climates. Bull Torrey Bot Club 86:367–373

Roth FJ, Orpurt PA, Ahearn DG (1964) Occurrence and distribution of fungi in a subtropical marine environment. Can J Bot 42:375–383

Steele CW (1967) Fungus populations in marine waters and coastal sands of the Hawaiian, Line, and Phoenic Islands. Pac Sci 21:317–331

Suberkropp K (1991) Relationships between growth and sporulation of aquatic *Hyphomycetes* on decomposing leaf litter. Mycol Res 95:843–850

Sugiyama J (1994) Fungal molecular systematics: towards a phylogenetic classification for the fungi. Nippon Nogeikagaku Kaishi 68:48–53

Talhinhas P, Sreenivasaprasad S, Neves-Martins J, Oliveira H (2002) Genetic and morphological characterization of *Colletotrichum acutatum* causing anthracnose of lupins. Phytopathology 92:986–996

Taylor JW, Jacobson DJ, Kroken S, Kasuga T, Geiser DM, Hibbett DS, Fisher MC (2000) Phylogenetic species recognition and species concepts in fungi. Fungal Genet Biol 31:21–32

Tillmann U, Hesse KJ, Tillmann A (1999) Large-scale parasitic infection of diatoms in the North Frisian Wadden Sea. J Sea Res 42:255–261

Tisdall JM, Oades JM (1982) Organic matter and water stable aggregates in soils. J Soil Sci 33:141–164

Vogel C, Rogerson A, Schatz S, Laubach H, Tallman A, Fell J (2007) Prevalence of yeasts in beach sand at three bathing beaches in South Florida. Water Res 41:1915–1920

Vrijmoed LLP (2000) Isolation and culture of higher filamentous fungi. In: Hyde KD, Pointing SB (eds) Marine mycology: a practical approach. Fungal Diversity Press, Hong Kong

Vrijmoed LLP, Kueh CSW, Shen HQ, Cai CH, Zhou YP (1993) A preliminary investigation of marine fungi in the South China Sea. In: Morton B (ed) Proceedings of the first international conference on the marine biology of Hong Kong and the South China Sea. Hong Kong University Press, Hong Kong

Walsh JJ (1983) Death in the sea – enigmatic phytoplankton losses. Prog Oceanogr 12:1–86

Wang G, Johnson ZI (2009) Impact of parasitic fungi on the diversity and functional ecology of marine phytoplankton. In: Kersey WT, Munger SP (eds) Marine phytoplankton. Nova Science Publisher, Hauppauge, NY

Wang G, Li Q, Zhu P (2008a) Phylogenetic diversity of culturable fungi associated with the Hawaiian sponges *Suberites zeteki* and *Gelliodes fibrosa*. Antonie Van Leeuwenhoek 93:163–174

Wang P, Xiao X, Zhang H, Wang F (2008b) Molecular survey of sulphate-reducing bacteria in the deep-sea sediments of the west Pacific Warm Pool. J Ocean Univ China 7:269–275

White TJ, Bruns T, Lee S, Taylor J (1990) Amplification and direct sequencing of fungal ribosomal RNA genes for phylogenetics. In: Innis MA, Gelfand DH, Sninsky JJ, White TJ (eds) PCR protocols: a guide to methods and application. Academic Press, San Diego

Zhang J, Sung X, Zhang J, Liu F, Song Q, Lu Y, Wang W (1989) Ecological distribution and genus composition of heterotrophic bacteria, yeasts and filamentous fungi in the ocean area of Northwest Antarctic Peninsula. In: Proceeding of China symposium on the Southern Ocean

Chapter 5
Fungi and Their Role in Corals and Coral Reef Ecosystems

Chandralata Raghukumar and J. Ravindran

Contents

Abstract Fungi in coral reefs exist as endoliths, endobionts, saprotrophs and as pathogens. Although algal and fungal endoliths in corals were described way back in 1973, their role in microboring, carbonate alteration, discoloration, density banding, symbiotic or parasitic association was postulated almost 25 years later. Fungi, as pathogens in corals, have become a much discussed topic in the last 10 years. It is either due to the availability of better tools for investigations or greater awareness among the research communities. Fungi which are exclusive as endoliths (endemic) in corals or ubiquitous forms seem to play a role in coral reef system. Fungi associated with sponges and their role in production or induction of

---

C. Raghukumar (✉)
National Institute of Oceanography, (Council for Scientific and Industrial Research), Dona Paula, Goa 403 004, India
e-mail: lata_raghukumar@rediffmail.com

C. Raghukumar (ed.), *Biology of Marine Fungi*,
Progress in Molecular and Subcellular Biology 53,
DOI 10.1007/978-3-642-23342-5_5,

secondary metabolites in their host is of primary interest to various pharmaceutical industries and funding agencies. Fungal enzymes in degradation of coral mucus, and plant detritus hold great promise in biotechnological applications. Unravelling fungal diversity in corals and associated reef organisms using culture and culture-independent approaches is a subject gaining attention from research community world over.

5.1 The Coral Reef Ecosystem

Coral reefs, oases in the blue deserts of the ocean, are home to colourful, hard and soft corals, sponges, a diverse population of fishes, holothurians, calciferous algae and other myriad communities (Connell 1978). They support tourism, food production and coastal protection from natural hazards. Their diverse life forms have vast biotechnological potentials. Corals fall in two categories: (1) the hard reef-building or hermatypic corals and (2) the soft or ahermatypic corals. The reef-building scleractinian corals (hard corals) possess calcium carbonate skeletons and are responsible for building reefs. They harbour symbiotic dinoflagellates *Symbiodinium*, also called zooxanthellae. The soft corals lack the calcium carbonate skeletons and lack symbiotic zooxanthellae.

Corals offer a diverse and organically rich environment for microorganisms. These are mucus layer on the coral, coral polyps, gastroderm cavity, gastrovascular spaces, organic matrix of the skeleton and calcium carbonate skeleton (Fig. 5.1). Corals with their microbial community of viruses, bacteria, archaea, cyanobacteria, fungi and endolithic algae form complex holobionts in the tropical nutrient-limited waters (Bourne et al. 2009). The symbiotic algae are a major source of crucial nutrients for scleractinian corals, while the microbes form mutualistic and

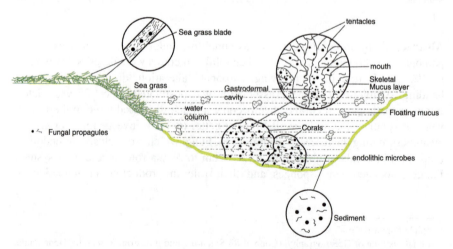

Fig. 5.1 Schematic diagram showing fungal distribution in corals and reef ecosystem in a coral reef lagoon

occasionally parasitic associations. Bacterial diversity associated with different coral species (Bourne and Munn 2005; Kvenefors et al. 2010), mucus (Ritchie 2006), coral reef sediments and water column (Van Duyl and Gast 2001) has been studied in detail. Its diversity in corals has been shown to differ spatially (Olson et al. 2009; Littman et al. 2009). Bacterial diversity has been studied using culture-dependent (Sabdono and Radjasa 2008) and culture-independent approaches (Rohwer et al. 2001; Cooney et al. 2002; Vega Thurber et al. 2009). A varied bacterial diversity in healthy, bleached (Bourne et al. 2008) and diseased corals (Sekar et al. 2008; Sunagawa et al. 2009), and as a response to various pollutants (Kuntz et al. 2005; Smith et al. 2008), has been reported. It is hypothesized that environmental stressors that affect the host physiology will have an impact on microbial community (Ainsworth et al. 2010). Archaeal diversity associated within a single coral species (Wegley et al. 2007), in three species of scleractinian corals from geographically different locations (Siboni et al. 2008; Lins-de-Barros et al. 2010), in healthy and stressed corals (Vega Thurber et al. 2009), in surface microlayer of corals (Kellogg 2004) has been documented in detail by sequencing the 16S rRNA gene. Diversity of cyanobacteria in healthy and diseased corals is also known to a certain extent (Frias-Lopez et al. 2003). After the initial study of fungal diversity in different species of corals (Kendrick et al. 1982) there are increasing reports of fungi in corals, both by culture-dependent and culture-independent methods (Ravindran et al. 2001; Wegley et al. 2007). We attempt to review the fungal diversity in corals and coral reefs and discuss their possible ecological role in this ecosystem.

5.2 Fungal Diversity in Coral Reef Ecosystems

5.2.1 Autochthonous (Indigenous or Native) Coral Fungi

The generally accepted broad ecological definition of marine fungi is that the obligate marine fungi grow and sporulate exclusively in a marine or estuarine habitat, and those from freshwater or terrestrial milieus that are able to grow and also sporulate in the marine environment are termed facultative marine fungi (Kohlmeyer and Kohlmeyer 1979). The best method to determine if a fungus is truly marine or not is to collect pieces of substrate from marine environment and inspect them for the presence of fungal hyphae or fruiting structures (Hyde and Pointing 2000). Based on this method, obligate marine fungi native to corals belonging to the class Ascomycetes were reported from corals (Kohlmeyer and Volkmann-Kohlmeyer 1992). Most of these fungi were reported from coral reefs off the coast of Belize (Central America). Among these, the genus *Koralionastes* with five species was placed in a new family Koralionastetaceae (Kohlmeyer and Volkmann-Kohlmeyer 1987). These five species were found attached to coral slabs in shallow water and were associated with crustaceous sponges. *Koralionastes angustus* and *K. giganteus* are reported from Belize, *K. ellipticus* and *K. ovalis* from Belize and Australia and *K. violaceus* from Australia and Fiji (Kohlmeyer and Volkmann-Kohlmeyer 1990). However, these fungi were not cultured and these authors suggested that they obtain their nutrition through close association with encrusting sponges (see Kohlmeyer and Volkmann-Kohlmeyer 2003). It is worth pursuing whether these fungi show any host specificity. There is a need for systematic survey and also to develop special culturing methods and media for obtaining such fungi in cultures.

5.2.2 Allochthonous (Non-native) Coral Fungi

In contrast to native fungi, several marine-derived terrestrial fungi find home in corals. The organic matrix of the carbonate skeletons is utilized by fungi to grow inside corals as endolithic saprotrophs. Kendrick et al. (1982) isolated saprotrophic fungi from several live hard corals from the Caribbean and the South Pacific using common mycological culture media (Table 5.1). These were identified to be common terrestrial fungi. The bioeroding and boring nature of two of these fungi, *Aspergillus versicolor* and *Penicillium stoloniferum*, was demonstrated by inoculating them in the coral *Siderastrea siderea*. The corals were incubated in sea water for 28 days and fungal boring was demonstrated by resin-cast method. In this technique, the dehydrated coral specimens are infiltrated with resin. After heat-induced polymerization, the carbonate is removed with HCl and the resin-cast borings are studied by scanning electron microscopy. Using the same technique, the presence of endolithic black septate fungi in skeletons of hard corals was demonstrated (Le Campion-Alsumard (1979), Le Campion-Alsumard et al. (1995a), Bentis et al. (2000) and Golubic et al. (2005). These fungi were neither cultured nor identified. *Aspergillus*-like conidiophores of endolithic fungi were observed in the scleractinian coral *Porites lobata* from Mayotte Island, Mozambique channel in the Indian Ocean (Priess et al. 2000) and in the same coral species on the island Moorea in the French Polynesia (Golubic et al. 2005). Bak and Laane (1987) observed a dark mycelial fungus with perithecia, probably belonging to Ascomycetes in black bands of *Porites* species in the eastern part of the Indonesian Archipelago. Partial dissolution of such black bands with EDTA (ethylenediamine tetracetic acid) revealed the presence of branched filaments. Sections of five species of massive corals from the Andaman Islands in the Bay of Bengal revealed the presence of a septate dark brown mycelial fungus on the subsurface of the dead patches (Raghukumar and Raghukumar 1991). The fungus formed a distinct dense brown to black zone of 0.5–1.5 cm width immediately below the surface layer of the corals. The fungus was isolated after partial dissolution of the coral skeleton with EDTA and a pure culture was obtained. It was identified to be a *Scolecobasidium* sp. It was isolated from several coral species showing necrotic patches in the Lakshadweep Island in the Arabian Sea as well (personal observation). Ravindran et al. (2001) also isolated

Table 5.1 Bioeroding saprotrophic fungi associated with various coral species (modified from Kendrick et al. 1982)

Host species	Location	Fungal species
Acropora sp.	Lizard islands Great Barrier Reef (GBR), Australia	*Aspergillus versicolor, Hormonema dematioides, Pithomyces chartarum,* sterile hyaline mycelium with swellings
	Heron islands, GBR	Sterile dark mycelium with swellings
A. hyacinthus	Heron islands, GBR	*A. versicolor, Humicola alopallonella*
A. palmata	Barbados, West Indies	*Aspergillus sydowii, Penicillium expansum, P. stoloniferum*
A. palifera	Lizard islands GBR	Sterile dark mycelium with swellings
	Heron islands, GBR	*Bipolaris rostrata*
Diploastrea heliopora	Lizard islands, GBR	*Aspergillus restrictus, Asteromella sp., Phialophora bubaki*
	Rarotonga, Cook islands	*Hormonema dematioide*
	Heron islands, GBR	Sterile hyaline mycelium with swellings
Diploria labyrinthiformis	Barbados, West Indies	*P. stoloniferum*
Goniastrea retiformis	Lizard islands, GBR	*A. versicolor, Phialophora bubaki*
	Rarotonga, Cook islands	*Hormonema dematioides*
G. australensis	Heron islands, GBR	*Paecilomyces lilacinus*, sterile hyaline mycelium with swellings
Meandrina meandrites	Barbados, West Indies	*Acremonium* sp.
Montastrea annularis	Barbados, West Indies	*A. sydowii, Cladosporium sphaerospermum*
M. cavernosa	Barbados, West Indies	*Paecilomyces godlewski*
Porites sp.	Rarotonga, Cook islands	*Cladosporium sphaerospermum, Wallemia ichthyophaga*, sterile dark mycelium with swellings
P. australensis	Lizard islands, GBR	*A. versicolor, Phialophora bubaki*
	Rarotonga, Cook islands	*Hormonema dematioides*
	Heron islands, GBR	*Penicillium restrictum*
P. porites	Barbados, West Indies	*A. sydowii, Penicillium avellaneum*
Stylophora pistillata	Rarotonga, Cook islands	*Hormonema dematioides*

common terrestrial fungi from healthy, pink-line syndrome-affected, bleached and dead (devoid of polyps) patch of *Porites lutea* from the Lakshadweep Island in the Arabian Sea (Table 5.2). The presence of two of these fungi in the coral skeleton was confirmed using polyclonal immunofluorescence probes. *Conidia*-like structures were observed within the coral skeleton by the authors. Besides isolation, they also observed fungal filaments directly in coral skeleton down to a depth of 3–7 cm in the coral sections of ~1 cm thickness. Fungal biomass constituted up to a 0.05% of the weight of the corals. Several ascomycetous fungi were isolated and obtained in pure culture from healthy and diseased branched coral, *Acropora formosa* (Yarden et al. 2007). These fungi were identified using internal transcribed spacer (ITS) sequences along with morphological features (Table 5.2). In both the

Table 5.2 Fungi in corals

Host	Location	Fungal species	State of the host/location in the host	References
Montastrea annularis	Venezuelan Reefs	Non-septate branched hyphae, possibly a lower marine fungus	Diseased showing black band, hyphae/epidermis, gastrodermis and mesoglea	Ramos-Flores 1983
Porites spp.	Indonesian Archipelago	Dark haphae with perithecia, probably belonging to Halosphaeriaceae (Ascomycota)	Healthy corals with black bands in coral skeleton	Bak and Laane 1987
Montipora tuberculosa, Porites lutea, Goniopora sp.	Andaman islands, Bay of Bengal, Indian Ocean	*Scolecobasidium* sp.	Necrotic patches in the corals/below the subsurface of corals	Raghukumar and Raghukumar 1991
Porites lobata	Moorea islands, French Polynesia	Unidentified septate mycelia	In the pore spaces emptied by the polyps	Le Campion-Alsumard et al. (1995a, b)
Porites lutea and *P. lobata*	Mayotte islands, Mozambique Channel and Moorea islands, French Polynesia	*Aspergillus*-like fungus	Within the coral skeletons attacking endolithic algae	Priess et al. 2000
Porites lutea	Lakshadweep islands, Arabian Sea	Dark, orange and hyaline non-sporulating mycelial forms, *Acremonium* sp., *Fusarium* sp., *Aspergillus* sp., *Cladosporium* sp., Mycelial yeasts, *Labyrinthula* sp., *Chaetomium* sp., *Aureobasidium* sp.	Skeleton of healthy and pink-line syndrome- affected corals and from dead patches	Ravindran et al. 2001
Acropora formosa	Wistari and Heron Reefs, GBR	*Phoma* sp., *Alternaria* sp.	Healthy coral	Yarden et al. 2007
		Phoma sp., *Humicola fuscoatra, Cladosporium* sp., *Penicillium citrinum*,	With brown band	
		Phoma sp., *Fusarium* sp., *Aureobasidium pullulans, Alternaria* sp.	Skeletal eroding	
		Phoma sp., *Cladosporium* sp.	Brown band and skeletal eroding	

Acopora sp., *A. secale*, *Porites* sp.	Lakshadweep islands, Arabian Sea	*Corallochytrium limacisporum*	Coral mucus	Raghukumar 1987
Coral, fresh coral mucus, floating and attached mucus detritus	Lakshadweep islands, Arabian Sea	*Corallochytrium limacisporum*, *Thraustochytrium motivum*, *Labyrinthuloides minuta*, *L. yorkensis*, *Ulkenia visurgensis*		Raghukumar and Balasubramanian 1991
Fungia granulosa	Gulf of Eilat, Israel	Thraustochytrid strain Fng1	Surface mucus	Harel et al. 2008
Various coral species	GBR, Australia	*Alternaria*, *Aspergillus* , *Beauveria* spp., *Dreschlera* spp., *Humicola* spp., *Monilia* spp., *Oididendron* spp., *Penicillium* spp., *Phialophora* spp., *Phoma* spp., *Torulomyces* spp., *Pestlotiopsis*-like sp., *Absidia* spp., unidentified fungus	Surface-sterilized segments	Morrison-Gardiner 2002

above instances (Ravindran et al. 2001; Yarden et al. 2007), alterations in fungal prevalence correlated with coral health. Report on fungi isolated from Australian Great Barrier Reef corals indicated the presence of several fungi, almost all of them being common terrestrial ones (Morrison-Gardiner 2002). There were 54 distinct fungal taxa indentified in addition to the presence of several non-sporulating fungi. Majority of the isolates were mitosporic fungi. Corals from near shore locations showed increased presence of fungi compared to that from offshore locations (Table 5.2). These results indicated that fungi, although terrestrial, are apparently an important component of corals.

5.2.3 Fungi in Other Reef Organisms

After the report of aspergillosis disease in the sea fan (soft coral) *Gorgonia ventalina* (Smith et al. 1996) caused by *Aspergillus sydowii*, a sudden spate of interest in mycoflora of this species is observed worldwide. Toledo-Hernández et al. (2007) isolated 15 species of terrestrial fungi from *Gorgonia ventalina* collected from Puerto Rico. These authors compared the methods of isolating fungi using two strategies: plating of tissue fragments of different sizes versus plating of homogenized tissue. The results showed that homogenization of small tissue fragments yielded better estimate of the fungal diversity. The fungi isolated were identified using ITS gene sequences and all of these were terrestrial fungi (Table 5.3). In a further study, Toledo-Hernández et al. (2008) compared fungal diversity in healthy and aspergillosis-affected *Gorgonia ventalina* collected from 15 reefs around Puerto Rico. These were identified by sequencing the nuclear ribosomal ITS region and by morphology. Thirty terrestrial fungal species belonging to 15 genera were isolated from 203 sea fan colonies (Table 5.3). Healthy sea fans harboured a greater fungal diversity than that of diseased ones. In another study, Koh et al. (2000) isolated 16 fungal genera and 51 species, including 2 yeasts, from 10 species of gorgonian corals in Singapore by culturing. Marine sponges collected from coral reefs of Palau and Bunaken Island, Indonesia harboured several filamentous fungi (Namikoshi et al. 2002). About 83% and 98% of the sponges collected from Palau and Bunaken Island, respectively, harboured fungi. Unfortunately, the authors have not provided the diversity of fungal species obtained in this study. These results indicate that fungi are residential flora within gorgonians too.

Seagrasses are major primary producers in coral reef ecosystems (Bakus et al. 1994), playing an important role in the food web. Seagrasses are angiosperms, capable of growing and flowering in submerged seawater of lagoons and intertidal regions. They form one of the major biological zones of coral reefs, responsible for nutrient recycling and creating sedimentary environment (Macintyre et al. 1987). *Thalassia hemprichii* is one of the dominant seagrasses in the Lakshadweep Islands

Table 5.3 Fungi associated with sea fan and sponges

Host	Location	Fungal genera	State of the host/location in the host	References
Several sponges ($n = 10$)	Fringing reef south of Singapore	*Acremonium strictum, A. aculeatus, A. cervinus, A. nutans, A. pulverulentus, A. flavus, A. ochraceus, A. ornatus, A. terricola, Cladosporium musae, C. sphaerospermum, Fusarium* sp., *Gliomastix cerealis, G. luzulae, G. murorum, Microascus triganosporus, Oidodendron griseum, Penicillium brevi-compactum, P. camemberti, P. canescens, P. citrinum, P. decumbens, P. frequentans, P. implicatum, P. janthinellum, P. lanoso, P. lilacinum, P. notatum, P. oxalicum, Phoma-like, Scolecobasidium humicola, Stibella* sp., *Trichoderma pseduokoningii, T. hamatum, T. harzianum, T. koningii, T. longibrachiatum, Tritirachium* sp., *Verticillium* sp., black yeast, white yeast, sterile isolates	Healthy sponges	Koh et al. (2000)
Several sponges ($n = 10$)	Fringing reef south of Singapore	*Acremonium butryi, A. furcatum, A. strictum, A. aculeatus, A. foetidus* var. *pallidus, A. kangawensis, A. nutans, A. ficuum, A. flavus, A. ornatus, A. terricola, A. wentii, Chaetophoma* sp., *Cladosporium musae, C. sphaerospermum, Fusarium* sp., *Gliomastix cerealis, G. luzulae,*	Unhealthy sponges	Koh et al. (2000)

(continued)

Table 5.3 (continued)

Host	Location	Fungal genera	State of the host/location in the host	References
		G. murorum, *Hymenula* sp., *Oidodendron griseum*, *Penicillium brevi-compactum*, *P. camemberti*, *P. canescens*, *P. citrinum*, *P. frequentans*, *P. implicatum*, *P. janthinellum*, *P. lilacinum*, *P. steckii*, *Scolecobasidium humicola*, *Sporotrichum* sp., *Trichoderma pseduokoningii*, *T. hamatum*, *T. harzianum*, *T. longibrachiatum*, *Tritirachium* sp., *Verticillium* sp., *Virgaria* sp., black yeast, white yeast, sterile isolates		
Several sponges	Coral reefs of Palau and Bunaken islands, Indonesia	Filamentous fungi (not identified)	Sponge tissue and fluid	Namikoshi et al. (2002)
Sea fan *Gorgonia ventalina*	San Juan, Puerto Rico	[a]*Aspergillus flavus/oryzae* (4,6), *A. ustus* (1,1), *A. niger* (0,1), *A. unguis* (1,2), *A. sydowii* (1,0), Xylariales (1,0), *Stachybortys chartarum* (1,0), *Cladosporium sphaerocarpum* (0,2), *Rhodotorula nympheae* (0,1), *Xylaria hypoxylon* (0,1), *Gloeotinia tremulenta* (1,0), *Penicillium citrinum* (0,3), *P. coffeae* (0,1), *P. citreonigrum* (1,1), *P. steckii* (0,1)	Healthy tissue	Toledo-Hernández et al. (2007)
Gorgonia ventalina	Puerto Ricoa	[b]*Aspergillus aculeatus* (2,1,0), *A. flavus/oryzae* (14,5,9), *A. melleus* (1,0,0), *A. niger* (3,0,2), *A. niger/foetidus/awamori* (0,1,0), *A. ochraceus* (0,0,1),	Healthy tissue from healthy *G. ventalina*, healthy tissue from diseased sea fan and diseased tissue from diseased colony	Toledo-Hernández et al. 2008

A. sydowii (10,0,0), *A. tamarii* (2,0,1), *A. terreus* (3,2,1), *A. unguis* (1,0,0), *A. ustus* (4,1,0), *A. versicolor* (2,0,1), *Candida* sp. (2,0,0), *Chalaropsis* sp. (0,0,1), *Cladosporium* sp. (0,0,1), *C. cladosporioides* (3,3,1), *C. sphaerosperma* (0,1,0), *Davidiella tassiana* (2,0,0), *Gloetinia temulenta* (4,1,2), Helotiaceae (0,1,0), *Hypocrea lixii* (0,0,1), *Nectria* sp.
Bionectria (0,0,1), *N. haematococca* (0,1,0), *Penicillium chrysogenum* (2,0,0), *P. citreonigrum* (0,1,0), *P. commune* (0,1,0), *P. minioluteum* (1,0,1), *P. citrinum* (23,1,4), *Pichia guillermondi* (1,0,0), *Rhodotorula nymphaeae* (1,0,0), *Stachybotrys chartarum* (1,0,0), *S. chlorohalonata* (0,1,0), *Trichoderma harzianum* (3,1,0), *Tritirachium* sp. (1,0,4), *Xylaria hypoxylon* (0,0,1), unknown (6,5,13)

[a]Numbers within parenthesis indicate observed frequency per colony by plating tissue fragments and homogenized tissue respectively

[b]Numbers within parenthesis indicate number of isolates obtained by plating healthy tissue from healthy *G. ventalina* colony, healthy tissue from a diseased colony and diseased tissue from diseased colony respectively

Fig. 5.2 (**a**) Seagrass beds exposed during low tide in Kavaratti island of the Lakshadweep Archipelago. (**b**) Sea grass beds and detritus on the same location

in the Arabian Sea (Jagtap 1998) and consequently it contributes enormously to detrital material in the intertidal and shallow lagoons of these islands (Fig. 5.2). Detrital material of this seagrass harboured nearly 3% of fungal biomass and 0.23% thraustochytrid biomass (Sathe and Raghukumar 1991) playing an important role in biogeochemical cycling of nutrients.

5.2.4 Straminopilan Fungi in Corals

Labyrinthulomycetes, comprising labyrinthulids, aplanochytrids and thraustochytrids, belong to the Kingdom Stramenopila (Dick 2001). Despite being not considered as true fungi, mycologists consider them as straminopilan fungi and include them in marine mycology (Raghukumar 2002). Therefore, they are included in this review. The presence of these truly aquatic "fungi" in corals and coral reef ecosystem is very sparsely recorded. Fresh coral mucus, floating and attached mucus detritus proved to be rich sources of these protists. Raghukumar and Balasubramanian (1991) isolated *Thraustochytrium motivum, Labyrinthuloides minuta, L. yorkensis* and *Ulkenia visurgensis* from these sources in coral reefs of the Lakshadweep Islands in the Arabian Sea. Mucus detritus attached to corals yielded 1.9×10^6 thraustochytrid g^{-1} mucus, whereas fresh mucus collected from corals yielded 20 thraustochytrid ml^{-1} mucus. Harel et al. (2008) reported the presence of a new unidentified thraustochytrid Fng1 in the coral mucus and surface of the hermatypic coral *Fungia granulosa* and *Favia* sp. from the Gulf of Eilat. A mention must be made here about the presence of a novel marine protist, *Corallochytrium limacisporum* in the coral lagoon (Raghukumar 1987). This single-cell marine protist, a choanozoan belonging to the supergroup Opisthokonta, forms a sister clade with fungi (Cavalier-Smith and Chao 2003). Sumathi et al. (2006) demonstrated the presence of fungal signatures, namely, ergosterol and gene sequence of alpha aminoadipate reductase (α-AAR), a key enzyme involved in lysine synthesis in fungi in *C. limacisporum*. Several isolates of this protist were obtained from coral mucus from a coral lagoon in the Lakshadweep group of islands of the Arabian Sea (Raghukumar and Balasubramanian 1991). The presence of *C. limacisporum* in three corals, *Acropora* sp., *A. secale* and *Porites* sp. was confirmed using an immunofluorescence probe (Raghukumar and Balasubramanian 1991). Such studies on the yet unknown marine eukaryotic diversity in coral reef ecosystems may be a key to understanding the origin of fungi.

5.2.5 *Culture-Independent Approaches for Assessment of Diversity*

Few studies have focussed on the diversity of coral-inhabiting fungi using culture-independent approach. Metagenomics or community genomics approach describes the taxonomic components, their relative abundance and metabolic potentials. The microbial metagenome of *Porites astreoides* collected from Bocas del Toro, Panama showed fungi to be the dominant community, contributing 38% to the genome of the coral (Wegley et al. 2007). These belonged to Ascomycota, Basidiomycota and Chytridiomycota divisions. The majority of coral-associated fungal sequences (93%) were most similar to Ascomycota, comprising Sordariomycetes, Euratiomycetes, Saccharomycetes and Schizosaccharomycetes classes. In another instance, sequences showing similarity with aquatic chytrids were reported from corals (Vega Thurber et al. 2009). This is the first report of a chytrid being reported from corals. However, the corals in this study were maintained under different stress conditions in the laboratory microcosm experiments. Among these, *Harpochytrium*, the species closely related to chytrid sequences, is a zoosporic fungus found as an epibiont in algae (Sparrow 1960).

5.3 The Role of Fungi in Corals

5.3.1 *Fungi as Carbonate Borers*

Phototrophic algae and heterotrophic fungi are the two most commonly occurring endoliths in calcium carbonate structures of corals. Very often, cyanobacterial filaments and conchocelis stages of red algae are also seen as endoliths (Bentis et al. 2000). Both the groups grow in the skeletal structure and are detected in early stages of coral growth (Le Campion-Alsumard et al. 1995a, b) producing tunnels of different diameters. These authors (Le Campion-Alsumard et al. 1995a; Golubic et al. 2005) further demonstrated that fungal hyphae are also part of the endolithic community and they parasitize algal filaments by producing haustoria. In the process they release a dark tannin-like substance that stains the hyphae, algal filaments and the coral skeleton. These result in black bands in corals called density-banding (Priess et al. 2000). Coral's defence response to fungal boring is by covering the perforations produced by fungal hyphae with dense fine-grained carbonate over the borehole. This results in a difference between skeletal carbonate and the repair carbonate, which is fine-grained aragonite, a unstable form of carbonate. Thus boring fungi play an important role in biomineralization of coral skeleton. These benign endolithic fungi can become opportunistic pathogens when the equilibrium is disturbed by environmental stressors. At present we do not know what effects these penetrating fungal hyphae have on corals, their mechanism of penetration and the role their enzymes play in this process.

5.3.2 Fungi as Pathogens in Reef Ecosystems

Besides natural disasters and climate warming, diseases have contributed to coral decline worldwide (Harvell et al. 2002). Disease outbreaks result in significant loss in coral cover and also cause a shift of the coral dominant community towards an algal-dominant ecosystem (Bourne et al. 2009). Characterization and establishing a disease must prove Koch's postulates (Andrews and Goff 1985). Several fungal associations with diseased corals were reported, such as a lower marine fungus associated with black line disease of the star corals, *Montastrea annularis* (Ramos-Flores 1983), endolithic fungi in hard corals *Porites lobata* (Le Campion-Alsumard et al. 1995a) and a *Scolecobasidium* sp. associated with scleractinian corals, *Porites lutea, Goniopora* sp. and *Montipora tuberculosa* (Raghukumar and Raghukumar 1991). About 600 km of the coral reefs was affected along the coasts from Tanzania to Kenya and fungal association has been reported in the affected corals (McClanahan et al. 2002). However, these fungi were not proven to be the actual pathogens.

The only well-characterized fungal disease is that of aspergillosis of gorgonians (sea fans) caused by *Aspergillus sydowii* (Smith et al. 1996). The sea fans *Gorgonia ventalina* and *G. flabellum* affected by the disease show lesions and necrotic patches. The disease is wide spread in the Caribbean (Nagelkerken et al. 1997a). Both tissue and skeleton disappear in the diseased sea fans. The necrotic tissue along the edges of the lesions is either lighter or dark purple in colour, in contrast to the coloration of the healthy tissue (Nagelkerken et al. 1997a). Smith et al. (1996) showed that disease could be established in healthy colonies by inoculating them with pure cultures of *A. sydowii* or by grafting infected tissue onto the healthy tissue. Recent report (Rypien and Baker 2009) shows that the spread of the disease can also be through a vector. In this case a snail, *Cyphoma gibbosum*, was demonstrated to transmit the pathogen by feeding on the gorgonians. Aspergillosis was highly prevalent in the Caribbean in 1995 where >90% of gorgonians were affected (Nagelkerken et al. 1997a). Temperature plays crucial role in the virulence of the fungus *A. Sydowii* to the sea fan coral *Gorgonia ventalina*. The fungus grows at an optimum temperature of 30°C suggesting that increased water temperature during the summer is likely to promote its pathogenicity (Alker et al. 2001). African dust has been shown to be the source of *A. sydowii* in Caribbean gorgonian corals (Shinn et al. 2000). The transport of dust during Saharan dust storms from Africa to the Atlantic and Caribbean coral reef communities carries nutrients, including iron (Hayes et al. 2001), organic pollutants (Garrison et al. 2006) and pathogenic organisms (Shinn et al. 2000). The presence of spores of *A. sydowii* in the African dust was shown as a proof of it being the primary source of the coral pathogen (Weir-Brush et al. 2004). However, a systematic study involving specific culture techniques to study the fungal flora of the dust collected from Africa and the Cape Verde Islands in the Atlantic could not detect *A. sydowii*, suggesting that African dust may not be a viable source of this coral pathogen (Rypien 2008). Genetic structure of global samples of *A. sydowii*, including isolates from diseased corals, diseased humans and environmental sources, revealed that disease-causing isolates are not genetically distinct from environmental isolates (Rypien et al. 2008). These authors further suggested that no specific virulence factor in the pathogen was responsible and that any isolate of *A. sydowii* could cause aspergillosis. Environmental stressors and host immune system may play a vital role in causing the disease. Absence of *A. sydowii* in diseased sea fans collected from 13 reefs in Puerto Rico and its presence in healthy samples by culture-dependent approach suggested that *A. sydowii* was not the pathogen in these samples (Toledo-Hernández et al. 2008). However, reports of *Aspergillus* species causing disease in another species of the sea fan, *Annella* sp., in Southeast Asia after the 2004 tsunami (Phongpaichit et al. 2006) and in a marine sponge *Spongia obscura* (Ein-Gill et al. 2009) keeps the *Aspergillus*-as-pathogen debate alive.

As the fungus can be cultured and the host is easy to maintain in laboratory, aspergillosis disease of the sea fan *Gorgonia ventalina* is taken as a model system to understand host–pathogen interactions in context with environmental stressors. Climate warming is one of the environmental drivers of diseases. Effects of these have been studied on the aspergillosis disease as well. Increased temperature (31.5°C), caused 176% increase in the activity of host-derived antifungal compounds, but also increased the pathogen's growth rate, providing an opportunity to the pathogen to establish itself before the host could defend itself (Ward et al. 2007). Increased temperature caused loss in zooxanthellae abundance indicating that zooxanthellae were not responsible for high level of anti-fungal activity. With increasing global warming it is predicted that aspergillosis will continue to play havoc in gorgonian sea fan communities. Immune response of the host against the pathogen included production of anti-fungal lipid metabolites (Kim et al. 2000) and melanin deposition as a physical barrier to prevent fungal expansion (Petes et al. 2003), production of peroxidases that show anti-fungal activity (Mydlarz and Harvell 2007) and production of exochitinase against the fungal pathogen *A. sydowii* (Douglas et al. 2007).

Fungi are the dominant pathogens of algae (Kohlmeyer and Kohlmeyer 1979). Several fungal pathogens of green, brown and red algae (Table 5.4) were reported from the Lakshadweep Islands in the Arabian Sea (see references in Raghukumar 2006). Most of these pathogens belonged to Chytridiomycota, comprising aquatic fungi that are pathogens of algae and phytoplankton. Littler and Littler (1998) described infection of crustose coralline algae by an unidentified fungus in American Samoa. The pathogen occurred in shallow reef habitat but was not restricted to calm waters. The spreading dense black fungal bands caused death of coralline algae. Sparse information available on fungal pathogens of algae from coral reefs indicates that this field is open for survey with modern molecular tools. Their occurrence in coral reef fauna, either as pathogens or saprotrophs would play a role in population dynamics and nutrient recycling, respectively.

Table 5.4 Fungi as algal pathogens in coral reef ecosystem

Locality	Host algae/class	Fungi/class	References
Lakshadweep islands Arabian Sea	*Cladophora repens, C. fascicularis, Rhizoclonium* sp./ Chlorophyta	*Coenomyces* sp./ Chytridiomycota	Raghukumar 1986
Lakshadweep islands, Arabian Sea	*Cladophora expansa, C. fascicularis, C. frescatii, Rhizoclonium* sp./ Chlorophyta	*Olpidium rostriferum*/ Chytridiomycota	Raghukumar 1986
Lakshadweep islands, Arabian Sea	*Cladophora frescatii, C. gracilaris*/ Chlorophyta	*Sirolpidium bryopsidis*/ Chytridiomycota	Raghukumar 1986
Lakshadweep islands, Arabian Sea	*Cladophora* sp., *C. frescatii, C. repens, Rhizoclonium* sp./ Chlorophyta	*Labyrinthula* spp./ Labyrinthulomycota	See Raghukumar 2006
Lakshadweep islands, Arabian Sea	*Sargassum* sp./ Phaeophyta	*Lindra thalassiae*/ Ascomycota	See Raghukumar 2006
Lakshadweep islands, Arabian Sea	*Centroceras clavulatum*/ Rhodophyta	*Chytridium polysiphoniae*/ Chytridiomycota	See Raghukumar 2006
Lakshadweep islands, Arabian Sea	*Thalassia hemprichii* Seagrass	*Acremonium, Chaetomium, Graphium, Humicola, Penicillium* spp., *Labyrinthuloides minuta, Thraustochytrium motivum*	Sathe and Raghukumar (1991)
American Samoa	Crustose coralline alga/ Chlorophyta	An undescribed fungal Pathogen	Littler and Littler (1998)

5.3.3 Secondary Metabolites from Coral Reef Fungi

Coral reefs are an important source of pharmaceutical, nutraceutical, enzymes, cosmetics, pesticides and other novel commercial products. Coral reefs contain diversity of sessile animals, such as corals, sponges, tunicates, molluscs, bryozoan and echinoderms. These sessile animals have developed unique chemical defence mechanisms against predation, biofouling, diseases, environmental perturbations and other stressors. These chemicals are either synthesized by the organisms themselves or their endobiontic microorganisms. If these valuable compounds are to be obtained from reef organisms, a system for sustainable management of harvesting these potentially valuable invertebrates has to be in place. On the other hand, endobiontic microorganisms such as bacteria, fungi and actinomycetes, which produce these compounds or induce the hosts to produce, can be cultured on a large scale for extraction and characterization of bioactive molecules. Several novel compounds with potential medical applications have been derived from fungi associated with various reef organisms (Table 5.5). A number of antimicrobial compounds have been reported from coral reef associated fungi (Namikoshi et al. 2000; Bugni and Ireland 2004).

Fungal presence in corals of different health may, to a limited extent, give insight into the type of fungal succession and their r- or K-strategies. Fungi with r-strategy may grow and reproduce very fast and utilize most of the available organic matter. Whereas, K-strategists would grow slowly, may produce a lot of antimicrobial compounds. Such ecological information is important in drug discovery and screening programmes.

Table 5.5 Secondary metabolites from coral reef organisms

Host/locality	Fungus	Compound	Activity	References
Unidentified soft coral	Unidentified fungus	Spiroxins A-E	DNA-cleaving antitumor antibiotic	McDonald et al. (1999)
Coral reef at Yap	*Paecilomyces* sp.	Paecilospirone	Curing effect on deformed fungal mycelia	Namikoshi et al. (2000)
Mangrove wood from coral reef in Pohnpei	*Phomopsis* sp.	Phomopsidin	Inhibitory activity in microtubule assembly assay using purified porcine brain microtubule protein	Namikoshi et al. (2000)
Coral reef at Pohnpei	*Penicillium* sp. *Fusarium* sp. *Colletotrichum* sp.	Griseofulvin Fusarielin Deoxyfusapyrone	Antifungal activity	Namikoshi et al. (2000)
Coral reef at Yap	*Acremonium* sp.	Fusapyrone	Antifungal activity	Namikoshi et al. (2000)
Sponge tissue, Okinawa	Unidentified strain # 97F95	Verrucarin	Unidentified strain # 97F95	Namikoshi et al. (2000)
Coral reef at Pohnpei	Unidentified strain # 95F137	Verrucarin L acetate	Unidentified strain # 97F95	Namikoshi et al. (2000)

5.3.4 Enzymes of Fungi from Coral Reefs

No information on extracellular enzymes of coral-associated fungi or bioeroding endolithic fungi is available. Coral mucus contains sugars, fatty acids and proteins (Meikle et al. 1988). Thraustochytrid protists displayed very good proteolytic enzyme activity at alkaline pH (Colaço et al. 2006; Raghukumar et al. 2008b). Therefore, thruastochytrids present in coral mucus might play a role in their degradation.

Little research has been aimed at understanding the fungal degradation of polysaccharides such as carrageenans, agar, alginate and laminarin of algae inhabiting coral reefs and their role in nutrient cycling. However, some information on degradative enzymes produced by fungi associated with seagrasses in coral reef lagoon in the Arabian Sea is known (see Raghukumar et al. 2008a). A basidiomycetous fungus NIOCC #312 isolated from decaying *Thalassia hemprichii* leaves from one of the islands in the Lakshadweep in the Arabian Sea displayed lignin-degrading activity (Raghukumar et al. 2008a). It mineralized about 24% of U-ring^{14}C-labelled lignin to $^{14}CO_2$ within 24 days when grown in seawater medium. The lignin-degrading enzymes, lignin peroxidase, manganese peroxidase and laccase were produced when the fungus was grown in seawater medium. It decolorized several synthetic dyes, bleach plant effluent (paper mill effluent) and molasses spent wash (molasses-based distillery effluent) effectively within 6 days. The fungus showed growth, lignin mineralization and decolourization of coloured industrial pollutants in sea water medium (Raghukumar 2008). Several seagrass species occurring in shallow coral reef lagoons, at different stages of decomposition, may harbour fungi with an array of lignocellulose degrading enzymes. These resources are still untapped. These fungi may prove to be a good source of salt-tolerant enzymes as the salinity in lagoons is 25–33 psu (practical salinity units).

5.3.5 Other Possible Roles

A recent metagenomic approach in coral holobiont revealed the presence of fungal genes involved in carbon and nitrogen metabolism, suggesting their role in conversion of nitrate and nitrite to ammonia, enabling fixed nitrogen to cycle within the coral holobiont (Wegley et al. 2007). Besides, genes encoding cell wall synthesis, chitin synthase and glutamine and nitrogen metabolism were also found during these studies on the coral *Porites astreoides*. To understand their role in nutrition, such studies need to be pursued further using radiolabelled substrates. Domart-Coulon et al. (2004) presented experimental evidence of interactions between coral cells and fungal resident, a *Cryptococcus* sp. strain F19-3-1 in *Pocillopora damicornis* coral skeleton. Coral cells established in primary culture were shown to obtain short-term extension of survival in the presence of the fungal strain F19-3-1, indicating that this fungus produced a cryoprotective effect which selectively enhanced the survival of skeletogenic cell types.

The photosynthetic dinoflagellates are rich in intracellular solute of dimethylsulfoniopropionate (DMSP). When DMSP is released from zooxanthellae, it is broken down to dimethyl sulphide (DMS) mostly by bacteria. DMS oxidation products are cloud condensation nuclei, causing cloud cover over the oceans (Sievert et al. 2007). Recently, the coral pathogen *A. sydowii* has been demonstrated to catabolically produce DMS from DMSP. It contained dddP gene, encoding the enzyme that releases DMS from DMSP (Kirkwood et al. 2010). The ability to catabolise DMSP may give such fungi selective access to an abundant substrate present in the ecosystem. The role of coral-associated fungi in DMS production needs to be assessed.

Corals in the tropical oceans protect their tissues from damaging doses of ultraviolet radiation. This protection is considered to be offered by the symbiotic zooxanthellae living in their tissue through mycosporine-like amino acids (MAA). Some members of MAAs are found in fungi (Dunlap and Shick 1998). It will be worthwhile to study how much of MAAs are contributed by the coral-associated fungi in corals and what role they play in absorbing and dissipating UV energy.

5.4 Future Directions

Studies on microbial diversity, especially bacteria and the symbiotic dinoflagellate of coral holobiont, have received a lot of attention recently. Although the presence of fungi in this holobiont has been recognized, not enough information on their diversity and role is known. Cultivation-based methods for fungi have limitations in obtaining slow-growing true marine fungi, which means that there is an urgent need to develop methods to cultivate the uncultured. Modifications of microencapsulation method used for microbial cultivation (Zengler et al. 2002) should be considered for culturing true marine fungi from corals. Low light intensities, oxygen concentrations, pH, carbon and nitrogen sources should be standardised for obtaining fungi from corals. For in situ analysis of residential mycota in corals, culture-independent approach needs to be used for a large-scale quantification on spatial and temporal scale. Methods such as fluorescence in situ hybridization (FISH) or CARD-FISH (catalysed reported deposition) or double-labelling of oligonucleotide probes for FISH (DOPE-FISH), used for the identification, quantification and characterization of phylogenetically distinct population of prokaryotes (Amann and Fuchs 2008; Stoecker et al. 2010), need to be developed for studying fungal diversity in corals. Such probes are being used in the detection of phylloplane mycoflora (Inácio et al. 2010), freshwater fungi in streams (Baschien et al. 2008) and environmental samples (Mitchell and Zuccaro 2006). A protocol for FISH developed by Sterflinger et al. (1998) for detecting hyphomycetes in stone monuments can be easily applied for detecting specific slow-growing fungi in corals. However, these authors encountered a problem of substrate autofluorescence or other associated fungi. To overcome this problem, they recommend the use of digoxigenin (DIG)-labelled probes which can be detected colorimetrically by antibodies specific for digoxigenin, which are coupled with alkaline phosphatase. After addition of the enzyme substrate (4-nitroblue tetrazolium chloride/5-bromo-4-chloro-3-indolyl-phosphate), the intracellular blue-purple precipitate is detected by light microscopy (Schröder et al. 2000). Thus, colorimetric *in situ* hybridization (CISH) is a promising method for the detection of specific fungi in healthy or diseased corals.

Filamentous fungi isolated from corals by several researchers may be mostly allochthonous, as coral skeletons are perforated and may harbour dormant propagules of terrestrial fungi. However, in cases where actively growing mycelia were observed in healthy corals, there is a possibility that they are autochthonous forms. Whether these coral-inhabiting fungi differ physiologically from their terrestrial counterparts is a much-debated topic (Kohlmeyer and Volkmann-Kohlmeyer 2003). It needs to be assessed whether they play any active role in coral holobiont as symbionts, or pathogens, or saprotrophs. The role of *A. sydowii*, a ubiquitous terrestrial fungus causing aspergillosis in sea fan, was cited as an example in this controversy about proving the role of marine-derived fungi. However, recent studies have caused serious concern about it being the sole pathogen (Toledo-Hernández et al. 2008), and is considered being an opportunistic pathogen (Rypien et al. 2008). Such stringent studies are required to be undertaken for proving the presence of marine-derived fungi in corals.

Assuming that some of the fungi may be symbiotic, at what stage the corals start harbouring fungi still is a total black box. In one instance it was demonstrated that endosymbiotic yeasts in the three species of the sponge *Chondrilla* were directly propagated by vertical transmission in the female (Maldonado et al. 2005). These authors reported that a large number of yeast cells (ca. 4.4 cells per 10 μm^2) were transmitted from the soma through the oocytes to the fertilized eggs. Is it the case with yeasts associated with some of the coral species at least? Using immunofluorescence probe, the presence of the thraustochytrid *Ulkenia visurgensis* in gut of hydrozoa was demonstrated (Raghukumar 1988). Do the coral polyps harbour yeasts or thraustochytrids within the coelenteron as shown in the case of hydrozoa?

Do fungi provide nutrition in the event of bleaching, when the zooxanthellae are expelled? Several workers have reported fungal presence in bleached and diseased corals (Ravindran et al 2001; Yarden et al. 2007). Straminopilan fungi found on the surface and in the mucus of corals may provide nutrition in the form of polyunsaturated fatty acids helping corals to survive during bleaching event (Harel et al. 2008).

Do the physiological changes occurring within the host coral in response to environmental stressors alter the endobiontic fungal populations or turn some of these benign fungi into opportunistic pathogens? Fungi associated with corals in healthy, bleached, stressed and diseased corals, succession of fungi at different stages of coral growth, in different species of corals, spatial and temporal variations are some of the issues which need to be addressed using polyphasic approaches of culture-dependent and culture-independent techniques, metagenomics and other modern molecular tools available.

Acknowledgements The first author wishes to thank Council for Scientific and Industrial Research, New Delhi for the grant of ES scheme No. 21 (0649)/06/EMR-II. This is NIO's contribution No. 4959.

References

Ainsworth TD, Thurber RV, Gates RD (2010) The future of coral reefs: a microbial perspective. Trends Ecol Evol 25:233–240

Alker AP, Smith GW, Kim K (2001) Characterization of *Aspergillus sydowii* (Thom et Church), a fungal pathogen of Caribbean sea fan corals. Hydrobiologia 460:105–111

Amann RI, Fuchs BM (2008) Single-cell identification in microbial communities by improved fluorescence *in situ* hybridization techniques. Nat Rev Microbiol 6:339–348

Andrews JH, Goff LJ (1985) Pathology. In: Littler M, Littler DS (eds) Handbook of phycological methods. Cambridge University Press, Cambridge

Bak RPM, Laane RWPM (1987) Annual black bands in skeletons of reef corals (Sceleractina). Mar Ecol Prog Ser 38:169–175

Bakus GJ, Wright M, Schulte B, Mofidi F, Yazdandoust M, Gulko D, Naqvi SWA, Jagtap TG, Goes J, Naik C (1994) Coral reef ecosystems. Oxford & IBH, New Delhi, India

Baschien C, Manz W, Neu TR, Marvanová L, Szewzyk U (2008) *In situ* detection of freshwater fungi in an alpine stream by new taxon-specific fluorescence *in situ* hybridization probes. Appl Environ Microbiol 74:6427–6436

Bentis CJ, Kaufman L, Golubic S (2000) Endolithic fungi in reef-building corals (order: Scleractinia) are common, cosmopolitan, and potentially pathogenic. Biol Bull 198:254–260

Bourne D, Munn CB (2005) Diversity of bacteria associated with the coral *Pocillopora damicornis* from the Great Barrier Reef. Environ Microbiol 7:1162–1174

Bourne D, Iida Y, Uthicke S, Smith-Keune C (2008) Changes in coral-associated microbial communities during a bleaching event. ISME J 2:350–363

Bourne DG, Garren M, Work TM, Rosenberg E, Smith GW, Harvell CD (2009) Microbial disease and the coral holobiont. Trends Microbiol 17:554–562

Bugni TS, Ireland CM (2004) Marine-derived fungi: a chemically and biologically diverse group of microorganisms. Nat Prod Rep 21:143–163

Cavalier-Smith T, Chao EE (2003) Phylogeny of Choanozoa, Apusozoa, and other protozoa and early eukaryote megaevolution. J Mol Evol 56:540–563

Colaço A, Raghukumar C, Mohandass C, Cardigos F, Santos RS (2006) Effect of shallow-water venting in Azores on a few marine biota. Cah Biol Mar 47:359–364

Connell JH (1978) Diversity in tropical rain forests and coral reefs. Science 199:1302–1310

Cooney RP, Pantos O, Le Tissier MDA, Barer MR, O'Donnell AG et al (2002) Characterization of the bacterial consortium associated with black band disease in coral using molecular microbiological techniques. Environ Microbiol 4:401–413

Dick MW (2001) Straminipilous fungi. Kluwer Academic Publishers, Doldrecht, Netherlands

Domart-Coulon LJ, Sinclair CS, Hill RT, Tambutté S, Puverel S, Ostrander GK (2004) A basidiomycete isolated from the skeleton of *Pocillopora damicornis* (Scleractinia) selectively stimulates short-term survival of coral skeletogenic cells. Mar Biol 144:583–592

Douglas NL, Mullen KM, Talmage SC, Harvell CD (2007) Exploring the role of chitinolytic enzymes in the sea fan coral, *Gorgonia ventalina*. Mar Biol 150:1137–1144

Dunlap WC, Shick JM (1998) Ultraviolet radiation-absorbing mycosporine-like amino acids in coral reef organisms: a biochemical and environmental perspective. J Phycol 34:418–430

Ein-Gill N, Ilan M, Carmeli S, Smith GW, Pawlik JR, Yarden O (2009) Presence of *Aspergillus sydowii*, a pathogen of gorgonian sea fans in the marine sponge *Spongia obscura*. ISME J 3:752–755

Frias-Lopez J, Bonheyo GT, Jin QS, Fouke BW (2003) Cyanobacteria associated with coral black band disease in Caribbean and Indo-Pacific Reefs. Appl Environ Microbiol 69:2409–2413

Garrison VH, Foreman WT, Genauldi S, Griffin DW et al (2006) Saharan dust – a carrier of persistent organic pollutants, metals and microbes to the Caribbean? Rev Biol Trop 54:9–21

Golubic S, Radke G, Le-Campoin Alsumard T (2005) Endolithic fungi in marine ecosystems. Trends Microbiol 12:229–235

Harel M, Ben-dov E, Rasoulouniriana D, Siboni N, Kramarsky-winter E, Loya Y, Barak Z, Weisman Z, Kushmaro A (2008) A new thraustochytrid, strain Fng1, isolated from the surface mucus of the hermatypic coral *Fungia granulosa*. FEMS Microbiol Ecol 64:378–387

Harvell CD, Mitchell CE, Ward JR, Altizer S, Dobson AP, Ostfeld RS, Samuel MD (2002) Climate warming and disease risks for terrestrial and marine biota. Science 296:2158–2162

Hayes ML, Bonaventura J, Mitchell T, Prospero JM, Shinn EA, Van Dolah F, Barber RT (2001) How are climate and marine biological outbreaks functionally linked? Hydrobiologia 460:213–220

Hyde KD, Pointing SB (2000) Marine mycology: a practical approach. Fungal Diversity Press, Hong Kong

Inácio J, Ludwig W, Spencer-Martins I, Fonseca A (2010) Assessment of phylloplane yeasts on selected Mediterranean plants by FISH with group-and species-specific oligonucleotide probes. FEMS Microbiol Ecol 71:61–72

Jagtap TG (1998) Structure of major seagrass beds from three coral reef atolls of Lakshadweep, Arabian Sea, India. Aquat Bot 60:397–408

Kellogg CA (2004) Tropical Archaea: diversity associated with the surface microlayer of corals. Mar Ecol Prog Ser 273:81–88

Kendrick B, Risk MJ, Michaelides J, Bergman K (1982) Amphibious microborers: bioeroding fungi isolated from live corals. Bull Mar Sci 32:862–867

Kim K, Kim PD, Alkar AP, Harvell CD (2000) Chemical resistance of gorgonian corals against fungal infections. Mar Biol 137:393–401

Kirkwood M, Todd JD, Rypien KL, Johnston AWB (2010) The opportunistic coral pathogen *Aspergillus sydowii* contains dddp and makes dimethyl sulphide from dimethylsulfoniopropionate. ISME J 4:147–150

Koh LI, Tan TK, Chou LM, Goh NKC (2000) Fungi associated with gorgonians in Singapore. Proceedings of the ninth international Coral Reef Symposium, vol 1, pp 521–526

Kohlmeyer J, Kohlmeyer E (1979) Marine mycology: the higher fungi. Academic Press, New York

Kohlmeyer J, Volkmann-Kohlmeyer B (1987) Koralionastetaceae fam. nov (Ascomycetes) from coral rock. Mycologia 79:764–778

Kohlmeyer J, Volkmann-Kohlmeyer B (1990) New species of *Koralionastes* (Ascomycotina) from the Caribbean and Australia. Can J Bot 68:1554–1559

Kohlmeyer J, Volkmann-Kohlmeyer B (1992) Two Ascomycotina from coral reefs in the Caribbean and Australia. Cryptogamie Bot 2:367–374

Kohlmeyer J, Volkmann-Kohlmeyer B (2003) Fungi from coral reefs: a commentary. Mycol Res 107:386–387

Kuntz NM, Kline DI, Sandin SA, Rohwer F (2005) Pathologies and mortality rates caused by organic carbon and nutrient stressors in three Caribbean coral species. Mar Ecol Prog Ser 294:173–180

Kvenefors EC, Sampaya E, Ridgway T, Barnes AC, Hoegh-Guldberg O (2010) Bacterial communities of two ubiquitous Great Barrier Reef corals reveals both site- and species-specificity of common bacterial associates. PLoS One 5:e10401–e10414

Le Campion-Alsumard R (1979) Les Cyanophycées endolithes marines. Systématique, ultrastructure, écologie et biodestruction. Ocean Acta 2:143–156

Le Campion-Alsumard T, Golubic S, Hutchings P (1995a) Microbial endoliths in skeletons of live and dead corals: *Porites lobata* (Moorea, French Polynesia). Mar Ecol Prog Ser 117:149–157

Le Campion-Alsumard T, Golubic S, Priess K (1995b) Fungi in corals: symbiosis or disease? Interaction between polyps and fungi causes pearl-like skeleton biomineralization. Mar Ecol Prog Ser 117:137–147

Lins-de-Barros MM, Vieira RP, Cardoso AM, Monteiro VA, Turque AS, Silveira CB, Albano RM, Clementino MM, Martins OB (2010) Archaea, bacteria and algal plastids associated with the reef-building corals *Siderastrea stellata* and *Mussismilia hispida* from Búzios, South Atlantic Ocean, Brazil. Microb Ecol 59:523–532

Littler MM, Littler DS (1998) An undescribed fungal pathogen of reef-forming crustose coralline algae discovered in American Samoa. Coral Reefs 17:144

Littman RA, Willis BL, Pfeffer C, Bourne DG (2009) Diversities of coral-associated bacteria differ with location, but not species for three acroporid corals on the Great Barrier Reef. FEMS Microbiol Ecol 68:152–163

Macintyre IG, Glans RR, Reinthal PN, Littler DS (1987) The barrier reef sediment apron: Tobacco Reef, Belize. Coral Reefs 6:1–2

Maldonado M, Cortadellas N, Trillas I, Rützler K (2005) Endosymbiotic yeast maternally transmitted in a marine sponge. Biol Bull 209:94–106

McClanahan TR, McLaughlin SM, Davy JE, Wilson WH, Peters EC, Price KL, Maina J (2002) Observations of a new source of coral mortality along the Kenyan Coast. Hydrobiologia 530:469–479

McDonald LA, Abbanat DR, Barbieri LR, Bernan VS, Discafani CM, Greenstein M, Janota K, Korshalla JD, Lassota P, Tischler M, Carter GT (1999) Spiroxins, DNA cleaving antitumor antibiotic from a marine-derived fungus. Tetrahedron Lett 40:2489–2492

Meikle P, Richards NG, Yellowlees D (1988) Structural investigations on the mucus from six species of coral. Mar Biol 99:187–193

Mitchell J, Zuccaro A (2006) Sequences, the environment and fungi. Mycologist 20:62–74

Morrison-Gardiner S (2002) Dominant fungi from Australian coral reefs. Fungal Divers 9:105–121

Mydlarz LD, Harvell CD (2007) Peroxidase activity and inducibility in the sea fan coral exposed to a fungal pathogen. Comp Biochem Physiol A 146:54–62

Nagelkerken I, Buchan K, Smith GW, Bonair K, Bush P, Garzón-Ferreira J, Botero L, Gayle P, Heberer C, Petrovic C, Pors L, Yoshioka P (1997b) Widespread disease in Caribbean sea fans: I. Spreading and general characteristics. Proceedings of the eighth international Coral Reef Symposium , vol 1, pp 679–682

Nagelkerken I, Buchan K, Smith GW, Bonair K, Bush P, Garzón-Ferreira J, Botero L, Gayle P, Harvell CD, Heberer C, Kim K, Petrovic C, Pors L, Yoshioka P (1997b) Widespread disease in Caribbean sea fans: II. Patterns of infection and tissue loss. Mar Ecol Prog Ser 160:255–263

Namikoshi M, Kobayashi H, Yoshimoto T, Meguro S, Akano K (2000) Isolation and characterization of bioactive metabolites from marine-derived filamentous fungi collected from tropical and sub-tropical coral reefs. Chem Pharm Bull 48:1452–1457

Namikoshi M, Akano K, Kobayashi H, Koike Y, Kitazawa A, Rondonuwu AB, Pratasik SB (2002) Distribution of marine filamentous fungi associated with marine sponges in coral reefs of Palau and Bunaken Island, Indonesia. J Tokyo Univ Fish 88:1–20

Olson ND et al (2009) Diazotrophic bacteria associated with Hawaiian *Montipora* corals: diversity and abundance in correlation with symbiotic dinoflagellates. J Exp Mar Biol Ecol 371:140–146

Petes l, Harvell CD, Peters EC, Webb MAH, Mullen KM (2003) Pathogens compromise reproduction and induce melanization in Caribbean sea fans. Mar Ecol Prog Ser 264:167–171

Phongpaichit S, Preedanan S, Rungiindama N, Sakayroj J, Benzies C, Chuaypat J, Plathong S (2006) Aspergillosis of the gorgonian sea fan *Annella* sp. after the tsunami at Mu Ko Similan National Park, Andaman Sea, Thailand. Coral Reefs 25:296

Priess K, Le Campion-Alsumard T, Golubic S, Gadel F, Thomassin BA (2000) Fungi in corals: black bands and density-banding of *Porites lutea* and *P. lobata* skeleton. Mar Biol 136:19–27

Raghukumar C (1986) Fungal parasites of the marine green algae *Cladophora* and *Rhizoclonium*. Bot Mar 29:289–297

Raghukumar S (1987) Occurrence of the thraustochytrid, *Corallochytrium limacisporum* gen. et. sp. nov. in the coral reef lagoons of the Lakshadweep islands in Arabian Sea. Bot Mar 30:83–89
Raghukumar S (1988) Detection of the thraustochytrid protist *Ulkenia visurgensis* in a hydroid, using immunofluorescence. Mar Biol 97:253–258
Raghukumar S (2002) Ecology of the marine protists, the Labyrinthulomycetes (Thraustochytrids and Labyrinthulids). Eur J Protistol 38:127–145
Raghukumar C (2006) Algal-fungal interactions in the marine ecosystem: symbiosis to parasitism. In: Tewari A (ed) Recent advances on applied aspects of Indian Marine algae with reference to global scenario, vol 1. Central Salt & Marine Chemicals Research Institute, India, pp 366–385
Raghukumar C (2008) Marine fungal biotechnology: an ecological perspective. Fungal Divers 31:19–35
Raghukumar S, Balasubramanian R (1991) Occurrence of thraustochytrid fungi in corals and coral mucus. Indian J Mar Sci 20:176–181
Raghukumar C, Raghukumar S (1991) Fungal invasion of massive corals. PSZNI Mar Ecol 12:251–260
Raghukumar C, D'Souza-Ticlo D, Verma AK (2008a) Treatment of colored effluents with lignin-degrading enzymes: an emerging role of marine-derived fungi. Crit Rev Microbiol 34:189–206
Raghukumar C, Mohandass C, Cardígos F, D'Costa PM, Santos RS, Colaço A (2008b) Assemblage of benthic diatoms and culturable heterotrophs in shallow-water hydrothermal vent of the D. João de Castro Seamount, Azores in the Atlantic Ocean. Curr Sci 95:1715–1723
Ramos-Flores T (1983) Lower marine fungus associated with black line disease in star corals (*Montasrea annularis* E & S). Biol Bull 165:429–435
Ravindran J, Raghukumar C, Raghukumar S (2001) Fungi in *Porites lutea*: association with healthy and diseased corals. Dis Aquat Org 47:219–228
Ritchie KB (2006) Regulation of microbial populations by coral surface mucus and mucus-associated bacteria. Mar Ecol Prog Ser 322:1–14
Rohwer F, Breitbart M, Jara J, Azam F, Knowlton N (2001) Diversity of bacteria associated with the Caribbean coral *Montastraea franksi*. Coral Reefs 20:85–91
Rypien KL (2008) African dust is an unlikely source of *Aspergillus sydowii*, the causative agent of sea fan disease. Mar Ecol Prog Ser 367:125–131
Rypien KL, Baker DM (2009) Isotopic labelling and antifungal resistance as tracers of gut passage of the sea fan pathogen *Aspergillus sydowii*. Dis Aquat Org 86:1–7
Rypien KL, Andras JP, Harwell CD (2008) Globally panmictic population structure in the opportunistic fungal pathogen *Aspergillus sydowii*. Mol Ecol 17:4068–4078
Sabdono A, Radjasa OK (2008) Phylogenetic diversity of organophosphorous pesticide-degrading coral bacteria from mid-west coast of Indonesia. Biotechnology 7:694–701
Sathe V, Raghukumar S (1991) Fungi and their biomass in the detritus of the seagrass *Thalassia hemprichii* (Ehrenberg) Ascherson. Bot Mar 34:271–277
Schröder S, Hain M, Sterflinger K (2000) Colorimetric *in situ* hybridization (CISH) with digoxigenin-labeled oligonucleotide probes in autofluorescent hyphomycetes. Int Microbiol 3:183–186
Sekar R, Kaczmarsky LT, Richardson LL (2008) Microbial community composition of black band disease on the coral host *Siderastrea siderea* from three regions of the wider Caribbean. Mar Ecol Prog Ser 362:85–98
Shinn EA, Smith GW, Prospero JM, Betzer P, Hayes ML, Garrison V, Barber RT (2000) African dust and the demise of Caribbean coral reefs. Geophy Res Lett 27:3029–3032
Siboni N, Ben-Dove E, Sivan A, Kushmaro A (2008) Global distribution and diversity of coral-associated Archaea and their possible role in the coral holobiont nitrogen cycle. Environ Microbiol 10:2979–2990
Sievert SM, Kiene RP, Schulz-Vogt HN (2007) The sulfur cycle. Oceanography 20:117–123
Smith GW, Ives LD, Nagelkerken IA, Ritchie KB (1996) Caribbean sea-fan mortalities. Nature 383:487

Smith TB, Nemeth RS, Blondeau J, Calnan JM, Kadison E, Herzlieb S (2008) Assessing coral reef health across onshore to offshore stress gradients in the US Virgin Islands. Mar Pollut Bull 56:1983–1991

Sparrow FK Jr (1960) Aquatic phycomycetes, 2nd edn. University of Michigan Press, Ann Arbor

Sterflinger K, Krumbein WE, Schwiertz A (1998) A protocol for PCR *in situ* hybridization of hyphomycetes. Int Microbiol 1:217–220

Stoecker K, Dominger C, Daims H, Wagner M (2010) Double labelling of oligonucleotide probes for fluorescence *in situ* hybridization (DOPE-FISH) improves signal intensity and increases rRNA accessibility. Appl Environ Microbiol 76:922–926

Sumathi JC, Raghukumar S, Kasbekar DP, Raghukumar C (2006) Molecular evidence of fungal signatures in the marine protist *Corallochytrium limacisporum* and its implications in the evolution of animals and fungi. Protist 157:363–376

Sunagawa S, DeSantis TZ, Piceno YM, Brodie EL, DeSalvo MK, Voolstra CR, Weil E, Andersen GL, Medina M (2009) Bacterial diversity and White Plague Disease-associated community changes in the Caribbean coral *Montastraea faveolata*. ISME J 3:512–521

Toledo-Hernández C, Bones-González A, Oritz-Vázquez OE, Sabat AM, Bayman P (2007) Fungi in the sea fan *Gorgonia ventalina*: diversity and sampling strategies. Coral Reefs 26:725–730

Toledo-Hernández C, Zuluaga-Montero A, Bones-González A, Rodríguez JA, Sabat AM, Bayman P (2008) Fungi in healthy and diseased sea fans (*Gorgonia ventalina*): is *Aspergillus sydowii* always the pathogen? Coral Reefs 27:707–714

Van Duyl FC, Gast GJ (2001) Linkage of small-scale spatial variations in DOC, inorganic nutrients and bacterioplankton growth with different coral reef water types. Aquat Microb Ecol 24:17–24

Vega Thurber R, Willner-Hall D, Rodriguez-Mueller B, Desnues C, Edwards RA, Angly F, Dinsdale E, Kelly L, Rohwer F (2009) Metagenomic analysis of stressed coral holobionts. Environ Microbiol 11:2148–2163

Ward JR, Kim K, Harwell CD (2007) Temperature affects coral disease resistance and pathogen growth. Mar Ecol Prog Ser 329:115–121

Wegley L, Edwards R, Rodriguez-Brito B, Liu H, Rohwer F (2007) Metagenomic analysis of the microbial community associated with the coral *Porites astreoides*. Environ Microbiol 9:2707–2719

Weir-Brush JR, Garrison VH, Smith GW, Shinn EA (2004) The relationship between gorgonian coral (Cnidaria:Gorgonacea) diseases and African dust storms. Aerobiologia 20:119–126

Yarden O, Ainsworth TD, Roff J, Leggat W, Fine M, Hoegh-Guldberg O (2007) Increased prevalence of ubiquitous Ascomycetes in an acroporid coral (*Acropora formosa*) exhibiting symptoms of brown band syndrome and skeletal eroding band diseases. Appl Environ Microbiol 73:2755–2757

Zengler K, Toledo G, Rappe M, Elkins J, Mathur EJ, Short JM, Keller M (2002) Cultivating the uncultured. Proc Natl Acad Sci U S A 26:15681–15686

Chapter 6
Fungal Life in the Dead Sea

Aharon Oren and Nina Gunde-Cimerman

Contents

Abstract The waters of the Dead Sea currently contain about 348 g/l salts (2 M Mg^{2+}, 0.5 M Ca^{2+}, 1.5 M Na^{+}, 0.2 M K^{+}, 6.5 M Cl^{-}, 0.1 M Br^{-}). The pH is about 6.0. After rainy winters the surface waters become diluted, triggering development of microbial blooms. The 1980 and 1992 blooms were dominated

---

The authors dedicate this chapter to Prof. Dr. (Emeritus) Hans Peter Molitoris (Regensburg) to thank him for his profound interest in the Dead Sea ecosystem and its fungi.

A. Oren (✉)
Department of Plant and Environmental Sciences, The Institute of Life Sciences, The Hebrew University of Jerusalem, Jerusalem 91904, Israel
e-mail: orena@cc.huji.ac.il

C. Raghukumar (ed.), *Biology of Marine Fungi*,
Progress in Molecular and Subcellular Biology 53,
DOI 10.1007/978-3-642-23342-5_6, © Springer-Verlag Berlin Heidelberg 2012

by the unicellular green alga *Dunaliella* and red Archaea. At least 70 species (in 26 genera) of Oomycota (Chromista), Mucoromycotina, Ascomycota, and Basidiomycota (Fungi) were isolated from near-shore localities and offshore stations, including from deep waters. *Aspergillus* and *Eurotium* were most often recovered. *Aspergillus terreus*, *A. sydowii*, *A. versicolor*, *Eurotium herbariorum*, *Penicillium westlingii*, *Cladosporium cladosporioides*, *C. sphaerospermum*, *C. ramnotellum*, and *C. halotolerans* probably form the stable core of the community. The species *Gymnascella marismortui* may be endemic. Mycelia of Dead Sea isolates of *A. versicolor* and *Chaetomium globosum* remained viable for up to 8 weeks in Dead Sea water; mycelia of other species survived for many weeks in 50% Dead Sea water. Many isolates showed a very high tolerance to magnesium salts. There is no direct proof that fungi contribute to the heterotrophic activity in the Dead Sea, but fungi may be present at least locally and temporarily, and their enzymatic activities such as amylase, protease, and cellulase may play a role in the lake's ecosystem.

6.1 Introduction: The Dead Sea Ecosystem

The Dead Sea (Fig. 6.1) is one of the most saline bodies of water on Earth. Currently its waters contain about 348 g/l total dissolved salts. The ionic composition is dominated by divalent cations (2 M Mg^{2+}, 0.5 M Ca^{2+}) rather than monovalent cations (1.5 M Na^{+}, 0.2 M K^{+}). Cl^{-} (6.5 M) and Br^{-} (0.1 M) are the main anions, and the water has a pH of about 6.0. The maximum depth is about 310 m, and the lake is aerobic down to the bottom. Surface water temperatures vary from about 18°C in winter to 35°C in summer. The lake is saturated with respect to sodium ions, and massive amounts of NaCl precipitate to the bottom as halite crystals. The result is a decrease in sodium content with an increase in the relative concentrations of divalent cations (Table 6.1). In summer, a seasonal thermocline develops and floodwaters that enter occasionally in winter may cause the formation of a less saline upper water layer. Still, in the past decade complete mixing occurred at least once every year (a holomictic regime) (Oren 2003a).

Currently the lake's waters are too hostile for life of even the best salt-adapted microorganisms. No algae are found in the water column, and only a small community of halophilic Archaea (family *Halobacteriaceae*) survives (Bodaker et al. 2010). However, occasionally the winter floods from the surrounding area and water inflow from the Jordan River can be so massive that a diluted upper water layer is formed that remains for several years (a meromictic regime). When this happened following the exceptionally rainy winters of 1979–1980 and 1991–1992, dense microbial blooms developed in the diluted epilimnion, consisting of the unicellular green alga *Dunaliella* as primary producer and red halophilic Archaea

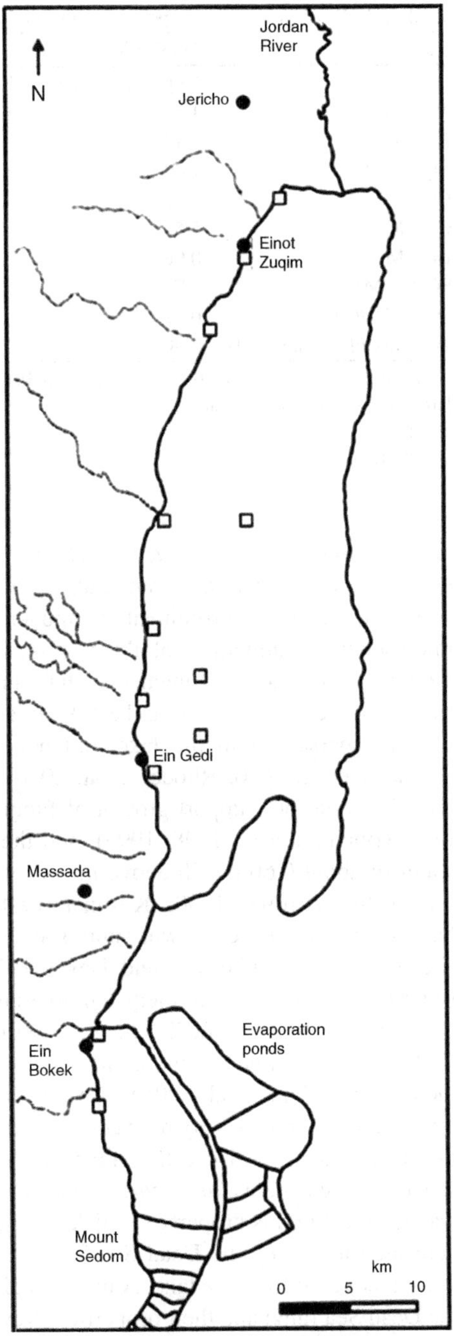

Fig. 6.1 Map of the Dead Sea, indicating the sites mentioned in the text. The sampling stations are indicated by *open squares*

Table 6.1 Dead Sea ionic composition and salt concentrations

Year	1959–1960		1977	1996	2007
	Upper water mass	Lower water mass	Average	–	–
Na^+ (M)	1.57	1.84	1.73	1.59	1.54
K^+ (M)	0.17	0.19	0.18	0.20	0.21
Mg^{2+} (M)	1.49	1.75	1.81	1.89	1.98
Ca^{2+} (M)	0.41	0.43	0.43	0.44	0.47
Cl^- (M)	5.56	6.18	6.34	6.34	6.48
Br^- (M)	0.06	0.07	0.07	0.07	0.08
SO_4^{2-} (M)	NR	NR	0.005	0.005	0.005
$(Na^++K^+)/(Mg^{2+}+Ca^{2+})$[a]	0.92	0.93	0.85	0.77	0.71
Total dissolved salts (g/l)	298	335	339	339	347

Data were derived from Beyth (1980), and unpublished data (I. Gavrieli, the Geological Survey of Israel, personal communication)
NR not reported
[a]Molar ratio

of the family *Halobacteriaceae* (Oren 1988, 1993, 1997, 1999, 2000, 2003b; Oren and Gurevich 1995; Oren et al. 1995, 2005).

Fungi are not a prominent component of the Dead Sea ecosystem: direct microscopic examination of the waters, including during periods of microbial blooms, never showed fungal mycelia, and metagenomic approaches, applied to samples collected both during the 1992 microbial bloom and during a recent period of very sparse life in the lake, did not provide evidence for fungal presence (Bodaker et al. 2010; Rhodes et al. 2010). Still, there is no a priori reason why the lake could not support growth of fungi. A relatively low pH generally favors fungi (Buchalo et al. 1998, 1999), and the water activity of undiluted Dead Sea water of about 0.66 is still above the a_w of 0.61 considered as the lower limit for fungal life (Brown 1990; Kushner 1978). Moreover, halophilic and highly halotolerant fungi are known: some species of *Aspergillus* and *Penicillium* grow above 25% NaCl (Tresner and Hayes 1971). Fungal species isolated from the salterns such as the extremely halotolerant black yeast *Hortaea werneckii* can grow from 0 to 32% NaCl (Gunde-Cimerman et al. 2000), while the halophilic *Wallemia ichthyophaga* cannot grow without the presence of at least 10% NaCl in the medium (Zalar et al. 2005a). Besides, recent evidence indicates that fungi are much more tolerant to very high concentrations of Mg^{2+} ions, particularly abundant in the waters of the Dead Sea, than previously expected. In particular, different species of the genera *Cladosporium* and *Wallemia*, both isolated from the Dead Sea waters, excel in this property (Sonjak et al. 2010). Therefore, research efforts were dedicated to survey the Dead Sea for fungi and to characterize the properties of those recovered from the lake. A monograph has been devoted to the description of the Dead Sea fungi and their properties (Nevo et al. 2003). This chapter reviews the result of these investigations and some not yet published investigations.

6.2 The Fungi of the Dead Sea

6.2.1 Approaches for the Isolation of Dead Sea Fungi

Most studies on Dead Sea fungi were performed with samples collected in the period 1995–2003, when the lake was holomictic. Different approaches were used: mixing of water samples with molten agar and incubation of the solidified agar in Petri dishes, filtration of water samples through 0.45-μm-pore-sized Millipore filters and incubation of the filters on agar plates, and also enrichment cultures in which Dead Sea water samples were added to portions of liquid media.

Media used were generally conventional media employed in mycological studies, such as glucose-yeast extract, malt extract agar, or Czapek agar, and these were prepared with different concentrations of Dead Sea water (10, 20, 40, or 50%, by volume) or different NaCl concentrations (17–32% NaCl). Media without added salt were used as well. To prevent overgrowth of prokaryotic microorganisms, different antibiotics were used: streptomycin and chloramphenicol or tetracycline, at a concentration of 100 μg/ml each. Cultures were incubated at 20, 26, and 37°C (Buchalo et al. 1999, 2000b; Butinar et al. 2005a; Kis-Papo et al. 2001, 2003c).

6.2.2 Fungal Diversity in the Dead Sea Water Column

The first fungal isolates were retrieved from Dead Sea surface water sampled in the vicinity of the freshwater springs of Einot Zuqim, and these included *Ulocladium chlamydosporum*, *Penicillium westlingii*, and *Gymnascella marismortui* (Buchalo et al. 1999, 2000b). Further studies of the spatial and temporal diversity in the fungal community of the lake were based on sampling of different near-shore localities and at a number of stations offshore, including collection of deep water samples down to the bottom (304 m) in the center of the lake (Fig. 6.1; Table 6.2) (Buchalo et al. 1999, 2000b; Butinar et al. 2005b; Kis-Papo et al. 2001, 2003a, c, d; Molitoris et al. 2000; Wasser et al. 2003).

The annotated list compiled by Wasser et al. (2003) encompasses 70 species of filamentous fungi, belonging to 26 genera, 10 orders, and 3 divisions. Ascomycota dominate with 66 species (only 12 of which have known teleomorphs); Oomycota and Mucoromycotina are represented with three and one species, respectively. Nearly half of the species are melanin-containing micromycetes. The Eurotiales are well represented with 33 species, including *Aspergillus* (19 species including *Emericella*) and *Penicillium* (13 species). Other genera frequently encountered were *Chaetomium* (Sordariales; 5 species) and *Cladosporium* (Davidiellaceae, Capnodiales; 5 species). These features can be considered as characteristic for mycobiota of a highly stressful environment. The list compiled by Wasser et al. (2003) was further enlarged by additional isolates obtained by Gunde-Cimerman et al. (unpublished data), and is shown in Tables 6.2 and 6.3.

Table 6.2 Systematic diversity of the Dead Sea microfungi (adapted from Wasser et al. 2003)

Division	Order or class	Number of taxa	
		Genera	Species
Oomycota	Pythiales	1	1
Mucoromycotina	Mucorales	2	3
Ascomycota	Capnodiales	1	13
	Dothideales	1	1
	Eurotiales	4	35
	Hypocreales	6	10
	Onygenales	2	2
	Ophiostomatales	1	1
	Pleosporales	5	12
	Sordariales	2	6
Saccharomycetes	Saccharomycetales	2	3
Basidiomycota	Russulales	1	1
	Sporidiobolales	1	1
	Tremellales	2	2
	Polyporales	2	2
	Wallemiales	1	1
Total		34	94

Many of the fungi recovered from Dead Sea water samples are non-halophilic, non-xerophilic, terrestrial species, and many species were also reported from soils in the surrounding lakes (Sect. 6.2.5). The genera most often recovered were related species with sexual stages in *Eurotium* and asexual ones in *Aspergillus*. *Eurotium* and its associated *Aspergillus* anamorph are known as xerophilic, adapted to matric stress, and is also often encountered in hypersaline waters (Tresner and Hayes 1971; Pitt and Hocking 1997). Thus, six species of the genus (*E. amstelodami*, *E. repens*, *E. herbariorum*, *E. rubrum*, *E. chevalieri*, and a potential new species "*E. halotolerans*") were repeatedly isolated from environments >17% salt, including coastal saltern ponds in Spain, Portugal, France, Slovenia, Israel, Namibia, and the Dominican Republic. *E. amstelodami* was also isolated from an industrial evaporation pond at the southern end of the lake (Butinar et al. 2005b, Gunde-Cimerman et al. 2005, 2009). However, *E. herbariorum* and *E. rubrum* grew poorly on agar media prepared with 50% Dead Sea water (Kis-Papo et al. 2003a). *Ulocladium chlamydosporum* is also poorly adapted to salinity of Dead Sea (Molitoris et al. 2000). *A. versicolor*, *E. herbariorum*, and *P. westlingii* were most abundant in winter; *A. sydowii*, *Cladosporium cladosporioides*, and *C. sphaerospermum* were found in comparable numbers the year round. Hypersaline waters of salterns on different continents most frequently harbor *A. niger*, *E. amstelodami*, and *P. chrysogenum* and to a lesser extent *A. sydowii*. *E. herbariorum*, and *A. versicolor* were isolated only sporadically, and they were considered only as temporal inhabitants, not adapted to long-term survival (Butinar et al. 2011). The diversity of *Cladosporium* species isolated appears to be higher than previously reported (Wasser et al. 2003). Besides *C. cladosporioides* and

Table 6.3 Fungal species reported from Dead Sea water

Group/class	Order	Genus	Species[a]
Oomycota	Pythiales	*Pythium*	*Pythium* sp.
Mucoromycotina	Mucorales	*Absidia*	*A. corymbifera* (Cohn) Sacc. et Trotter
			A. glauca Hagem, 1908
	Mucorales	*Circinella*	*C. rigida* G. Sm.
Ascomycota	Hypocreales	*Acremonium*	*A. implicatum* (J.C. Gilman et E.V. Abbott (W. Gams)
			A. persicinum (Nicot) W. Gams
			A. rutilum W. Gams
	Pleosporales	*Alternaria*	*A. alternata* (Fr.) Kessler
			A. pluriseptata (R. Karst., Har. et Peck) Jorst.
			Alternaria sp.
	Eurotiales	*Aspergillus*	*A. caespitosus* Raper et Thom
			A. carneus (Tiegh.) Blochwitz
			A. flavus Link
			A. fumigatus Fresen.
			A. niger Van Tiegh.
			A. penicilloides Speg.
			A. phoenicis (Corda) Thom de Currie (*A. niger* (nom. cons.), or *A. tubingensis*)
			A. ochraceus G.Wilh.
			A. proliferans G.Sm.
			A. sclerotiorum G.A. Huber
			A. sydowii Bainier et Sartory (Thom et Church)
			A. terreus Thom
			A. tubingensis Mosseray
			A. ustus (Bainier) Thom et Church
			A. versicolor (Vuill.) Tirab.
			A. wentii Wehmer
	Dothideales	*Aureobasidium*	*A. pullulans* (de Bary) Arnaud
	Saccharomycetales	*Candida*	*C. glabrata* (H.W. Anderson) S.A. Mey. et Yarrow
			C. parapsilosis (Ashford) Langeron & Talice)
	Sordariales	*Chaetomium*	*C. aureum* Chivers
			C. flavigenum van Warmelo
			C. funicola Cooke
			C. globosum Kunze
			C. nigricolor L.M. Ames
	Onygenales	*Chrysosporium*	*Chrysosporium* sp.
	Capnodiales	*Cladosporium*	*C. cladosporioides* (Fresen.) G.A. de Vries
			C. aff. *cladosporioides*
			C. halotolerans Zalar, de Hoog & Gunde-Cim.
			C. herbarum (Pers.) Link

(continued)

Table 6.3 (continued)

Group/class	Order	Genus	Species[a]
			C. aff. *herbarum* (Pers.) Link
			C. aff. *inversicolor* Bensch, Crous & U. Braun
			C. macrocarpum Preuss
			C. oxysporum Berk. et M.A. Curtis
			C. ramotenellum K. Schub., Zalar, Crous & U. Braun 2007
			C. sphaerospermum Penz.
			C. aff. *sphaerospermum.*
			C. tenellum K. Schub., Zalar, Crous & U. Braun 2007
	Cladosporium sp.		
	Pleosporales	*Curvularia*	*C. protuberata* R.R. Nelson et Hodges
	Eurotiales	*Emericella*	*E. nidulans* (Eidam) Vuill.
			E. purpurea Samson et Mouchacca
			E. neopurpurea (a new species under description)
	Hypocreales	*Engyodontium*	*E. album* (Limber) de Hoog
	Eurotiales	*Eurotium*	*E. amstelodami* L. Mangin
			E. herbariorum (F.H. Wigg.) Link
			E. rubrum W. Bremer
	Onygenales	*Gymnascella*	*G. marismortui* Buchalo et al.
	Hypocreales	*Isaria*	*I. farinosa* (Holmsk.) Fr.
	Saccharomycetales	*Meyerozyma*	*M. guillermondii* (Wick.) Kurtzman & M. Suzuki
	Hypocreales	*Myrothecium*	*M. roridum* Tode
	Eurotiales	*Penicillium*	*P. brevicompactum* Dierckx
			P. chrysogenum Thom
			P. citrinum Thom
			P. commune Thom
			P. corylophilum Dierckx
			P. crustosum Thom
			P. fellutanum Biourge
			P. glabrum (Wehmer) Westling
			P. implicatum Biourge
			P. restrictum J.C. Gilman et E.V. Abbott
			P. steckii K.M. Zalessky
			P. variabile Sopp
			P. westlingii K.M. Zalessky
	Pleosporales	*Phoma*	*P. leveillei* Boerema et G.J. Bollen
	Ophiostomatales	*Sporothrix*	*S. guttuliformis* de Hoog
	Hypocreales	*Stachybotrys*	*S. chartarum* (Ehrenb.) S. Hughes
	Pleosporales	*Stemphylium*	*Stemphylium herbarum* E.G. Simmons
	Sordariales	*Thielavia*	*T. terricola* (J.C. Gilman et E.V. Abbott) C.W. Emmons
	Hypocreales	*Trichoderma*	*T. koningii* Oudem.
			T. aff. *atroviridae.*
			Trichoderma sp.

(continued)

Table 6.3 (continued)

Group/class	Order	Genus	Species[a]
	Pleosporales	*Ulocladium*	*U. alternariae* (Cooke) E.G. Simmons
			U. atrum Preuss
			U. chlamydosporum Mouchacca
			U. oudemansii E.G. Simmons
			U. tuberculatum E.G. Simmons
Ulocladium sp.			
Basidiomycota	Polyporales	*Bjerkandera*	*Bjerkandera* sp.
	Tremelalles	*Cryptococcus*	*C. albidus* var. *kuetzingii* (Fell & Phaff) Fonseca, Scorzetti & Fell
	Sporidiobolales	*Rhodotorula*	*R. laryngis* Reiersöl
	Russulales	*Stereum*	*S. hirsutum* (Willd.) Pers.
	Polyporales	*Trametetes*	*T. versicolor* (L.) Lloyd
	Tremellales	*Trichosporon*	*T. mucoides* E. Guého et M.T. Sm.
	Wallemiales	*Wallemia*	*W. sebi* (Fr.) Arx

Data were derived from Buchalo et al. (1998, 1999, 2000a, b), Butinar et al. (2005a, b), Kis-Papo et al. (2001, 2003c), Molitoris et al. (2000), and Wasser et al. (2003, unpublished results)

[a] Lists of synonyms can be found in the Index Fungorum (http://www.indexfungorum.org)

C. sphaerospermum, *C. halotolerans* and *C. ramnotellum* also represent the core fungal community (Zalar et al. 2007).

Fungal diversity increased near the outlets of less saline springs near the shore. Some of the Dead Sea fungi could therefore have entered the lake with river or spring water or from the air. Species richness was highest in winter, and varied greatly between sampling localities. This implies an unstable and variable community with very few constant species. More than 40% of the diversity was isolated >0.5 km offshore. Most species recovered were isolated during one season only and from a limited number of sampling sites. No single species had a temporal or spatial frequency of occurrence of >75%.

Only a few species were consistently isolated from the water column, and these may constitute the core of the Dead Sea mycobiota. These are *Aspergillus sydowii*, *A. versicolor*, *Eurotium herbariorum*, *Cladosporium cladosporoides*, and *C. sphaerospermum*. Moreover, at least for some of these species the spores and mycelia survived prolonged suspension in Dead Sea water (Sect. 6.3.1). Surprisingly, halotolerant black yeasts, the dominant representatives of fungi in hypersaline waters of salterns worldwide, were not isolated from the lake (Butinar et al. 2005c; Gunde-Cimerman et al. 2000), but only from the surface of halophytic plants growing on the shores (Gunde-Cimerman et al., unpublished data).

6.2.3 Gymnascella marismortui: *An Endemic Dead Sea Fungus?*

One of the Ascomycota recovered from Dead Sea surface water near the Einot Zuqim springs in January–November 1995 was a new species of the genus *Gymnascella*, now described as a new species, *Gymnascella marismortui*

(Buchalo et al. 1998, 1999, 2000a). At the time seven isolates were obtained, but to our knowledge it was not found in later studies. It is an obligate halophile that grows optimally between 0.5- and 2 M NaCl, or in media containing 10–30% by volume of Dead Sea water, and some growth was even obtained in 50% Dead Sea water media (Molitoris et al. 2000). High sucrose concentrations instead of salt supported only a slow growth. Mycelia retained their viability for 4 weeks in undiluted Dead Sea water and 12 weeks in Dead Sea water diluted to 80% (Kis-Papo et al. 2003a). Also its spores survived prolonged suspension in Dead Sea water better than the spores of other species (Kis-Papo et al. 2003d) (see also Sect. 6.3.1).

This species might thus be well adapted to life in the Dead Sea. Thus far it was not reported from other sites, and until further notice it may be considered endemic to the lake (Kis-Papo et al. 2003d). Further studies should ascertain the temporal and spatial distribution of *G. marismortui* to decide whether it is a true inhabitant of the Dead Sea, or whether its habitat is restricted to the areas of lowered salinity where freshwater from the Einot Zuqim springs mixes with the brines, and/or other areas with a reduced salinity due to freshwater influx.

G. marismortui displays several enzymatic activities of interest: urease (optimal production at 27°C at 17.5% salt), cellulase (best at lower temperatures), and amylase, but in contrast to other Dead Sea fungi tested (Sect. 6.3.2) it did not display any dye decolorization activity (Molitoris et al. 2000).

6.2.4 Yeasts in the Dead Sea

In an unpublished M.Sc. thesis, Kritzman (1973) reported the isolation of an osmophilic yeast from the Dead Sea that grew in a medium with 15% glucose and 12% salt. Unfortunately no cultures have been preserved. It should be noted that in the early 1970s the lake was stably meromictic, and the conditions were less extreme than today (Table 6.1).

The list of 70 Dead Sea fungi given by Wasser et al. (2003) included a single yeast species (*Candida glabrata*). Butinar et al. (2005a) reported the isolation of a *Candida glabrata*-like isolate, *Candida parapsilosis*, *Rhodotorula laryngis*, and *Trichosporon mucoides*. In addition, *Cryptococcus albidus* and *Meyerozyma guillermondii* were isolated (Gunde-Cimerman, unpublished results). *T. mucoides* was found also in the salterns of Eilat.

6.2.5 Comparison of the Fungal Community of the Dead Sea Waters with that of the Surrounding Soils

Of the 70 species reported from Dead Sea water (Wasser et al. 2003), 46 were previously isolated from the terrestrial shore, including the commonly found *Eurotium herbariorum*, *Aspergillus versicolor*, and *Cladosporium cladosporoides*.

The fungal diversity of soils in the Dead Sea area has been investigated by a number of groups (Guiraud et al. 1995; Steiman et al. 1995, 1997; Grishkan et al. 2003, 2004; Volz and Wasser 1995). Grishkan et al. (2003) found 78 micromycete species from 40 genera in soils from western shore of the lake. The majority contained melanin, and most isolates did not show special halophilic or thermophilic properties. The same was even true for the mycoflora of the saline Arubotaim Cave, a Holocene karst cave within the halite mountain of Mount Sedom (Grishkan et al. 2004). Soils from the Dead Sea shore and from oases in the area do not have a highly unusual fungal flora (Steiman et al. 1994, 1995, 1997). Many species encountered have a worldwide distribution. A number of new fungal species isolated from the area were described: *Aspergillus homomorphus*, *A. pseudoheteromorphus* (Steiman et al. 1994), *Exserohilum sodomii* (Guiraud et al. 1997), and *Bipolaris israeli* (Steiman et al. 1996).

6.3 Special Adaptations and Properties of Dead Sea Fungi

6.3.1 Survival of Fungal Spores in Dead Sea Water

The procedures used to isolate fungi from the Dead Sea (Sect. 6.2.1) do not discriminate between vegetative mycelia and spores, and therefore do not provide information whether the fungi may actually be growing in the Dead Sea water or may have entered the lake as resistant conidiospores, ascospores, basidiospores, or chlamydospores, depending on the species, and had survived for prolonged periods in the lake as spores. Therefore, the survival of spores and vegetative mycelia was examined after suspension in the undiluted and diluted Dead Sea water (80%, 50%, 10%, by volume), comparing the result with the survival of strains of the same species or closely related ones isolated from the Dead Sea terrestrial shore and from a control Mediterranean environment (Kis-Papo et al. 2003a, d). Mycelia were grown in media containing 10% Dead Sea water, checked microscopically for absence of spores, and then transferred serially to increasing concentrations of the Dead Sea water (30–50–80%) to prevent severe osmotic shock. Spores were added to Dead Sea water or its dilutions to a density of 200/ml. After incubation periods of up to 12 weeks the number of viable spores and the presence of viable mycelium were tested by plating on suitable media (for *Gymnascella* and *Eurotium* supplemented with 30% Dead Sea water).

Mycelia of *Aspergillus versicolor* and *Chaetomium globosum* remained viable for up to 8 weeks in undiluted Dead Sea water; four isolates (*A. versicolor*, *Eurotium herbariorum*, *Gymnascella marismortui*, and *C. globosum*) retained their viability for 12 weeks in water diluted to 80%. In water diluted to 50% of its original salinity, mycelia of all species survived for many weeks.

Ascospores of *G. marismortui* survived very well, with 70% viable spores in undiluted Dead Sea water and nearly 100% in 80% Dead Sea water at the termination of the experiment. Ascospores of an *Emericella nidulans* isolated from a non-saline environment proved less resistant than conidia: after 14 days, 40% of the conidia were still viable, as compared to 5% of the ascospores. This is contrary to the common finding that ascospores have improved stress resistance compared with conidia; the tolerance of sexual spores is one of the reasons why they are more frequently encountered in stressful environments (Jin and Nevo 2003). Spores of *Aspergillus sydowii* and *Eurotium herbariorum* (a xerophilic genus that can grow in 50% Dead Sea water) survived for 3 months in undiluted and diluted water; however, spores of these species obtained from the shore of the lake proved less tolerant (10–20% survival only). Spores isolated from a site with low salt concentrations survived for less than 2 months (Kis-Papo et al. 2003a, d).

When spores and mycelium of six different species of the genus *Eurotium*, most frequently isolated from salterns, were tested for viability in hypersaline waters, at the highest salinity (30% NaCl) spore viability of most species decreased drastically within the first week and then with a slower pace. Viability of *E. herbariorum* spores remained unchanged after the initial decrease of 65%, while only up to 15% of the spores of the other species remained viable. The highest adaptive ability of survival in hypersaline water was shown for *E. rubrum* spores, since almost half of them survived. Surprisingly, mycelia of all tested species survived 10 weeks of incubation even at the highest tested salinities. Mycelia of *E. amstelodami*, *E. herbariorum*, *E. repens*, and *E. chevalieri* remained viable throughout the test period of 12 weeks (Butinar et al. 2005b).

6.3.2 Enzyme Production and Dye Decoloration by Dead Sea Fungi

Many fungi possess exoenzymes and display interesting metabolic properties such as decolorization of dyes and degradation of other toxic chemicals. Therefore, selected fungal isolates from the Dead Sea were tested for different enzymatic activities: amylase, caseinase, urease, and cellulase (Buchalo et al. 1999; Molitoris et al. 2000). *P. westlingii* and *U. chlamydosporum* were the best urease producers, whereas *G. marismortui* was the best cellulase producer. Enzyme production generally decreased with increasing temperature and salinity, but in some cases (*G. marismortui* amylase, cellulase, urease, *U. chlamydosporum* amylase, and caseinase) optimal production was observed at intermediate salinities and temperatures.

Dye decolorization was tested as a model system to evaluate the potential of Dead Sea fungi to degrade pollutants. Dyes of different chemical groups were included in the tests: anthraquinones (remazol brilliant blue R, disperse blue), heterocyclic compounds (fluorescein, methylene blue, bromophenol blue), azo dyes (methyl orange, methyl red, reactive orange), and diazo dyes (congo red, sudan black B, reactive black). The anthraquinone and the monoazo dye methyl red were most rapidly decolorized, while the diazo dyes were only poorly degraded. Among the mitosporic taxa, *U. chlamydosporum* was the best dye degrader, an unnamed *Acremonium* species was the weakest. Among the ascomycetes *E. nidulans* and *A. phoenicis* were the best decolorizers (Buchalo et al. 1999; Molitoris et al. 2000).

6.3.3 Genetic Diversity of Dead Sea Fungi and Comparison with Related Isolates from Other Environments

To further examine whether fungi isolated from the Dead Sea differ from strains of the same species found in less salt-stressed ecosystems, the genetic diversity was investigated using fluorescent amplified fragment length polymorphism (AFLP) markers as a high-resolution fingerprinting tool (Kis-Papo et al. 2003b). Three species were included in these studies: the cosmopolitan *Aspergillus versicolor*, the halophilic *Eurotium amstelodami*, and *E. herbariorum*. *A versicolor* showed the highest level of gene diversity with 78% of polymorphic loci from the 605 bands scored. *E. herbariorum* showed the lowest percentage of polymorphic loci (31%), and all isolates including those from the lake shore clustered in one group. *E. amstelodami* had two significant genotypes with no apparent correlation to distribution. In both *E. amstelodami* and *A. versicolor* the water column population was significantly differentiated from populations from the Jordan River and a non-saline environment in the north of Israel (Nahal Keziv), but not from the populations of the Dead Sea shore. For the halophilic *E. amstelodami*, the freshwater of the Jordan River is no less stressful than the saline water of the Dead Sea, and accordingly low polymorphism levels were found in the Jordan River samples. In the three species the shore isolates were not differentiated from strains from Dead Sea surface water, which may be explained by migration between the Dead Sea and the shore populations.

Genetic diversity showed an increase from the mild north to the stressful Dead Sea area culminating in the sea surface, but dropped to very low genetic diversity in the water column collected from 30 to 300 m depth, where only a few adapted ecotypes survive. Clearly the highly stressful Dead Sea environment lowers the diversity of its fungi with only a few adapted ecotypes surviving. In addition, the percentage sexually reproducing species increases with stress, but drops dramatically at the highest stress levels, which select for few well-adapted but homozygous genotypes (Kis-Papo et al. 2003b, e).

Further genetic studies were made of the EhHOG gene of a Dead Sea isolate of *Eurotium herbariorum*, encoding a key component of a signal transduction cascade that plays an essential role in the osmoregulatory pathway. When EhHOG was expressed in a *S. cerevisiae hog1*Δ mutant, the growth and aberrant morphology of the *hog1*Δ mutant were restored under conditions of high osmotic stress. Moreover, intracellular glycerol content in the transformant increased to a higher level than that in the mutant during salt-stress conditions. Such genes from Dead Sea fungi may be a promising resource to improve salt tolerance of plants. Preliminary results showed that indeed the EhHOG gene may provide increased stress tolerance in transgenic *Arabidopsis* (Jin et al. 2005).

6.4 How Important Are Fungi in the Dead Sea Ecosystem?

6.4.1 *Occurrence of Fungi on Wood Submerged in Dead Sea Water*

A promising place to look for development of fungal mycelia is submerged wood and wooden structures. Thus, *Cladosporium* sp. was found on submerged pine wood in Great Salt Lake at a salt concentration of 290–360 g/l salt (Cronin and Post 1977). Zalar et al. (2005b) reported that the black yeasts *Hortaea werneckii* and *Trimmatostroma salinum* degrade wood immersed in hypersaline waters of the salterns. Isolates of both species showed xylanolytic and lignolytic activity under hypersaline and non-saline conditions; *T. salinum* displayed cellulolytic activity as well. These results suggested an active lignicolous saprobic role of halophilic fungi in hypersaline water environments.

The finding of fungal mycelia on near-shore wooden structures submerged in the Dead Sea was mentioned in several publications (Buchalo et al. 1998, 1999, 2000a, b; Kis-Papo et al. 2001, 2003b; Molitoris et al. 2000). However, nowhere were the findings documented in-depth. Kis-Papo et al. (2003b) stated that "live mycelium of nine fungal species, including *Eurotium amstelodami*, *E. herbariorum*, and *E. rubrum*, were found growing on driftwood submerged near the aquatic shore," and cited papers by Kis-Papo et al. (2003c) and Buchalo (2003), but these sources do not provide any additional information. Buchalo et al. (2000b) described an experiment in which 1–2 cm long pieces of wood on which fungal hyphae were observed under a binocular microscope were placed in Petri dishes on filter paper wetted with sterile Dead Sea water and incubated at 28°C for 30–40 days, the filter paper being regularly wetted with sterile distilled water. However, the results of this experiment were not reported.

6.4.2 *Isolation of Vegetative Fungi in the Dead Sea by Baiting*

Baiting experiments to isolate fungi, using media containing starch and cellulose were set up in the summer of 2000 at the lake surface at an offshore station and again in the spring of 2003 in near-shore areas. No fungal mycelia were detected after 6 weeks of incubation (Kis-Papo et al. 2003c). Possibly the baits selected were not appropriate, since most halotolerant and halophilic fungi display strong proteolytic and lipolytic activities and less cellulose and starch degrading ones (Gunde-Cimerman, unpublished results).

6.4.3 Final Comments: Are Fungi Important in the Dead Sea Ecosystem?

A great variety of fungal species were isolated from the lake, and a stable core of species was detected in different seasons and depths, including species such as *A. sydowii*, *A. versicolor*, *E. herbariorum*, *P. westlingii*, *C. cladosporioides*, and *C. sphaerospermum*. These species were also found in deep samples in summer, when the thermocline is expected to act as a barrier preventing migration of spores or mycelium from the surface to the deeper water. New fungi may be expected to enter the lake continuously as spores by rain, runoff, and wind, but due to the harsh conditions there is little chance for intruding fungal species originating from a non-saline environment to survive for prolonged times in Dead Sea brines.

Proof of vegetative growth of mycelia in the Dead Sea is still missing, and growth experiments also showed that even the best-adapted species do not grow in undiluted Dead Sea water. We do not know whether indeed fungi did grow in the upper water layers of the Dead Sea at the time of the blooms of *Dunaliella* and red halophilic Archaea that developed in 1980 and in 1992 following massive rain floods diluting the upper meters of the water column by 15% and 28%, respectively. However, based on the information reviewed above, it appears that a potential for fungal growth in the Dead Sea does exist. Development of mycelium may be possible only during rare episodes in which the salinity of the upper water layers becomes reduced as a result of massive freshwater inflow. In addition, areas near the mouth of the Jordan River and freshwater springs around the lake may be dilute havens where fungi may get the chance to germinate, grow, develop mycelia, and mature conidiophores in the diluted brines and complete their life cycle.

Acknowledgments The authors thank Prof. Dr. (Emeritus) Hans Peter Molitoris for his valuable comments and Dr. Polona Zalar for identifications of isolates that were not previously reported elsewhere.

References

Beyth M (1980) Recent evolution and present stage of Dead Sea brines. In: Nissenbaum A (ed) Hypersaline brines and evaporitic environments. Elsevier, Amsterdam

Bodaker I, Sharon I, Suzuki MT, Reingersch R, Shmoish M, Andreishcheva E, Sogin ML, Rosenberg M, Belkin S, Oren A, Béjà O (2010) The dying Dead Sea: comparative community genomics in an increasingly extreme environment. ISME J 4:399–407

Brown AD (1990) Microbial water stress physiology. Principles and perspectives. Wiley, Chichester

Buchalo AS (2003) Fungi in saline water bodies. In: Nevo E, Oren A, Wasser SP (eds) Fungal life in the Dead Sea. A.R.G. Gantner Verlag, Ruggell

Buchalo AS, Nevo E, Wasser SP, Oren A, Molitoris HP (1998) Fungal life in the extremely hypersaline water of the Dead Sea: first records. Proc R Soc Lond B 265:1461–1465

Buchalo AS, Wasser SP, Molitoris HP, Volz PA, Kurchenko I, Lauer I, Rawal B (1999) Species diversity and biology of fungi isolated from the Dead Sea. In: Wasser SP (ed) Evolutionary theory and processes: modern perspectives. Papers in honour of Eviatar Nevo. Kluwer Academic Publishers, Dordrecht

Buchalo AS, Nevo E, Wasser SP, Molitoris HP, Oren A, Volz PA (2000a) Fungi discovered in the Dead Sea. Mycol Res 104:132–133

Buchalo AS, Nevo E, Wasser SP, Volz PA (2000b) Newly discovered halophilic fungi in the Dead Sea. In: Seckbach J (ed) Journey to diverse microbial worlds. Kluwer Academic Publishers, Dordrecht

Butinar L, Santos S, Spencer-Martins I, Oren A, Gunde-Cimerman N (2005a) Yeast diversity in hypersaline habitats. FEMS Microbiol Lett 244:229–234

Butinar L, Zalar P, Frisvad JC, Gunde-Cimerman N (2005b) The genus *Eurotium* – members of indigenous fungal community in hypersaline waters of salterns. FEMS Microbiol Ecol 51:155–166

Butinar L, Sonjak S, Zalar P, Plemenitaš A, Gunde-Cimerman N (2005c) Melanized halophilic fungi are eukaryotic members of microbial communities in hypersaline waters of solar salterns. Bot Mar 1:73–79

Butinar L, Frisvad JC, Gunde-Cimerman N (2011) Hypersaline waters – a potential source of foodborne toxigenic Aspergilli and Penicillia. FEMS Microbiol Ecol 77:186–199

Cronin AD, Post FJ (1977) Report of a dematiaceous hyphomycete from the Great Salt Lake, Utah. Mycologia 69:846–847

Grishkan I, Nevo E, Wasser SP (2003) Soil micromycete diversity in the hypersaline Dead Sea coastal area, Israel. Mycol Progr 2:19–28

Grishkan I, Nevo E, Wasser SP (2004) Micromycetes from the saline Arubotaim Cave: Mount Sedom, The Dead Sea southwestern shore, Israel. J Arid Environ 57:431–443

Guiraud P, Steiman R, Seigle-Murandi F, Sage L (1995) Mycoflora of soil around the Dead Sea. II. Deuteromycetes (except *Aspergillus* and *Penicillium*). Syst Appl Microbiol 18:318–322

Guiraud P, Steiman R, Seigle-Murandi F, Sage L (1997) *Exserohilum sodomii*, a new species isolated from soil near the Dead Sea (Israel). Antonie van Leeuwenhoek 72:317–325

Gunde-Cimerman N, Zalar P, de Hoog GS, Plemenitaš A (2000) Hypersaline waters in salterns: natural ecological niches for halophilic black yeasts. FEMS Microbiol Ecol 32:235–240

Gunde-Cimerman N, Frisvad JC, Zalar P, Plemenitaš A (2005) Halotolerant and halophilic fungi. In: Desmukh SK, Rai MK (eds) Biodiversity of fungi: their role in human life. Science Publishers, Enfield, NH

Gunde-Cimerman N, Ramos J, Plemenitaš A (2009) Halotolerant and halophilic fungi. Mycol Res 113:1231–1241

Jin Y, Nevo E (2003) Osmoadaptation strategies of the Dead Sea fungi to hypersaline stress. In: Nevo E, Oren A, Wasser SP (eds) Fungal life in the Dead Sea. A.R.G. Gantner Verlag, Ruggell

Jin Y, Weining S, Nevo E (2005) A MARK gene from Dead Sea fungus confers stress tolerance to lithium salt and freezing-thawing: prospects for saline agriculture. Proc Natl Acad Sci U S A 102:18992–18997

Kis-Papo T, Grishkan I, Oren A, Wasser SP, Nevo E (2001) Spatiotemporal diversity of filamentous fungi in the hypersaline Dead Sea. Mycol Res 105:749–756

Kis-Papo T, Oren A, Wasser SP, Nevo E (2003a) Survival of spores of filamentous fungi in Dead Sea water. Microb Ecol 45:183–190

Kis-Papo T, Kirzhner V, Wasser SP, Nevo E (2003b) Evolution of genomic diversity and sex at extreme environments: fungal life under hypersaline Dead Sea stress. Proc Natl Acad Sci U S A 100:14970–14975

Kis-Papo T, Grishkan I, Gunde-Cimerman N, Oren A, Wasser SP, Nevo E (2003c) Spatiotemporal patterns of filamentous fungi in the Dead Sea. In: Nevo E, Oren A, Wasser SP (eds) Fungal life in the Dead Sea. A.R.G. Gantner Verlag, Ruggell

Kis-Papo T, Oren A, Wasser S, Nevo E (2003d) Physiological adaptations of Dead Sea fungi to the Dead Sea: survival of spores and vegetative cells. In: Nevo E, Oren A, Wasser SP (eds) Fungal life in the Dead Sea. A.R.G. Gantner Verlag, Ruggell

Kis-Papo T, Oren A, Nevo E (2003e) Genetic diversity of filamentous fungi under hypersaline stress in the Dead Sea. In: Nevo E, Oren A, Wasser SP (eds) Fungal life in the Dead Sea. A.R. G. Gantner Verlag, Ruggell

Kritzman G (1973) Observations on the microorganisms in the Dead Sea. M.Sc. Dissertation, The Hebrew University of Jerusalem (in Hebrew)

Kushner DJ (1978) Life at high salt and solute concentrations: halophilic bacteria. In: Kushner DJ (ed) Microbial life in extreme environments. Academic Press, New York

Molitoris HP, Buchalo AS, Kurchenko I, Nevo E, Rawal BS, Wasser SP, Oren A (2000) Physiological diversity of the first filamentous fungi isolated from the hypersaline Dead Sea. In: Hyde DK, Ho WH, Pointing SB (eds) Aquatic mycology across the millennium, vol 5. Fungal Diversity Press, Hong Kong, pp 55–70

Nevo E, Oren A, Wasser SP (eds) (2003) Fungal life in the Dead Sea. A.R.G. Gantner Verlag, Ruggell

Oren A (1988) The microbial ecology of the Dead Sea. In: Marshall KC (ed) Advances in microbial ecology, vol 10. Plenum Publishing Company, New York

Oren A (1993) The Dead Sea – alive again. Experientia 49:518–522

Oren A (1997) Microbiological studies in the Dead Sea: 1892–1992. In: Niemi T, Ben-Avraham Z, Gat JR (eds) The Dead Sea – the lake and its setting. Oxford University Press, New York

Oren A (1999) Microbiological studies in the Dead Sea: future challenges toward the understanding of life at the limit of salt concentrations. Hydrobiologia 405:1–9

Oren A (2000) Biological processes in the Dead Sea as influenced by short-term and long-term salinity changes. Arch Hydrobiol Spec Issues Adv Limnol 55:531–542

Oren A (2003a) Physical and chemical limnology of the Dead Sea. In: Nevo E, Oren A, Wasser SP (eds) Fungal life in the Dead Sea. A.R.G. Gantner Verlag, Ruggell

Oren A (2003b) Biodiversity and community dynamics in the Dead Sea: Archaea, bacteria and eucaryotic algae. In: Nevo E, Oren A, Wasser SP (eds) Fungal life in the Dead Sea. A.R.G. Gantner Verlag, Ruggell

Oren A, Gurevich P (1995) Dynamics of a bloom of halophilic archaea in the Dead Sea. Hydrobiologia 315:149–158

Oren A, Gurevich P, Anati DA, Barkan E, Luz B (1995) A bloom of *Dunaliella parva* in the Dead Sea in 1992: biological and biogeochemical aspects. Hydrobiologia 297:173–185

Oren A, Gavrieli I, Gavrieli J, Kohen M, Lati J, Aharoni M (2005) Microbial communities in the Dead Sea – past, present and future. In: Gunde-Cimerman N, Oren A, Plemenitaš A (eds) Adaptation to life at high salt concentrations in Archaea, Bacteria, and Eukarya. Springer, Dordrecht

Pitt J, Hocking AD (1997) Fungi and food spoilage, 2nd edn. Blackie Academic & Professional, London

Rhodes ME, Fitz-Gibbon S, Oren A, House CH (2010) Amino acid signatures of salinity on an environmental scale with a focus on the Dead Sea. Environ Microbiol, in press.

Sonjak S, Bukay YG, Gunde-Cimerman N (2010) Magnesium tolerant fungi from the bitterns. Meeting of the ISHAM Working groups on Black Yeasts and Chromoblastomycosis, Ljubljana, May 14–16, 2010. http://blackyeast2010.bf.uni-lj.si/lectures. Accessed 18 Nov 2010

Steiman R, Guiraud P, Sage L, Seigle-Murandi F (1994) New strains from Israel in the *Aspergillus niger* group. Syst Appl Microbiol 17:620–624

Steiman R, Guiraud P, Sage L, Seigle-Murandi F, Lafond JL (1995) Mycoflora of soil around the Dead Sea. I. Ascomycetes (including *Aspergillus* and *Penicillium*), Basidiomycetes, Zygomycetes. Syst Appl Microbiol 18:310–317

Steiman R, Guiraud P, Seigle-Murandi F, Sage L (1996) *Bipolaris israeli* sp. nov. from Israel: description and physiological features. Syst Appl Microbiol 19:182–190

Steiman R, Guiraud P, Sage L, Seigle-Murandi F, Lafond JL (1997) Soil mycoflora from the Dead Sea oases of Ein Gedi and Einot Zuqim (Israel). Antonie van Leeuwenhoek 72:261–270

Tresner HD, Hayes JA (1971) Sodium chloride tolerance of terrestrial fungi. Appl Microbiol 22:210–213

Volz PA, Wasser SP (1995) Soil micromycetes from selected areas of Israel. Israel J Plant Sci 43:281–290

Wasser SP, Grishkan I, Kis-Papo T, Buchalo AS, Volz PA, Gunde-Cimerman N, Zalar P, Nevo E (2003) Species diversity of the Dead Sea fungi. In: Nevo E, Oren A, Wasser SP (eds) Fungal life in the Dead Sea. A.R.G. Gantner Verlag, Ruggell

Zalar P, de Hoog GS, Schroers H-J, Frank JM, Gunde-Cimerman N (2005a) Taxonomy and phylogeny of the xerophilic genus *Wallemia* (Wallemiomycetes and Wallemiales, cl. et ord. nov.). Antonie van Leeuwenhoek 87:311–328

Zalar P, Kocuvan MA, Plemenitaš A, Gunde-Cimerman N (2005b) Halophilic black yeasts colonize wood immersed in hypersaline water. Bot Mar 48:323–326

Zalar P, de Hoog GS, Schroers H-J, Crous J, Groenewald JZ, Gunde-Cimerman N (2007) Phylogeny and ecology of the ubiquitous saprobe *Cladosporium sphaerospermum*, with descriptions of seven new species from hypersaline environments. Stud Mycol 58:157–183

Chapter 7
The Mycobiota of the Salterns

Janja Zajc, Polona Zalar, Ana Plemenitaš, and Nina Gunde-Cimerman

Contents

Abstract Solar salterns are constructed as shallow multi-pond systems for the production of halite through evaporation of seawater. The main feature of salterns is the discontinuous salinity gradient that provides a range of well-defined habitats with increasing salinities, from moderate to hypersaline. These present one of the most extreme environments, because of the low levels of biologically available water and the toxic concentrations of ions. Up to the year 2000, hypersaline environments were considered to be populated almost exclusively by prokaryotic

N. Gunde-Cimerman (✉)
Biology Department, University of Ljubljana, Večna pot 111, Ljubljana SI-1000, Slovenia

Centre of Excellence for Integrated Approaches in Chemistry and Biology of Proteins (CIPKeBiP), Jamova 39, Ljubljana SI-1000, Slovenia
e-mail: nina.gunde-cimerman@bf.uni-lj.si

C. Raghukumar (ed.), *Biology of Marine Fungi*,
Progress in Molecular and Subcellular Biology 53,
DOI 10.1007/978-3-642-23342-5_7,

microorganisms till fungi were reported to be active inhabitants of solar salterns. Since then, numerous fungal species have been described in hypersaline waters around the world. The mycobiota of salterns is represented by different species of the genus *Cladosporium* and the related meristematic melanized black yeasts, of non-melanized yeasts, of the filamentous genera *Penicillium* and *Aspergillus* and their teleomorphic forms (*Eurotium* and *Emericella*), and of the basidiomycetous genus *Wallemia*. Among these, two species became new model organisms for studying the mechanisms of extreme salt tolerance: the extremely halotolerant ascomycetous black yeast *Hortaea werneckii* and the obligate halophilic basidiomycete *Wallemia ichthyophaga*.

7.1 Introduction

7.1.1 The Saltern Ecosystem

Thalassohaline hypersaline environments are generally considered to be those originating by evaporation of sea water and with halite (NaCl) concentrations of greater than 10% (m/w) (Oren 2002). These provide some of the most extreme habitats in the World. They are common all around the globe, and include, for example, marine ponds and salt marshes that are subjected to evaporation, salt or soda lakes, and sea-salt and man-made salterns (Trüper and Galinski 1986). Solar salterns, the focus of this chapter, are composed of multiple shallow ponds that are located in tropical, subtropical, and temperate parts of the World. In these ponds, the NaCl is gradually concentrated as the seawater evaporates.

Salt production in salterns usually begins with seawater as the initial source of brine, which is evaporated through a series of ponds, to the final pond where the NaCl and other salts precipitate out of the saturated brine. The bittern that remains after this crystallization of the halite is rich in magnesium chloride and provides a special ecological habitat within the salterns. Variable water activities (a_w) because of the increasing concentrations of NaCl, as well as the low oxygen concentrations and high light intensity, present life-limiting parameters in salterns (Brock 1979).

The Sečovlje solar salterns (Fig. 7.1a) were established in the ninth century as a man-made system of ponds, and they are located in the northern Adriatic Sea, in the Gulf of Trieste in Slovenia. The sub-Mediterranean climate and unstable weather conditions allow the production of salt only during the summer, when climate is arid. The strong local winds enhance evaporation of water and keep the temperatures of the water in brines moderate (18–32°C). When the water reaches a suitable concentration of salt, it is directed over the wooden barriers to the next of the 16 evaporative ponds separated by canals. These are the only salterns along the eastern Adriatic coast where salt is produced according to the traditional procedures, with the daily gathering of the saturated brine on the cultivated microbial mat covering the salterns, known as the petola. The role of the petola is dual; it prevents the mixing of the crystallized halite with the mud at the bottom of

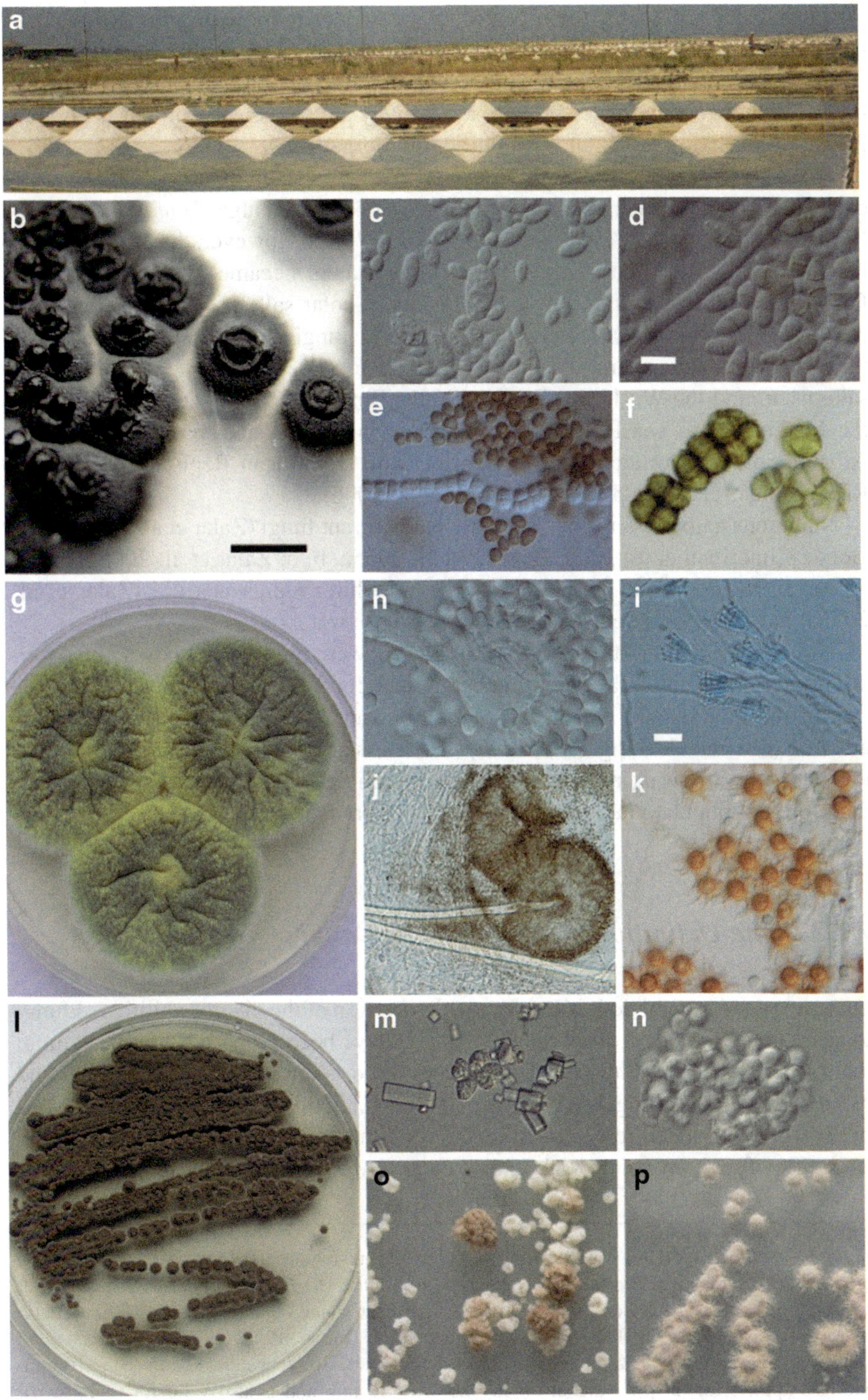

Fig. 7.1 Representatives of the saltern mycobiota. (**a**) salterns Sečovlje; (**b**) *Hortaea werneckii* colonies; (**c**) *Aureobasidium pullulans* budding cells; (**d**) *Hortaea werneckii* budding cells;

the ponds, and it prevents the incorporation of undesired ions of iron and manganese into the halite crystals (Pahor and Poberaj 1963).

The solar salterns Sečovlje with their successive evaporation ponds, the salt-pan mud, and the wooden fences provide a relatively simple ecosystem that is popular for studies of halotolerant and halophilic microorganisms.

These hypersaline waters in salterns were at first believed to be populated almost exclusively by archaea, bacteria, and the eukaryotic alga *Dunaliella salina* (Rodriguez-Valera et al. 1981; Javor 1989; Oren 2005). However, the high diversity of eukaryotic microorganisms in hypersaline waters became evident soon after fungi were first reported as active inhabitants of solar salterns (Gunde-Cimerman et al. 2000). Numerous halotolerant and halophilic fungi were initially isolated from the Sečovlje solar salterns (Gunde-Cimerman et al. 2000), and later also from salterns around the World: La Trinidad in the Ebro River Delta and Santa Pola on the Mediterranean coast of Spain, Camargue in France, and the salterns on the Atlantic coast in Portugal, and in Namibia, the Dominican Republic and Puerto Rico (Butinar et al. 2005a, b, c; Cantrell et al. 2006).

Numerous halotolerant and extremely halotolerant fungi (Zalar et al. 1999a, b, c; Gunde-Cimerman et al. 2000; Butinar et al. 2005a, b, c; Zalar et al. 2005b, 2007, 2008a, b; Butinar et al. 2011) and a few halophilic representatives (Zalar et al. 2005a) have been isolated from these hypersaline waters.

The xerophilic fungi that can grow at low a_w were previously known only from domestic environments, as they can contaminate food preserved through the reduction of the biologically available water, by means of drying, freezing, or solute addition (Pitt and Hocking 1977, 1997). It was assumed that these food-borne fungi that can grow at low a_w reflect a general xerophilic phenotype, and that they would therefore not populate natural hypersaline environments.

7.1.2 *The Definition of Halophily in Fungi*

Over the years, several definitions for the description of the diverse abilities of fungi to adapt to a wide range of salt concentrations have been proposed. For these fungi that prefer reduced a_w, the most commonly used adjectives were xerophilic and osmophilic, as these fungi can grow at a_w below 0.85; this corresponds to 17% NaCl or 50% glucose added to a growth medium (Gunde-Cimerman et al. 2005). However, these fungi are not only able to grow at low a_w; some also show preferences

---

Fig. 7.1 (continued) (**e**) *Phaeotheca triangularis* conidia and hyphae; (**f**) *Trimmatostroma salinum* meristematic clumps; (**g**) *Eurotium amstelodami* in culture; (**h**) *Eurotium chevalieri* anamorph; (**i**) *Penicillium crustosum* conidiophores; (**j**) *Aspergillus niger* conidiophores; (**k**) *Emericella stella-maris* ascospores; (**l**) *Wallemia muriae* in culture; (**m**, **n**) *W. ichthyophaga meristematic clumps;* (**o**) *W. ichthyophaga* colonies; (**p**) *W. sebi* colonies. Scalebar indicated in (**b**) is 10 mm, and also applies for (**o**, **p**); scalebar on (**d**) indicates 10 μm, and is also valid for (**c**–**h**, **k**, **n**); scalebar on (**i**) indicates 20 μm and is valid for (**i**, **j**, **m**) for certain chemical natures of the solutes that lower the a_w (de Hoog et al. 2005; Gunde-Cimerman and Plemenitaš 2006). Hence these osmotolerant/osmophilic fungi can live in environments that are rich in sugar, whereas those that are halotolerant/halophilic can live in environments that are rich in salt. In contrast to obligate halophilic archaea and bacteria, in the fungal kingdom, no evidence of obligate requirements for salt was reported until only recently. When fungi were found to constitute active communities in the hypersaline water of solar salterns, they were considered halophilic if they were regularly isolated from water at 17–32% NaCl, primarily on saline selective media, and if they were able to grow *in vitro* on 17% NaCl (Gunde-Cimerman et al. 2000). These fungi were then shown to sustain a range of different salt concentrations, right across the whole salinity range. Later, a few fungal species that show superior growth on media with NaCl as the controlling solute and necessarily require lowered a_w for growth were first reported (Zalar et al. 2005a); hence the term halophilic fungi was challenged again. Thus, halotolerant and extreme halotolerant are now the terms used to describe fungi that can grow across a range of different salt concentrations, even from fresh water to NaCl saturation (Gunde-Cimerman and Plemenitaš 2006), and the term halophily remains reserved for those that require salt for growth.

7.2 The Mycobiota of the Salterns

7.2.1 Methods for the Isolation of the Fungi from the Salterns

To avoid selection of certain mycobiota that are favored by a particular method, four different methods of isolation have often been used in the studies of fungal diversity in hypersaline brines: (a) filtration of the saline and placing the membrane filter on different selective media; (b) agar baits in dialysis tubing and glass tubes that were left for 5 months in crystallization ponds and then placed on media; (c) enrichment of saline with glucose and yeast extract; and (d) spreading of biofilms on selective media (Gunde-Cimerman et al. 2000).

The selective media have low a_w (at least 0.89, and lower) because of the supplements of either salt or sugar. In the salt-based selective media, 10–32% NaCl is added to malt-extract yeast agar media, and for the sugar-selective media, these are supplemented with 50–70% glucose, or with 18% glycerol (Gunde-Cimerman et al. 2000; Butinar et al. 2005a). The enumeration of the colony-forming units (CFU) per 100 ml hypersaline water can be performed with a general-purpose medium (dichloran rose bengal chloramphenicol [DRBC]; a_w 1.0) (King et al. 1979) and a medium for the detection of moderate xerophiles (Hocking and Pitt 1980). Bacterial overgrowth is prevented by adding different antibiotics to all of the selective and enumeration media, like chloramphenicol (100 mg l^{-1}). The cultures are incubated for 1–10 weeks at 25°C, and examined every few days (Gunde-Cimerman et al. 2000).

7.2.2 *Diversity of the Saltern Mycobiota*

After a decade of research into the fungal diversity in salterns, together with new taxa, a number of fungal genera with high diversities of halotolerant and halophilic species have been described. The melanized fungi isolated are represented by meristematic black yeast-like hypomycetes of the following species: *Hortaea werneckii*, *Phaeotheca triangularis*, *Trimmatostroma salinum*, *Aureobasidium pullulans*, and the phylogenetically related genus *Cladosporium* (Gunde-Cimerman et al. 2000).

A number of nonmelanized yeast species have been described for the different salterns and salt lakes worldwide. The most abundant among these are *Pichia guilliermondii, Debaryomyces hansenii*, and *Candida parapsilosis*, to name but a few (Butinar et al. 2005a). Among the filamentous fungi that appear with the highest frequencies there are different species of the genera *Aspergillus* and *Penicillium* (Butinar et al. 2011). Representatives of the newly described basidiomycetous class Wallemiomycetes cover the order *Wallemiales* with the genus *Wallemia*, and these were recently isolated from hypersaline waters of man-made salterns on different continents in Europe, Asia, Africa, and North America. Interestingly, the genus *Wallemia* represents one of the most xerophilic and halophilic fungal taxa known to date (Zalar et al. 2005a). Morphology and growth features of some of these fungi are shown in Fig. 7.1b–p. Their geographical distribution and salinity range for growth are listed in Table 7.1.

7.2.3 *The Black Yeasts*

Black yeasts are a group of dark pigmented (melanized) polymorphic hyphomycetes that besides filamentous have also the ability of yeast-like growth. Their slowly expanding melanized colonies have been detected in different extreme environments, such as on rock surfaces in arid, semi-arid, hot, and cold desserts that were previously considered to lack eukaryotic extremophiles (Gorbushina et al. 1996; Sterflinger and Krumbein 1997; Sterflinger et al. 1999).

These black yeasts are characterized by thick, melanized cell walls, with slow, often meristematic, growth and proliferation with endoconidiation. A similar morphology, which is regarded as an "extremophilic ecotype," was observed also with fungi isolated from the hypersaline waters of salterns (Gunde-Cimerman et al. 2000). Because of their morphological plasticity and polymorphic anamorphic stages, their certain identification can only be correctly performed by complementing morphological criteria with molecular methods and physiological tests of carbon and nitrogen metabolism without or with high NaCl concentrations. The morphological types of black yeasts isolated from the Sečovlje salterns were speculated to be different species, as they showed different growth patterns in the presence of salt. This was confirmed by additional molecular data that were obtained by sequencing the internal

Table 7.1 The most frequent fungal species isolated from hypersaline water of salterns across the globe

Group/class	Order	Species	Salinity range (NaCl) (%)		Geographical distribution: Slovenia	Spain		France	Portugal	Namibia	Dominican Republic		Puerto Rico	Israel	Bosnia and Herzegovina	USA (Utah)
					Sečovlje	La Trinidad	Santa Pola	Camargue	Samouco	Skeleton coast	Salterns	Lake Enriquillio	Cabo Rojo	Eilat	Ston	Great Salt Lake
Ascomycota	Capnodiales	*Hortaea werneckii*	0–32	L	X	X	X	X	X	X		X	X			
		Phaeotheca triangularis	0–26	L	X	X	X	X	X	X		X	X			
	Helotiales	*Trimmatostroma salinum*	0–24	L	X											
	Dothideales	*Aureobasidium pullulans*	0–18	L	X	X	X	X	X	X		X	X			
	Capnodiales	*Cladosporium cladosporioides*	0–17[a]	S	X								X		X	
		Cladosporium herbarum	nd													
		Cladosporium oxysporum	nd								X	X	X			
		Cladosporium sphaerospermum	0–20	S	X						X		X	X		
		Cladosporium halotolerans	0–20	S	X		X			X		X			X	
		Cladosporium dominicanum	0–20	S							X	X				
		Cladosporium velox	0–20	S	X											
		Cladosporium psychrotolerans	0–17	S	X						X					
		Cladosporium spinulosum	0–17	S	X											
		Cladosporium salinae	0–17	S	X		X									
		Cladosporium fusiforme	0–17	S	X											
		Cladosporium ramotenellum	nd		X											
		Cladosporium subinflatum	nd		X											
		Cladosporium tenuissimum	nd		X					X						
		Cladosporium tenellum	nd											X		
		Cladosporium herbaroides	nd											X		
		Cladosporium macrocarpum	nd		X											
		Cladosporium subtilissimum	nd		X											
		Cladosporium bruhnei	nd		X		X									
	Saccharomycetales	*Candida parapsilosis*	0–17[a]	S	X				X							
		Debaryomyces hansenii	0–17[a]	S						X						X
		Metschnikowia bicuspidata	nd													X
		Pichia guilliermondii	0–17[a]	S	X			X								
		Yarrowia lipolytica	nd											X		

(continued)

Table 7.1 (continued)

Group/class	Order	Species	Salinity range (NaCl) (%)		Geographical distribution											
					Slovenia	Spain		France	Portugal	Namibia	Dominican Republic		Puerto Rico	Israel	Bosnia and Herzegovina	USA (Utah)
					Sečovlje	La Trinidad	Santa Pola	Camargue	Samouco	Skeleton coast	Salterns	Lake Enriquillio	Cabo Rojo	Eilat	Ston	Great Salt Lake
	Eurotiales	*Eurotium amstelodami*	0–27.5	S	X	X	X			X	X	X		X		X
		Eurotium chevalieri	0–22.5	S	X	X	X									
		Eurotium herbariorum	0–25	S	X	X								X		
		Eurotium rubrum	0–20	S	X											
		Eurotium repens	0–25	S	X	X										
		Eurotium halotolerans	0–27.5	S												
		Emericella filifera	0–17	S	X											
		Emericella stella-maris	0–17	S	X											
		Aspergillus niger	0–17[a]	S	X	X	X	X			X		X			
		Aspergillus flocculosus	0–17[a]	S	X						X	X				
		Aspergillus caesiellus	0–17[a]	S	X											
		Aspergillus ochraceus	0–17[a]	S	X	X	X						X			
		Aspergillus flavus	0–17[a]	S	X	X							X			X
		Aspergillus roseoglobulosus	0–17[a]	S							X					
		Aspergillus tubingensis	0–17[a]	S	X									X		
		Aspergillus melleus	0–17[a]	S									X			
		Aspergillus sclerotium	0–17[a]	S	X											
		Aspergillus versicolor	0–17[a]	S	X									X		
		Aspergillus sydowii	0–17[a]	S	X					X	X			X		
		Aspergillus wentii	0–17[a]	S	X	X					X					
		Aspergillus flavipes	0–17[a]	S	X								X			
		Aspergillus terreus	0–17[a]	S	X											
		Aspergillus candidus	0–17[a]	S	X	X							X			
		Aspergillus penicillioides	0–17[a]	S	X								X			
		Aspergillus restrictus	0–17[a]	S	X											
		Aspergillus proliferans	0–17[a]	S	X											
		Aspergillus nidulans	0–17[a]	S									X			
		Aspergillus fumigates	0–17[a]	S	X											
		Penicillium chrysogenum	0–17[a]	S	X	X	X							X		

		Penicillium brevicompactum	0–17[a]	S								
		Penicillium citrinum	0–17[a]	S	X				X		X	
		Penicillium oxalicum	0–17[a]	S						X		
		Penicillium steckii	0–17[a]	S	X							X
		Penicillium sizovae	0–17[a]	S	X							X
		Penicillium westlingii	0–17[a]	S	X							
		Penicillium freii	0–17[a]	S	X							
		Penicillium cyclopium	0–17[a]	S	X							
		Penicillium solitum	0–17[a]	S	X							
		Penicillium albocoremium	0–17[a]	S	X							
		Penicillium crustosum	0–17[a]	S		X						
Basidiomycota	Wallemiales	*Wallemia sebi*	0–27	L	X		X		X		X	
		Wallemia muriae	0–25	L	X					X		X
		Wallemia ichthyophaga	9–32	L	X				X			
	Tremellales	*Trichosporon mucoides*	nd									X
	Sporidiobolales	*Rhodosporidium babjevae*	0–13[a]	S	X			X				
		Rhodosporidium sphaerocarpum	0–17[a]	S	X			X				

Data were adopted from Gunde-Cimerman et al. (2000); Butinar et al. 2005a, b, c, 2011; Diaz-Munoz and Montalvo-Rodriguez 2005; Zalar et al. 2005a, 2007, 2008a; Cantrell et al. 2006; Schubert et al. 2007, and unpublished data (Gostinčar, personal communication)

nd no data; *X* the presence of the species in the salterns; *L* salinity range determined in the liquid media; *S* salinity range determined in solid media

[a]Maximum tolerated NaCl concentration not determined

transcribed spacer (ITS) rDNA and by restriction fragment length polymorphism (RFLP) of small-subunit (SSU) and ITS rDNA. The representatives of the black yeasts identified include: *H. werneckii* (Zalar et al. 1999c) and *P. triangularis* (Zalar et al. 1999a; c), both comprise the order *Capnodiales*; *A. pullulans* (Zalar et al. 1999c), from the order *Dothideales*; and *T. salinum*, a newly described species from the order *Helotiales* (Zalar et al. 1999b).

Initially, these black yeasts were isolated from the Adriatic salterns, but they were also later identified in hypersaline waters of six salterns on three continents. These were La Trinidad in the Ebro River Delta and Santa Pola on the Mediterranean coast of Spain, Camargue in France, the Atlantic coast in Portugal, and in Namibia and the Dominican Republic. Exceptionally, *T. salinum* was detected only in the Adriatic salterns (Zalar et al. 1999b). *H. werneckii* represents 70–80% of all of the isolates, followed in abundance by *T. salinum* and *P. triangularis*, while *A. pullulans* appears mainly at lower salinities (Butinar et al. 2005b). The extremely halotolerant nature of the black yeasts was first noted when they were isolated from the saline media in the highest numbers according to two peaks, one at the beginning of the sampling season (May), and the other and during the crystallization (August) (Gunde-Cimerman et al. 2000); this was later confirmed by *in vitro* studies.

Hortaea werneckii has also been named as *Cladosporium werneckii, Exophiala werneckii, Dematium werneckii, Pullularia werneckii, Aureobasidium werneckii, A. mansonii, Sarcinomyces crustaceous*, and *Pheoannellomyces werneckii* in the past, but has been because of special conidiogenesis described in a new genus (Nishimura and Miyaji 1983). It was long known as the primary etiological agent of a human surface skin problem called tinea nigra, a superficial fungal infection of the human hand that is frequent in warmer areas of the world. The ecology of *H. werneckii* was linked to the presence of salt, as it was isolated from seawater (Iwatsu and Udagawa 1988), marine fish (Todaro et al. 1983), salted freshwater fish (Mok et al. 1981), and beach soil (de Hoog and Guého 2010). Its ecological niche has been suggested to be intermittently drying salty pools (de Hoog and Gerrits van den Ende 1992) or highly saline water of the crystallization ponds in the solar salterns (Gunde-Cimerman et al. 2000). Among all of the melanized fungi, *H. werneckii* is dominant in the hypersaline waters of the Adriatic salterns and of all of the other salterns that have environmental salinities above 20% (Gunde-Cimerman et al. 2000; Butinar et al. 2005b; Diaz-Munoz and Montalvo-Rodriguez 2005; Cantrell et al. 2006). *H. werneckii* is most abundant (1,400 CFU l^{-1}) at 23% salinity, and remains present with low CFU also after the end of the salt production season. *H. werneckii* is the only black yeast that can grow across the whole range of NaCl concentrations, from 0% to NaCl saturation, with a broad optimum from 6 to 14% NaCl (Butinar et al. 2005b), and it is thus considered to be the most halotolerant fungus so far described.

Trimmatostroma salinum is the only black yeast-like fungus from hypersaline waters that occurs only in the Adriatic salterns, and almost exclusively in one pond, where it was discovered growing on the wooden fence immersed in hypersaline water (Zalar et al. 1999b, 2005b). *T. salinum* peaks at 25% salinity (700 CFU l^{-1}) and it can appear within broad environmental salinities, as 8–25% NaCl (Butinar et al. 2005b). The broad salinity range for in-vitro growth is from 0 to 24% NaCl, with the optimal values at 2–8% NaCl.

Phaeotheca triangularis was detected in all of the ponds of the Adriatic salterns throughout the year, as well as in crystallizer ponds of other salterns that have been sampled (Butinar et al. 2005b). It shows high adaptability to saline conditions, as it appeared most abundantly (100–340 CFU l^{-1}) in the range of 18–25% NaCl; at lower salinities, its CFUs decrease to less than 50 CFU l^{-1}. In contrast to *H. werneckii*, *P. triangularis* cannot grow *in vitro* on 32% salt, with 24% (or at times also 26%) NaCl reported as its maximum. Its optimal concentration is much narrower, as 6–12% NaCl. *P. triangularis* has been most often isolated from storage ponds with relatively constant salinity. It has also been considered to be an oligotroph, as it can grow on nutritionally poor agar media, and it has been most frequently isolated from low-nutrient storage ponds (Butinar et al. 2005b). *P. triangularis* can form biofilms on solid and liquid saline media, and it frequently appears in sampled microbial biofilms (Gunde-Cimerman et al. 2000; Butinar et al. 2005b). *P. triangularis* is thus an extremely halotolerant species with a narrow ecological amplitude (Gunde-Cimerman et al. 2000).

Aureobasidium pullulans is a cosmopolitan species that can be found in different environments that have fluctuating water activities, such as the phyllosphere (Andrews et al. 1994), foods, feedstuff, bathrooms (Samson et al. 2002), polluted water (Vadkertiova and Slavikova 1995), and others. It has also been found in osmotically stressed environments, such as on rocks and monuments (Urzi et al. 1999), on surface sediments and detritus in salt marshes (Torzilli et al. 1985), and in the hypersaline waters of the Adriatic salterns (Gunde-Cimerman et al. 2000; Butinar et al. 2005b). According to multilocus molecular analysis, the strains of *A. pullulans* from the hypersaline water of the Sečovlje salterns have segregated mostly to the globally ubiquitous variety (var.) *pullulans* (Zalar et al. 2008b). In all of the ponds of the Adriatic salterns and of other sampled salterns, *A. pullulans* occurs with low CFU (up to 50 CFU l^{-1}), and generally at environmental salinities below 8% NaCl. The counts are the highest (800 CFU l^{-1}) at 5% salinity, before or after the salt production season. *A. pullulans* is regarded as a halotolerant species, rather than halophilic, as it can tolerate up to 18% NaCl in vitro and grows optimally on medium without NaCl (Butinar et al. 2005b).

The sampling and inspection of wooden boards that support the walls of the crystallization ponds that are immersed in the hypersaline waters of the active solar salterns of Sečovlje have demonstrated active lignicolous saprobic roles of the halophilic fungi in the hypersaline water. Melanized hyphae in the black-stained parts of the wood from these walls were recognized as belonging to *H. werneckii* and *T. salinum*. These show xylanolytic and lignolytic activities under hypersaline and nonsaline conditions; and *T. salinum* alone shows cellulolytic activity. This suggests a complementary role of these two black yeasts for the invasion of wood in hypersaline environments (Zalar et al. 2005b).

7.2.4 The Genus *Cladosporium*

The cosmopolitan genus *Cladosporium* is currently in revision process (Crous et al. 2007; Bensch et al. 2010), but is accommodating over 772 different species names (Dugan et al. 2004). It was phylogenetically placed into the order Capnodiales, after the discovery of a teleomorph stage *Davidiella* (Braun et al. 2003). A few of these species, namely *C. herbarum, C. cladosporioides, C. sphaerospermum, C. tenuissimum,* and *C. oxysporum*, have often been isolated from habitats characterized by low a_w, such as sugary and salty foods (Samson et al. 2002), Egyptian salt marshes, the rhizosphere of halophytic plants, and the phylloplane of Mediterranean plants (Abdel-Hafez et al. 1978). The minimal a_w for growth are 0.82, 0.85, and 0.86 for *C. sphaerospermum, C. herbarum,* and *C. cladosporioides*, respectively. In the hypersaline waters of the Sečovlje solar salterns, *Cladosporium* strains are among the most frequently found and are the most abundant of the melanized fungi (Gunde-Cimerman et al. 2000; Butinar et al. 2005b). They have the broadest occurrence among all of the melanized fungi throughout the year in the ponds of the Adriatic salterns, as well as in other hypersaline environments such as Cabo Rojo solar salterns in Puerto Rico (Cantrell et al. 2006). Interestingly, they occur with the highest CFU (1,000–3,600 CFU l^{-1}) between 15 and 25% NaCl, and as the NaCl concentrations increase, they seem to be gradually replaced by different species of black yeasts. As these were mainly obtained from sugar-based media, they are considered as xerotolerant or xerophilic. First, *C. herbarum, C. cladosporioides, C. sphaerospermum,* and *C. oxysporum* were identified (Butinar et al. 2005b; Cantrell et al. 2006). Then with the different taxonomic methods that were performed later, *C. ramotenellum, C. tenellum, C. subinflatum,* and *C. herbaroides* were also shown to be part of the hypersaline mycobiota (Schubert et al. 2007).

Strains at that time identified as *C. sphaerospermum*, which was later recognized as a species complex of cosmopolitan as well as specialized air-borne species, have been reported from a wide range of habitats, including osmotically non-stressed niches, and they show pronounced osmotolerant behavior as they can grow at very low a_w (0.82) (Hocking et al. 1994).

In a wide ecological and phylogenetical study of isolates from the hypersaline waters from Mediterranean salterns, from different coastal areas along the Atlantic Ocean, and from the Red Sea, the Dead Sea, and the salt Lake Enriquillio (Dominican Republic), seven more species of the *Cladosporium* genus were newly described for these hypersaline water: *C. halotolerans, C. dominicanum, C. velox, C. psychrotolerans, C. spinulosum, C. salinae,* and *C. fusiforme*. *C. psychrotolerans* and *C. spinulosum* are currently known only from hypersaline water. Interestingly, a pathogen on the human skin, *Cladosporium langeronii* (= *Hormodendrum langeronii*), is halotolerant, although it has not yet been recorded from hypersaline environments. The maximum NaCl concentration in the growth media for the development of colonies of these various halotolerant *Cladosporium* representatives has been reported to be 17–20% (Zalar et al. 2007).

7.2.5 The Nonmelanized Yeasts

Only a few reports on the existence of the osmophilic yeasts in natural hypersaline waters have appeared in the literature, although these yeasts have been recognized for their tolerance to high concentrations of sugars or salt. A few species, such as *Torulopsis famata, Rhodotorula rubra, Pichia etchelsii, C. parapsilosis*, and *D. hansenii*, are known to be able to grow above 10–15% NaCl (Samson et al. 2002). Most strains of *Metschnikowia bicuspidata* var. *bicuspidata* have been found to be associated with diseased brine shrimps (*Artemia salina*) from salt lakes and ponds with 10% NaCl, and they appear to require 2% NaCl to grow in vitro (Blackwell 2001). Lahav et al. (2002) described two yeasts, *Pichia guilliermondii* and *Rhodotorula mucilaginosa*, that can survive the extremely high salinity (between 3% and saturation) and pH (2.0–10.0) fluctuations.

Until 2005, there was no evidence that natural hypersaline brines contain nonmelanized yeast populations. Then, yeast diversity was reported for hypersaline waters of eight different salterns and three salt lakes worldwide. These were the Dead Sea, Enriquillo Lake in the Dominican Republic, and the Great Salt Lake in Utah, USA (Butinar et al. 2005a). This diversity included *P. guilliermondii, D. hansenii, Yarrowia lipolytica, M. bicuspidata, C. parapsilosis, Rhodosporidium sphaerocarpum, Rhodosporidium babjevae*, and *Trichosporon mucoides*. In bittern (water rich in magnesium chloride) from the La Trinidad salterns (Spain), two new species were discovered and provisionally named as *Candida atmosphaerica*-like and *Pichia philogaea*-like. Among the isolates obtained, *P. guilliermondii, D. hansenii, Y. lipolytica*, and *C. parapsilosis* were already known contaminants of low a_w food, whereas *R. sphaerocarpum, R. babjevae*, and *T. mucoides* were identified for the first time in hypersaline habitats. Moreover, the ascomycetous yeast *M. bicuspidata*, which is known to be a parasite of the brine shrimp, was isolated as a free-living form from Great Salt Lake brine. The frequency of its occurrence was low, as the counts only occasionally reached several hundred cells per liter across all of sampled salterns (on average, between 0 and 300 CFU l^{-1}). The most frequently occurring species in the Adriatic salterns were identified as *P. guilliermondii* (up to 270 CFU l^{-1}) and *C. parapsilosis*. They have also been sporadically isolated in other salterns (Butinar et al. 2005a).

In contrast to the melanized fungi that appear in the highest numbers during the crystallization in the solar salterns (Gunde-Cimerman et al. 2000; Butinar et al. 2005b), nonmelanized yeasts were isolated primarily outside the salt production season (Adriatic salterns) or in waters with NaCl concentrations below 20% (Butinar et al. 2005a). In addition, these yeasts were never isolated on 32% NaCl medium, but mainly on medium with 10% NaCl, and to a lesser extent on media with 17–25% NaCl (Butinar et al. 2005a).

7.2.6 The Order Eurotiales

The group of filamentous fungi that have been isolated from different salterns around the World is mainly represented by the order *Eurotiales*, or more precisely, by the teleomorphic genera *Eurotium* and *Emericella* and the anamorphic *Aspergillus,* and *Penicillium.* Tolerance for high salt concentrations has for a long time been recognized for many species of the ubiquitous food-borne genera *Aspergillus* and *Penicillium* (Tresner and Hayes 1971), but only recently have their biodiversity, together with their teleomophic forms, been investigated in the hypersaline waters of salterns from different geographical locations (Cantrell et al. 2006; Butinar et al. 2011).

The mycotoxin-producing genus *Eurotium* was for a long time known to comprise species that are contaminants of foods and feedstuffs that are preserved by high concentrations of sugar or NaCl. They can grow at a_w as low as 0.7, and are therefore considered xerophilic (Pitt and Hockering 1997). Occasionally, different *Eurotium* species have been isolated from natural habitats with low a_w, including saline soils and waters (Abdel-Hafez et al. 1978) and the Dead Sea (Kis-Papo et al. 2001). In a study of the mycodiversity of hypersaline water of different salterns from Europe, Asia, Africa, and North America, six different *Eurotium* species were identified, namely *E. amstelodami, E. chevalieri, E. herbariorum, E. rubrum, E. repens*, and *Eurotium* sp. The last one is probably a new species, and it has tentatively been named "*E. halotolerans.*" Most (74%) of the 208 *Eurotium* spp. isolates obtained belonged to *E. amstelodami*, followed by *E. repens* and *E. herbariorum* (10% of the isolates). These three species are probably part of the indigenous mycobiota of salterns, while other *Eurotium* species are believed to be only temporal inhabitants or occasional air-borne contaminants of low-salinity brines. During the salt production season, two pronounced peaks of occurrence appear: first, at the 10–15% NaCl range, with counts up to 5,000 CFU l^{-1}, and then in the 18–25% NaCl range, with the highest counts (up to 30,000 CFU l^{-1}); outside the season of salt production, the counts for all of the *Eurotium* species remain below 100 CFU l^{-1}. The salinity growth ranges determined in vitro are broad, ranging from 0 up to 27.5% NaCl, and the spores and mycelia of these species can survive long-term exposure to solutions of up to 30% NaCl, with growth stimulated up to 10% NaCl for *E. rubrum, E. chevalieri*, and *E. amstelodami* and up to 12.5% NaCl for "*E. halotolerans,*" *E. repens*, and *E. herbariorum. E. amstelodami* and "*E. halotolerans*" have the broadest salinity range, growing up to 27.5% NaCl, followed by *E. repens* and *E. herbariorum* (up to 25%), and *E. chevalieri* (22.5%) and *E. rubrum* up to 20% (Butinar et al. 2005c). Some of the *Erotium* species (e.g., *E. amstelodami* and *E. rubrum*) prefer media with sugar to media with NaCl, indicating that they have a xerophilic, rather than a halophilic, character (Wheeler and Hocking 1988).

The representatives of genus *Emericella*, which are recognizable by hülle cells in the chleistothecial walls and ornamented ascospore, have frequently been isolated from dry substrata in hot and arid areas worldwide. These appear to be well adapted to dry and warm climates (Samson and Mouchacca 1974) and low a_w (Zalar et al. 2008a). The soil representative *E. nidulans* was isolated also from desert saline soil

(Samson and Mouchacca 1974), while from the hypersaline water of the Sečovlje salterns in Slovenia, two newly described halotolerant species, *E. filifera* and *E. stella-maris*, were reported. The ascospores of *E. filifera* form long appendages that emerge radially from narrow stellate crests, and the ascospores of *E. stella-maris* have star-shaped equatorial crests (Zalar et al. 2008a). *E. striata* was obtained from Lake Enriquillo in Dominican Republic (Butinar et al. 2011).

The genus *Aspergillus* inhabits hypersaline waters that cover the greatest global diversity. *A. niger* and *A. caesiellus* contribute to the stable fungal communities in natural hypersaline waters, while *A. ochraceus*, *A. flavus*, *A. roseoglobulosus*, and *A. tubingensis* are primarily or exclusively present in hypersaline localities at higher environmental temperatures. *A. melleus*, *A. sclerotiorum*, and holomorphic species *Petromyces alliaceus* have been recognized within these taxonomic groups, although they have appeared only locally. *A. versicolor* and *A. sydowii* are both common in marine environments and in dry foods, and these have also been identified as part of the fungal communities in the hypersaline environments. *A. wentii*, *A. flavipes*, *A. terreus*, and particularly *A. candidus* have been repeatedly isolated from Adriatic salterns, whereas *A. penicillioides*, *A. proliferans*, and *A. restrictus* have been found only sporadically at salinities below 10% NaCl. *A. fumigates* is common in arid environments (deserts) at high temperatures, and has been found consistently in solar salterns, although it is also most abundant at salinities below 10% NaCl (Butinar et al. 2011).

Although many species in the subgenus *Penicillium* grow well in salted foods, only five species have been recognized as part of the indigenous fungal communities here. *P. chrysogenum* is a common, widely distributed species, which appears regularly in saline lakes and salterns worldwide, while *P. brevicompactum* is particularly common in Adriatic salterns. The other three species, two of which are most probably new, are seen consistently at low counts or at low salinities, or appear in hypersaline waters only sporadically. A series of soil-borne *Penicillium* species has also been identified as part of the hypersaline mycobiota, namely *P. citrinum*, *P. oxalicum*, and *P. steckii*. Although *P. sizovae* and *P. westlingii* are also common, their counts decrease with increased salinity and they are therefore considered as temporal inhabitants. Four Penicillia isolates identified as *P. nordicum* were isolated from salt used for human consumption and from the salting of foods, at counts of 5 CFU g^{-1} salt crystals (Butinar et al. 2011).

To conclude, in hypersaline environments, the pan-global stable mycobiota of the order *Eurotiales* are represented by *A. niger*, *E. amstelodami*, and *P. chrysogenum*, and possibly also by *A. sydowii*, *A. candidus*, and *E. herbariorum*, which are also quite abundant, although more locally distributed (Butinar et al. 2011).

7.2.7 The Genus *Wallemia*

Fungi from the genus *Wallemia* are xerophilic food- and air-borne fungi (Samson et al. 2002) that have also been isolated from soil (Domsch et al. 1990), sea salt (DasSarma et al. 2010; Butinar et al. 2011), and hypersaline water (Zalar et al. 2005a).

Until recently (Zalar et al. 2005a), a single cosmopolitan species *Wallemia sebi* was recognized in the genus which was known as the causative agent of allergological problems arising in farmer's lung disease (Lappalainen et al. 1998; Roussel et al. 2004). Its unresolved phylogenetic position was worked out when a large group of strains were collected globally from environments with low a_w (e.g., foods, hypersaline waters from different part of the World) and the sequence data of their different genes were analyzed (Zalar et al. 2005a; Matheny et al. 2006). On the basis of their unique morphology, xerotolerance, and isolated phylogenetic position, a new basidiomycetous class of Wallemiomycetes and the order *Walleniales* were proposed for the genus *Wallemia*.

The genus *Wallemia* is an early diverging lineage of Basidiomycota and it segregates into at least three species that have been identified so far: *W. ichthyophaga, W. sebi*, and *W. muriae*. These species have been isolated consistently, although with low counts (maximally, 40 CFU l^{-1}), from hypersaline waters of salterns in Slovenia, Spain, the Dominican Republic, Israel, and Namibia. This is particularly extraordinary, because basidiomycetous fungi rarely show xerophilic or halophilic characteristics, as they generally appear outstandingly intolerant to NaCl (Tresner and Hayes 1971). Also within the Ascomycota there are only a few known fungi, such as *Basipetospora halophila*, *Polypaecilum pisce* (Wheeler et al. 1988), and *H. werneckii* (Gunde-Cimerman and Plemenitaš 2006), which are stimulated by NaCl, although with no obligate requirements.

Interestingly, tests on xerotolerance of the *Wallemia* spp. have shown that it represents one of the most xerophilic fungal taxa (Zalar et al. 2005a). The two halophilic *Wallemia* species, *W. muriae* and *W. ichthyophaga,* necessarily require media with low a_w, whereas *W. sebi* grows also on media without additional solutes. The narrow a_w ranges for growth of *W. muriae* and *W. ichthyophaga* are 0.984–0.805 and 0.959–0.771, respectively. Moreover, *W. ichthyophaga* shows preference for certain solutes for the lowered a_w, as it shows poor growth that is characterized by prolonged growth phases and smaller colonies on media with high concentrations of glucose, compared with media with high concentrations of salt (Zalar et al. 2005a; Kralj Kunčič et al. 2010). As it can thrive in media with NaCl necessarily above 1.7 M and up to saturation (5.3 M NaCl), this makes it the most halophilic fungi known to date. *W. sebi* can grow over a wider range of a_w (0.997–0.690) in glucose/fructose media (Pitt and Hocking 1977), but in media with NaCl as the major solute, its lowest a_w for growth has been reported to be 0.80 (Pitt and Hocking 1977; Zalar et al. 2005a), corresponding to 4.5 M NaCl.

7.3 Special Adaptations and Properties of the Fungi in the Salterns

The responses of eukaryotic cells to low a_w and high concentrations of toxic ions involves complex alterations in gene expression that lead to metabolic changes and subsequent adaptation to the new conditions (Yale and Bohnert 2001; Petrovič et al. 2002; Vaupotič and Plemenitaš 2007). The mechanisms of salt tolerance in fungi have been mostly studied in salt-sensitive *Saccharomyces cerevisiae* (Blomberg and Adler 1992; Blomberg 2000; Hohmann 2002), the halotolerant yeast *D. hansenii* (Larsson and Gustafsson 1987; Andre et al. 1988; Larsson et al. 1990; Larsson and Gustafsson 1993; Prista et al. 1997; Almagro et al. 2000) and the extremely halotolerant black yeast *H. werneckii* (Turk et al. 2004, 2007a; Kogej et al. 2005, 2006a, b, 2007; Plemenitaš et al. 2008). The last species has been proposed as a model organism for eukaryotic halophily studies. Physiological and molecular adaptations of the true halophilic representative known so far, *W. ichthyophaga*, are only at an early stage.

The pathway for the sensing of osmolarity changes is of vital importance for the survival of the cell. In *S. cerevisiae*, this pathway is known as the high-osmolarity glycerol (HOG) signaling pathway, which is also part of one of the best understood mitogen-activated protein kinase (MAPK) cascades (Hohmann 2002). The existence of a similar signaling pathway has been demonstrated and extensively studied in *H. werneckii* (Turk and Plemenitaš 2002; Plemenitaš et al. 2003; Lenassi and Plemenitaš 2005, 2007; Plemenitaš et al. 2008; Fettich et al. 2011). An important level of adaptation to hypersaline environment is the need to balance the osmotic pressure of the medium by accumulating and/or synthesizing compatible organic solutes (Oren 1999) and maintaining low salt concentrations within the cytoplasm. Additional adaptations at the level of the plasma membrane composition (Petrovič et al. 1999; Turk et al. 2004; Gostinčar et al. 2008, 2009) and the cell wall structure (Kralj Kunčič et al. 2010), which represent the first line of defense against environmental stress (Mager and Siderius 2002), are also required to prevent damage of the cells in such environments.

Numerous morphological adaptations that are reflected in the extremophilic ecotype characterized by meristematic growth, pigmentation, and changes in colony morphology are believed to have important roles for successful growth under extremely saline conditions (Kogej et al. 2006a; Kralj Kunčič et al. 2010). The morphology of black yeasts uniquely depends on the environment, as isodiametrical cells are typical in water and hyphal growth is shown only on solid media. This optimization of the volume-to-surface ratio, together with meristematic growth, thick melanized cell walls, extracellular polysaccharides, and adhesion and propagation with endoconidiation, is interpreted as a response to stress factors such as low a_w, low nutrition, and high temperatures (Sterflinger et al. 1999). A study of the morphological adaptations to moderate and high NaCl concentrations of *Wallemia* spp. using different microscopy approaches revealed an impact of high concentrations of NaCl on the cell morphology of the *Wallemia* spp. (Kralj Kunčič et al. 2010). Hyphal compartments of *W. sebi* and *W. muriae* are thicker and shorter, and their mycelial pellets are larger at high salinity. The phylogenetically distinct *W. ichthyophaga* differs from the other two *Wallemia* spp. at the level of its morphology also, as it forms sarcina-like multicellular clumps that are composed of compactly packed spherical cells. The size of the cells does not respond to increased salinity, whereas these multicellular clumps become significantly large. Clustered growth allows the sheltering of the cells in the interior and minimizes the number of cells that are directly in contact with the hostile environment. Therefore, the ability to grow meristematically or in the form of multicellular clumps is believed to greatly enhance their survival in stressful environments (Wollenzien et al. 1995; Palkova and Vachova 2006).

The presence of extracellular polysaccharides is part of the protection against desiccation in rock-inhabiting fungi (Selbmann et al. 2005), and it might have a protective function at high sanities, as a pronounced extracellular polysaccharide layer has been observed in all the three *Wallemia* spp. (Kralj Kunčič et al. 2010) and for *T. salinum* (Kogej et al. 2006a). Furthermore, an increase in the thickness of the multilayered cell wall occurs at higher salinities in all the three *Wallemia* spp., although it is especially pronounced in *W. ichthyophaga*. The extremely thick cell wall of *W. ichthyophaga* is however an exception in the fungal responses to extremely saline conditions known to date (Kralj Kunčič et al. 2010).

A common strategy of osmoadaptation in response to increased salinity in the environment in most eukaryotic microorganisms, as well as in halotolerant and halophilic fungi, is based on the synthesis and cytoplasmic accumulation of a mixture of polyols as compatible solutes. Among these, glycerol is the most significant, and it has been measured in the highest amounts in the extremely halotolerant black yeast *H. werneckii* (Petrovič et al. 2002; Kogej et al. 2007) as well as in the halophilic *W. ichthyophaga* (our unpublished data). Besides glycerol, *H. werneckii* accumulates erythritol, arabitol, and mannitol (Kogej et al. 2007), and interestingly, mycosporine-glutaminol-glucoside also, as complementary compatible solutes (Kogej et al. 2006b). *H. werneckii* and *A. pullulans* keep their intracellular concentrations of sodium and potassium cations low in environments with high concentrations of salt, and they are therefore considered as Na^+-excluders (Kogej et al. 2005). So far, only the halotolerant yeasts *D. hansenii* (Prista et al. 1997) and *P. guilliermondii* (Lahav et al. 2002) have been shown to maintain relatively high internal concentrations of sodium when coping with salt stress, together with the production and intracellular retention of compatible solutes, particularly glycerol (Prista et al. 1997; Lahav et al. 2002).

Glycerol can easily pass through lipid bilayers because of its small molecular mass. Therefore, eukaryotic cells using glycerol as a compatible solute either accumulate lost glycerol using energetically costly transport systems or change their membrane structure by an increased sterol content or reduced membrane fluidity (Oren 1999). Interestingly, in the black yeasts *H. werneckii* and *P. triangularis*, the total sterol content remains mainly unchanged with increased salinity (Turk et al. 2004), while the plasma membrane is significantly more fluid over a wide range of salinities, in comparison with the membranes of salt-sensitive and halotolerant fungi (Turk et al. 2004, 2007a). Higher plasma-membrane fluidity results from an increase in the unsaturated fatty acid content and length, because of the salt stress (Turk et al. 2004, 2007a). One of the mechanisms that allows the precise regulation of membrane fluidity in *H. werneckii* and *A. pullulans* is the change in the expression of fatty-acid-modifying enzymes, such as desaturases and elongase (Gostinčar et al. 2008, 2009). Similar to halophilic/halotolerant black yeasts, the membrane fluidity in the osmotolerant *D. hansenii* is not significantly affected at high levels of NaCl, although the sterol-to-phospholipid ratio and the fatty acid unsaturation increase (Turk et al. 2007b). As high salt tolerance correlates well with higher membrane fluidity, this is of crucial importance for tolerance to salt stress.

Hortaea werneckii can grow at very high salinities, which require large amounts of intracellular glycerol, while at the same time it maintains a very fluid membrane and a constant sterol content. Instead of modifying its membrane properties *H. werneckii* uses its melanized cell wall to reduce glycerol leakage from cells at the optimal salinities (Kogej et al. 2007). The outer part of the melanized cell wall has a continuous layer of melanin granules that minimizes glycerol loss from the cells, as this layer creates a mechanical permeability barrier for glycerol by reducing the size of the pores in the cell wall (Jacobson and Ikeda 2005). At higher salinities, melanization is diminished, and glycerol retention is less effective; therefore, the growth rates and biomass yields of *H. werneckii* are reduced (Kogej et al. 2007).

7.4 The Ecological Types of Fungi in the Salterns

Fungal species can be classified into three ecological groups according to their abilities to inhabit extreme environments. First, there are the mesophiles, which predominantly inhabit environments without extreme conditions. The second group is represented by the generalists, which tolerate a variety of moderately stressful conditions, but not the extreme ones, and they have their growth optimum under moderate conditions. As a result of their limited competition with mesophiles and their inability to survive the most extreme conditions, they are often predominantly found under moderately stressful conditions. In hypersaline environments, *A. pullulans* is a good example of a generalist, as it can tolerate up to 18% NaCl, but not in saturated NaCl solution, and it grows optimally without NaCl. The third group comprises the specialists, which are extremely halo tolerant or even halophilic, and their growth optima are shifted toward extreme conditions. They predominantly inhabit extreme habitats, as they cannot compete with species in moderate environments, or are simply not able to survive moderate conditions. Examples of the specialists among the hypersaline mycobiota are *H. werneckii*, which grows under extreme (up to saturation) salt conditions as well as without salt, and *W. ichthyophaga*, which necessarily requires 10% NaCl and also grows in saturated NaCl solution. Therefore, adaptive extremophiles with a broad ecological amplitude (e.g., *H. werneckii*) and obligate extremophiles with a narrow ecological amplitude (e.g., *W. ichthyophaga*) have been distinguished. Both *H. werneckii* and *W. ichthyophaga* are found in hypersaline environments, and they are characterized by growth optima at extreme salinity. This narrow-amplitude strategy is more of an exception among the fungal kingdom, because *W. ichthyophaga* is so far the only representative of halophilic fungi with an obligate salt requirement (Gostinčar et al. 2010).

7.5 The Importance of the Saltern Mycobiota

Over the last few years, fungi thriving under conditions that are extreme from an anthropocentric point of view, and which thus live at the so-called ecological periphery, have deserved and received increasing scientific attention. Some halotolerant and halophilic fungi have possible important biotechnological applications. These will arise from studies of their basic characteristics and adaptation mechanisms, at the level of their secondary metabolites, cell membranes, intracellular and extracellular enzymes, genetic transfer systems and intracellular osmolytes, and especially of their compatible solutes, which have a wide range of applications because of their ability to stabilize proteins and nucleic acids (Arakawa and Timasheff 1985; Kurz 2008). The salt-tolerant black yeast *A. pullulans* produces exopolymer pullulan that has a broad spectrum of use in the food and pharmaceutical industry (Leathers 2003; Singh and Saini 2008), and the extremely halotolerant black yeasts *T. salinum* and *H. werneckii* have been shown to produce extracellular hydrolytic enzymes that are active at high salt concentrations and that could therefore have important roles in different industries (Zalar et al. 2005b). *H. werneckii* also produces antibiotic compounds that remain to be commercially exploited (Brauers et al. 2001). Indeed, many halophilic and halotolerant fungi synthesize specific bioactive metabolites under stress conditions, and particularly at increased salt concentrations, when higher hemolytic activities have been seen (Sepčić et al. 2011).

Progressive salinization represents a serious agricultural problem worldwide, as 10 million hectares of arable land is lost in this way annually. Genetic manipulations of crops that have increased salt tolerance have still not yielded satisfactory results. The use of halophilic and halotolerant fungal genetic sources might provide the desired improvement in the breeding of such crops in the future. The halotolerant *H. werneckii* has been shown to be a promising source of salt-tolerant transgenes for agriculture. In yeast, as well as in plants, Hal2 is a sodium- and lithium-sensitive 3′-phosphoadenosine-5′-phosphatase, which is an important determinant for halotolerance (Gläser et al. 1993). Thus, overexpression of novel isoenzymes or of Hal2-like proteins from *H. werneckii* can remarkably increase the halotolerance of *S. cerevisiae* (Vaupotič et al. 2007).

Microorganisms and their metabolites can affect the salt production in the evaporating ponds of salterns, as they can physically affect the evaporation process and as their by-products can chemically modify or bind with the dissolved ions (Javor 2002). Moreover, the biological systems in the salterns can also "contaminate" the salt that is used for food preservation. It has been known for a long time that haloarchaea can be introduced into food via salt and can spoil heavily salted proteinaceous products (Norton and Grant 1988; Grant 2004). Recently, different fungi have been isolated from salt used for human consumption and for food salting (Butinar et al. 2011).

As many fungi from salt and hypersaline water produce the same mycotoxins as those found on the salted meat products, the salt is most probably the contamination source for some toxinogenic fungi (Andersen 1995; Larsen et al. 2001; Butinar et al. 2011). Precautionary measures should therefore be considered seriously, such as heat treatment of salt prior to adding it to the meat products, to kill the fungal conidia (Butinar et al. 2011).

7.6 Conclusion

The taxonomic and physiological characterization of halotolerant/halophilic fungi isolated from different salterns and hypersaline lakes on three continents have shown a surprising diversity of fungi. The reports of existance of fungi along with other complex microbial communities in the salterns has improved our understanding of interlinked microbial and chemical processes. As evaporation and mineral precipitation are intimately linked to microbial communities and their products, microbial activity, including fungal, can affect both the quality and quantity of salt. The mycobiota can thus be "imported" in natural hypersaline environments through air, but it can also be "exported" as living cells bounded to salt crystals, and in this way contaminating our foods and homes.

Acknowledgments The scientific studies integral to this report were financed partly through the "Centre of Excellence for Integrated Approaches in Chemistry and Biology of Proteins" (No. OP13.1.1.2.02.0005) of the European Regional Development (30%), partly by the Slovenian Ministry of Higher Education, Science and Technology (35%), and partly by the Slovenian Research Agency (35%).

References

Abdel-Hafez S, Maubasher A, Abdel-Fattah H (1978) Cellulose-decomposing fungi of salt marshes in Egypt. Folia Microbiol (Praha) 23(1):37–44

Almagro A, Prista C, Castro S, Quintas C, Madeira-Lopes A, Ramos J, Loureiro-Dias MC (2000) Effects of salts on *Debaryomyces hansenii* and *Saccharomyces cerevisiae* under stress conditions. Int J Food Microbiol 56(2–3):191–197

Andersen SJ (1995) Compositional changes in surface mycoflora during ripening of naturally fermented sausages. J Food Prot 58:426–429

Andre L, Nilsson A, Adler L (1988) The role of glycerol in osmotolerance of the yeast *Debaryomyces hansenii*. J Gen Microbiol 134:669–677

Andrews JH, Harris RF, Spear RN, Lau GW, Nordheim EV (1994) Morphogenesis and adhesion of *Aureobasidium pullulans*. Can J Microbiol 40(1):6–17

Arakawa T, Timasheff SN (1985) The stabilization of proteins by osmolytes. Biophys J 47(3):411–414

Bensch K, Groenewald JZ, Dijksterhuis J, Starink-Willemse M, Andersen B, Summerell BA, Shin HD, Dugan FM, Schroers HJ, Braun U, Crous PW (2010) Species and ecological diversity within the *Cladosporium cladosporioides* complex (Davidiellaceae, Capnodiales). Stud Mycol 67:1–94

Blackwell M (2001) The yeasts, a taxonomic study. In: Kurtzman CP, Fell JW (eds) Mycopathologia, vol 149(3). Springer, Netherlands, pp 157–158

Blomberg A (2000) Metabolic surprises in *Saccharomyces cerevisiae* during adaptation to saline conditions: questions, some answers and a model. FEMS Microbiol Lett 182(1):1–8

Blomberg A, Adler L (1992) Physiology of osmotolerance in fungi. Adv Microb Physiol 33:145–212

Brauers G, Ebel R, Edrada R, Wray V, Berg A, Grafe U, Proksch P (2001) Hortein, a new natural product from the fungus *Hortaea werneckii* associated with the sponge *Aplysina aerophoba*. J Nat Prod 64(5):651–652

Braun U, Crous P, Dugan F, Groenewald J, Sybren De Hoog G (2003) Philogeny and taxonomy of *Cladosporium*-like hyphomycetes, including *Davidiella* gen. nov., the teleomorph of *Cladosporium s. str.* Mycol Prog 2(1):3–18

Brock TD (1979) Ecology of saline lakes. In: Shilo M (ed) Strategies of microbial life in extreme environments. Dahlem Konferenzen, Berlin, pp 29–47

Butinar L, Santos S, Spencer-Martins I, Oren A, Gunde-Cimerman N (2005a) Yeast diversity in hypersaline habitats. FEMS Microbiol Lett 244(2):229–234

Butinar L, Sonjak S, Zalar P, Plemenitaš A, Gunde-Cimerman N (2005b) Melanized halophilic fungi are eukaryotic members of microbial communities in hypersaline waters of solar salterns. Bot Mar 48(1):73–79

Butinar L, Zalar P, Frisvad JC, Gunde-Cimerman N (2005c) The genus *Eurotium* – members of indigenous fungal community in hypersaline waters of salterns. FEMS Microbiol Ecol 51(2):155–166

Butinar L, Frisvad JC, Gunde-Cimerman N (2011) Hypersaline waters – a potential source of foodborne toxigenic *aspergilli* and *penicillia*. FEMS Microbiol Ecol 77:186–199

Cantrell SA, Casillas-Martinez L, Molina M (2006) Characterization of fungi from hypersaline environments of solar salterns using morphological and molecular techniques. Mycol Res 110:962–970

Crous PW, Braun U, Schubert K, Groenewald JZ (2007) Delimiting *Cladosporium* from morphologically similar genera. Stud Mycol 58:33–56

DasSarma P, Klebahn G, Klebahn H (2010) Translation of Henrich Klebahn's 'Damaging agents of the klippfish – a contribution to the knowledge of the salt-loving organisms'. Saline Systems 6(1):7

de Hoog GS, Gerrits van den Ende AH (1992) Nutritional pattern and eco-physiology of *Hortaea werneckii,* agent of human tinea nigra. Antonie Van Leeuwenhoek 62(4):321–329

de Hoog GS, Guého E (2010) White piedra, black piedra, and tinea nigra. Topley and Wilson's microbiology and microbial infections. Wiley, New York

de Hoog GS, Zalar P, van den Ende BG, Gunde-Cimerman N (2005) Relation of halotolerance to human-pathogenicity in the fungal tree of life: an overview of ecology and evolution under stress. In: Gunde-Cimerman N, Oren A, Plemenitas A (eds) Adaptation to life at high salt-concentration in Archaea, Bacteria and Eukarya, vol 9. Springer, Dordrecht, pp 373–395

Diaz-Munoz G, Montalvo-Rodriguez R (2005) Halophilic black yeast *Hortaea werneckii* in the Cabo Rojo Solar Salterns: its first record for this extreme environment in Puerto Rico. Caribb J Sci 41(2):360–365

Domsch KH, Gams W, Anderson TH (1990) Compendium of soil fungi. Academic, London

Dugan FM, Schubert K, Braun U (2004) Check-list of *Cladosporium* names. Schlechtendalia 11:1–103

Fettich M, Lenassi M, Veranič P, Gunde-Cimerman N, Plemenitaš A (2011) Identification and characterization of putative osmosensors, HwSho1A and HwSho1B, from the extremely halotolerant black yeast *Hortaea werneckii*. Fungal Genet Biol 48(5):475–484

Gläser HU, Thomas D, Gaxiola R, Montrichard F, Surdinkerjan Y, Serrano R (1993) Salt tolerance and methionine biosynthesis in *Saccharomyces cerevisiae* involve a putative phosphatase gene. EMBO J 12(8):3105–3110

Gorbushina AA, Panina LK, Vlasov DY, Krumbein WE (1996) Fungi deteriorating marble in Chersonessus. Mikologiya I Fitopatologiya 30(4):23–27

Gostinčar C, Turk M, Trbuha T, Vaupotič T, Plemenitaš A, Gunde-Cimerman N (2008) Expression of fatty-acid-modifying enzymes in the halotolerant black yeast *Aureobasidium pullulans* (de Bary) G. Arnaud under salt stress. Stud Mycol 61:51–59

Gostinčar C, Turk M, Plemenitaš A, Gunde-Cimerman N (2009) The expressions of Delta(9)-, Delta(12)-desaturases and an elongase by the extremely halotolerant black yeast *Hortaea werneckii* are salt dependent. FEMS Yeast Res 9(2):247–256

Gostinčar C, Grube M, de Hoog S, Zalar P, Gunde-Cimerman N (2010) Extremotolerance in fungi: evolution on the edge. FEMS Microbiol Ecol 71(1):2–11

Grant WD (2004) Life at low water activity. Philos Trans R Soc Lond B Biol Sci 359 (1448):1249–1266

Gunde-Cimerman N, Plemenitaš A (2006) Ecology and molecular adaptations of the halophilic black yeast *Hortaea werneckii*. Rev Environ Sci Biotechnol 5(2):323–331

Gunde-Cimerman N, Zalar P, de Hoog S, Plemenitaš A (2000) Hypersaline waters in salterns – natural ecological niches for halophilic black yeasts. FEMS Microbiol Ecol 32(3):235–240

Gunde-Cimerman N, Oren A, Plemenitaš A, Butinar L, Sonjak S, Turk M, Uršič V, Zalar P (2005) Halotolerant and halophilic fungi from coastal environments in the Arctics. In: Seckbach J (ed) Adaptation to life at high salt concentrations in Archaea, Bacteria, and Eukarya, vol 9, Cellular origin, life in extreme habitats and astrobiology. Springer, Netherlands, pp 397–423

Hocking AD, Pitt JI (1980) Dichloran-glycerol medium for enumeration of xerophilic fungi from low-moisture foods. Appl Environ Microbiol 39(3):488–492

Hocking AD, Miscamble BF, Pitt JI (1994) Water relations of *Alternaria alternata, Cladosporium cladosporioides, Cladosporium sphaerospermum, Curvularia lunata* and *Curvularia pallescens*. Mycol Res 98(1):91–94

Hohmann S (2002) Osmotic stress signaling and osmoadaptation in yeasts. Microbiol Mol Biol Rev 66(2):300–372

Iwatsu TU, Udagawa SI (1988) *Hortaea werneckii* isolated from sea-water. Jpn J Med Mycol 29:142–145

Jacobson ES, Ikeda R (2005) Effect of melanization upon porosity of the cryptococcal cell wall. Med Mycol 43(4):327–333

Javor BJ (1989) Hypersaline environments. In: Schiewer U (ed) Microbiology and biogeochemistry, vol 76(2). Springer, Berlin, p 287

Javor BJ (2002) Industrial microbiology of solar salt production. J Ind Microbiol Biotechnol 28(1):42–47

King AD, Hocking AD, Pitt JI (1979) Dichloran-rose bengal medium for enumeration and isolation of molds from foods. Appl Environ Microbiol 37(5):959–964

Kis-Papo T, Grishkan I, Oren A, Wasser SP, Nevo E (2001) Spatiotemporal diversity of filamentous fungi in the hypersaline Dead Sea. Mycol Res 105(6):749–756

Kogej T, Ramos J, Plemenitaš A, Gunde-Cimerman N (2005) The halophilic fungus *Hortaea werneckii* and the halotolerant fungus *Aureobasidium pullulans* maintain low intracellular cation concentrations in hypersaline environments. Appl Environ Microbiol 71(11):6600–6605

Kogej T, Gorbushina AA, Gunde-Cimerman N (2006a) Hypersaline conditions induce changes in cell-wall melanization and colony structure in a halophilic and a xerophilic black yeast species of the genus *Trimmatostroma*. Mycol Res 110(Pt 6):713–724

Kogej T, Gostinčar C, Volkmann M, Gorbushina AA, Gunde-Cimerman N (2006b) Mycosporines in extremophilic fungi – novel complementary osmolytes? Environ Chem 3(2):105–110

Kogej T, Stein M, Volkmann M, Gorbushina AA, Galinski EA, Gunde-Cimerman N (2007) Osmotic adaptation of the halophilic fungus *Hortaea werneckii*: role of osmolytes and melanization. Microbiology 153(Pt 12):4261–4273

Kralj Kunčič M, Kogej T, Drobne D, Gunde-Cimerman N (2010) Morphological response of the halophilic fungal genus *Wallemia* to high salinity. Appl Environ Microbiol 76 (1):329–337

Kurz M (2008) Compatible solute influence on nucleic acids: many questions but few answers. Saline Systems 4:6

Lahav R, Fareleira P, Nejidat A, Abeliovich A (2002) The identification and characterization of osmotolerant yeast isolates from chemical wastewater evaporation ponds. Microb Ecol 43(3):388–396

Lappalainen S, Pasanen AL, Reiman M, Kalliokoski P (1998) Serum IgG antibodies against *Wallemia sebi* and *Fusarium* species in Finnish farmers. Ann Allergy Asthma Immunol 81(6):585–592

Larsen TO, Svendsen A, Smedsgaard J (2001) Biochemical characterization of ochratoxin A-producing strains of the genus *Penicillium*. Appl Environ Microbiol 67(8):3630–3635

Larsson C, Gustafsson L (1987) Glycerol production in relation to the ATP pool and heat-production rate of the yeasts *Debaryomyces hansenii* and *Saccharomyces cerevisiae* during salt stress. Arch Microbiol 147(4):358–363

Larsson C, Gustafsson L (1993) The role of physiological-state in osmotolerance of the salt-tolerant yeast *Debaryomyces hansenii*. Can J Microbiol 39(6):603–609

Larsson C, Morales C, Gustafsson L, Adler L (1990) Osmoregulation of the salt-tolerant yeast *Debaryomyces hansenii* grown in a chemostat at different salinities. J Bacteriol 172(4):1769–1774

Leathers TD (2003) Biotechnological production and applications of pullulan. Appl Microbiol Biotechnol 62(5–6):468–473

Lenassi M, Plemenitaš A (2005) HwSln1p, a putative sensor protein of the HOG signalling pathway in the halophilic black yeast *Hortaea werneckii*. FEBS J 272:309

Lenassi M, Plemenitaš A (2007) Novel group VII histidine kinase HwHhk7B from the halophilic fungi *Hortaea werneckii* has a putative role in osmosensing. Curr Genet 51(6):393–405

Mager WH, Siderius M (2002) Novel insights into the osmotic stress response of yeast. FEMS Yeast Res 2(3):251–257

Matheny PB, Gossmann JA, Zalar P, Kumar TKA, Hibbett DS (2006) Resolving the phylogenetic position of the Wallemiomycetes: an enigmatic major lineage of Basidiomycota. Canadian Journal of Botany-Revue Canadienne de Botanique 84(12):1794–1805

Mok WY, Castelo FP, Dasilva MSB (1981) Occurrence of *Exophiala werneckii* on salted fresh-water fish *Osteoglossum bicirrhosum*. J Food Technol 16(5):505–512

Nishimura K, Miyaji M (1983) Studies on the phylogenesis of pathogenic black yeasts. Mycopathologia 81(3):135–144

Norton CF, Grant WD (1988) Survival of halobacteria within fluid inclusions in salt crystals. J Gen Microbiol 134(5):1365–1373

Oren A (1999) Bioenergetic aspects of halophilism. Microbiol Mol Biol Rev 63(2):334–348

Oren A (2002) Hypersaline environment and their biota. In: Oren A (ed) Halophilic microorganisms and their environments, vol 5, Cellular origin, life in extreme habitats and astrobiology. Kluwer Academic Publishers, Dordrecht, p 575

Oren A (2005) A hundred years of *Dunaliella* research: 1905–2005. Saline Syst 1:2

Pahor M, Poberaj T (1963) Stare Piranske Soline. Mladinska knjiga, Ljubljana

Palkova Z, Vachova L (2006) Life within a community: benefit to yeast long-term survival. FEMS Microbiol Rev 30(5):806–824

Petrovič U, Gunde-Cimerman N, Plemenitaš A (1999) Salt stress affects sterol biosynthesis in the halophilic black yeast *Hortaea werneckii*. FEMS Microbiol Lett 180(2):325–330

Petrovič U, Gunde-Cimerman N, Plemenitaš A (2002) Cellular responses to environmental salinity in the halophilic black yeast *Hortaea werneckii*. Mol Microbiol 45(3):665–672

Pitt JI, Hocking AD (1977) Influence of solute and hydrogen-ion concentration on water relations of some xerophilic fungi. J Gen Microbiol 101:35–40

Pitt JI, Hocking AD (1997) Fungi and food spoilage, 2nd edn. Blackie Academic & Professional, London

Plemenitaš A, Gorjan A, Gunde-Cimerman N, Turk M (2003) HOG signaling pathway in halophilic black yeast *Hortaea werneckii*. Yeast 20:S207

Plemenitaš A, Vaupotič T, Lenassi M, Kogej T, Gunde-Cimerman N (2008) Adaptation of extremely halotolerant black yeast *Hortaea werneckii* to increased osmolarity: a molecular perspective at a glance. Stud Mycol 61:67–75

Prista C, Almagro A, Loureiro-Dias MC, Ramos J (1997) Physiological basis for the high salt tolerance of *Debaryomyces hansenii*. Appl Environ Microbiol 63(10):4005–4009

Rodriguez-Valera F, Ruiz-Berraquero F, Ramos-Cormenzana A (1981) Characteristics of the heterotrophic bacterial populations in hypersaline environments of different salt concentrations. Microb Ecol 7(3):235–243

Roussel S, Reboux G, Dalphin JC, Bardonnet K, Millon L, Piarroux R (2004) Microbiological evolution of hay and relapse in patients with farmer's lung. Occup Environ Med 61(1):e3

Samson RA, Mouchacca J (1974) Some interesting species of *Emericella* and *Aspergillus* from Egyptian desert soil. Antonie Van Leeuwenhoek 40(1):121–131

Samson RA, Hoekstra ES, Frisvad JC, Filtenborg O (2002) Introduction to food- and airborne fungi, 6th edn. Centraalbureau voor Schimmelcultures, Baarn

Schubert K, Groenewald JZ, Braun U, Dijksterhuis J, Starink M, Hill CF, Zalar P, de Hoog GS, Crous PW (2007) Biodiversity in the *Cladosporium herbarum complex* (Davidiellaceae, Capnodiales), with standardisation of methods for *Cladosporium* taxonomy and diagnostics. Stud Mycol 58:105–156

Selbmann L, de Hoog GS, Mazzaglia A, Friedmann EI, Onofri S (2005) Fungi at the edge of life: cryptoendolithic black fungi from Antarctic desert. Stud Mycol 51:1–32

Sepčić K, Zalar P, Gunde-Cimerman N (2011) Low water activity induces the production of bioactive metabolites in halophilic and halotolerant fungi. Mar Drugs 9(1):59–70

Singh RS, Saini GK (2008) Pullulan-hyperproducing color variant strain of *Aureobasidium pullulans* FB-1 newly isolated from phylloplane of *Ficus* sp. Bioresour Technol 99(9):3896–3899

Sterflinger K, Krumbein WE (1997) Dematiaceous fungi as a major agent for biopitting on Mediterranean marbles and limestones. Geomicrobiol J 14(3):219–230

Sterflinger K, de Hoog GS, Haase G (1999) Phylogeny and ecology of meristematic ascomycetes. Stud Mycol 43:5–22

Todaro F, Berdar A, Cavaliere A, Criseo G, Pernice L (1983) Gasophthalmus in Black-Sea Bream (*Spondyliosoma cantharus*) caused by *Sarcinomyces crustaceus* Lindner. Mycopathologia 81(2):95–97

Torzilli AP, Vinroot S, West C (1985) Interactive effect of temperature and salinity on growth and activity of a salt-marsh isolate of *Aureobasidium pullulans*. Mycologia 77(2):278–284

Tresner HD, Hayes JA (1971) Sodium chloride tolerance of terrestrial fungi. Appl Microbiol 22(2):210–213

Trüper HG, Galinski EA (1986) Concentrated brines as habitats for microorganisms. Experientia 42(11–12):1182–1187

Turk M, Plemenitaš A (2002) The HOG pathway in the halophilic black yeast *Hortaea werneckii*: isolation of the HOG1 homolog gene and activation of HwHog1p. FEMS Microbiol Lett 216(2):193–199

Turk M, Mejanelle L, Šentjurc M, Grimalt JO, Gunde-Cimerman N, Plemenitaš A (2004) Salt-induced changes in lipid composition and membrane fluidity of halophilic yeast-like melanized fungi. Extremophiles 8(1):53–61

Turk M, Abramović Z, Plemenitaš A, Gunde-Cimerman N (2007a) Salt stress and plasma-membrane fluidity in selected extremophilic yeasts and yeast-like fungi. FEMS Yeast Res 7(4):550–557

Turk M, Montiel V, Zigon D, Plemenitaš A, Ramos J (2007b) Plasma membrane composition of *Debaryomyces hansenii* adapts to changes in pH and external salinity. Microbiol (Soc Gen Microbiol) 153:3586–3592

Urzi C, De Leo F, Lo Passo C, Criseo G (1999) Intra-specific diversity of *Aureobasidium pullulans* strains isolated from rocks and other habitats assessed by physiological methods and by random amplified polymorphic DNA (RAPD). J Microbiol Methods 36(1–2):95–105

Vadkertiova R, Slavikova E (1995) Killer activity of yeasts isolated from the water environment. Can J Microbiol 41:759–766

Vaupotič T, Plemenitaš A (2007) Differential gene expression and HogI interaction with osmoresponsive genes in the extremely halotolerant black yeast *Hortaea werneckii*. BMC Genomics 8:280

Vaupotič T, Gunde-Cimerman N, Plemenitaš A (2007) Novel 3'-phosphoadenosine-5'-phosphatases from extremely halotolerant *Hortaea werneckii* reveal insight into molecular determinants of salt tolerance of black yeasts. Fungal Genet Biol 44(11):1109–1122

Wheeler KA, Hocking AD (1988) Water relations of *Paecilomyces variotii, Eurotium amstelodami, Aspergillus candidus* and *Aspergillus sydowii*, xerophilic fungi isolated from Indonesian dried fish. Int J Food Microbiol 7(1):73–78

Wheeler KA, Hocking AD, Pitt JI (1988) Influence of temperature on the water relations of *Polypaecilum pisce* and *Basipetospora halophila,* 2 halophilic fungi. J Gen Microbiol 134:2255–2260

Wollenzien U, de Hoog GS, Krumbein WE, Urzí C (1995) On the isolation of microcolonial fungi occurring on and in marble and other calcareous rocks. Sci Total Environ 167(1–3):287–294

Yale J, Bohnert HJ (2001) Transcript expression in *Saccharomyces cerevisiae* at high salinity. J Biol Chem 276(19):15996–16007

Zalar P, de Hoog GS, Gunde-Cimerman N (1999a) Taxonomy of the endoconidial black yeast genera *Phaeotheca* and *Hyphospora*. Stud Mycol 43:49–56

Zalar P, de Hoog GS, Gunde-Cimerman N (1999b) *Trimmatostroma salinum*, a new species from hypersaline water. Stud Mycol 43:57–62

Zalar P, de Hoog GS, Gunde-Cimerman N (1999c) Ecology of halotolerant dothideaceous black yeasts. Stud Mycol 43:38–48

Zalar P, de Hoog GS, Schroers HJ, Frank JM, Gunde-Cimerman N (2005a) Taxonomy and phylogeny of the xerophilic genus *Wallemia* (Wallemiomycetes and Wallemiales, cl. et ord. nov.). Antonie Van Leeuwenhoek 87(4):311–328

Zalar P, Kocuvan MA, Plemenitaš A, Gunde-Cimerman N (2005b) Halophilic black yeasts colonize wood immersed in hypersaline water. Bot Mar 48(4):323–326

Zalar P, de Hoog GS, Schroers HJ, Crous PW, Groenewald JZ, Gunde-Cimerman N (2007) Phylogeny and ecology of the ubiquitous saprobe Cladosporium sphaerospermum, with descriptions of seven new species from hypersaline environments. Stud Mycol 58:157–183

Zalar P, Frisvad JC, Gunde-Cimerman N, Varga J, Samson RA (2008a) Four new species of Emericella from the Mediterranean region of Europe. Mycologia 100(5):779–795

Zalar P, Gostinčar C, de Hoog GS, Uršič V, Sudhadham M, Gunde-Cimerman N (2008b) Redefinition of *Aureobasidium pullulans* and its varieties. Stud Mycol 61:21–38

Chapter 8
Morphological Evaluation of Peridial Wall, Ascus and Ascospore Characteristics in the Delineation of Genera with Unfurling Ascospore Appendages (Halosphaeriaceae)

Ka-Lai Pang, Wai-Lun Chiang, and Jen-Sheng Jheng

Contents

Abstract In the Halosphaeriaceae, taxa with unfurling ascospore appendages and related species constitute 61 species (in 21 genera). Recent phylogenetic analyses of the rRNA genes have advanced our knowledge on the relationships between genera in the family, especially the group with unfurling ascospore appendages. However, many new genera resulting from these studies lack distinctive morphological characteristics from closely related taxa. In this chapter, peridial wall layers of the ascomata and morphology of asci and ascospores are re-examined to determine if these structures offer useful information for the delineation of genera. In particular, shape parameters (aspect ratio, convexity, elongation, shape factor, sphericity, area, perimeter, diameter max, diameter mean and diameter min) of ascospores were calculated to determine if these parameters can provide extra characters for the delineation of taxa. Results suggest that peridial wall structure alone is insufficient to separate genera in the Halosphaeriaceae. Shape parameters of ascospores can provide additional characters but more taxa are required to test their efficacy. Ascus shape and length of stalk are further characters that should be calculated for

K.-L. Pang (✉)
Institute of Marine Biology, National Taiwan Ocean University, 2 Pei-Ning Road, Keelung, 20224, Taiwan, ROC
e-mail: klpang@ntou.edu.tw

C. Raghukumar (ed.), *Biology of Marine Fungi*,
Progress in Molecular and Subcellular Biology 53,
DOI 10.1007/978-3-642-23342-5_8,

taxonomical consideration. Morphology of the ascomatal wall and shape of asci and ascospores in genera with unfurling ascospore appendages in the Halosphaeriaceae are partially concordant with their phylogeny, suggesting a more thorough examination of these characters for the delineation of taxa in the family.

8.1 Introduction

In the book "Biology of Marine Fungi," edited by the late Steve Moss, Jones et al. (1986) evaluated the criteria that could be used to delineate genera in the Halosphaeriaceae and Farrant (1986) studied the morphology of two genera with unfurling ascospore appendages, *Aniptodera* and *Halosarpheia*, based on ultrastructural observations. This chapter re-evaluates the morphology of a larger number of genera with unfurling ascospore appendages. Jones et al. (2009) documented a total of 530 marine fungi, with 142 species (in 56 genera) belonging to the Halosphaeriaceae. Genera with unfurling ascospore appendages include *Aniptodera*, *Ascosacculus*, *Cucullosporella*, *Halosarpheia*, *Magnisphaera*, *Moana*, *Nais*, *Natantispora*, *Oceanitis*, *Ophiodeira*, *Panorbis*, *Phaeonectriella*, *Saagaromyces*, *Tirispora*, *Trichomaris* and *Tunicatispora*. The genera *Anisostagma*, *Iwilsoniella*, *Lignincola*, *Neptunella* and *Thalassogena*, although without ascospore appendages, are morphologically related, in terms of morphology of asci and ascospores (Pang et al. 2003a). A total of 61 species are included in these 21 genera.

For the last decades, sequence analysis of the nuclear ribosomal RNA genes has advanced our knowledge on the phylogenetic relationships between genera in the Halosphaeriaceae, especially the group with unfurling ascospore appendages and resulted in a number of new genera. *Halosarpheia* was split into a number of genera, including: *Ascosacculus* (*A. aquatic*, *A. heteroguttulata*), *Magnisphaera* (*M. spartinae*), *Natantispora* (*N. lotica*, *N. retorquens*), *Oceanitis* (*O. cincinnatula*, *O. unicaudata*, *O. viscidula*), *Panorbis* (*P. viscosus*), *Saagaromyces* (*S. abonnis*, *S. ratnagiriensis*) (Campbell et al. 2003, Pang et al. 2003a, b, Dupont et al. 2009). Unfurling ascospore appendages were inferred to be an unimportant character in the delineation of genera but can be used to delimit species (Pang et al. 2003a). *Nais glitra*, although without appendages, formed a monophyletic group with *Halosarpheia abonnis* and *H. ratnagiriensis* and consequently, *Saagaromyces* was established for these three species (Pang et al. 2003a). After these revisions, *Halosarpheia sensu stricto* currently includes *H. fibrosa*, *H. trullifera* and *H. unicellularis*, while further collection and investigation are required to clarify the taxonomical position of the remaining *Halosarpheia* species. Morphologically, many of the newly established genera are well circumscribed. *Magnisphaera* differs from *Halosarpheia* in having multi-septate ascospores with a rough ascospore wall (Jones 1962, Campbell et al. 2003). Ascospores of *Oceanitis* species are elongate-fusiform, many times septate which differ from all other *Halosarpheia* species (Dupont et al. 2009). *Saagaromyces* is characterized by large ascomata, large asci with a long stalk and large ellipsoidal ascospores (Pang et al. 2003a). On the other hand, *Ascosacculus*, *Natantispora* and *Panorbis* are morphologically very similar to species assigned to *H. sensu stricto*. Is sequence analysis the only way to separate these genera in the Halosphaeriaceae? Have we carefully examined all the morphological traits for the delineation of the above genera?

The morphological characters defining the Halosphaeriaceae include perithecial ascomata, clavate asci which are mostly deliquescing, presence of catenophyses, ellipsoidal/fusiform, mostly hyaline ascospores with appendages (Spatafora et al. 1998). Taxonomic reviews of the family have been undertaken by many authors, including Kohlmeyer (1972), Jones et al. (1986), Jones (1995) and Pang (2002). Sakayaroj et al. (2011) recently constructed a phylogeny of the Halosphaeriaceae, using all available sequences of the small subunit (SSU) and large subunit (LSU) ribosomal RNA and RNA polymerase II subunit (RPB2) genes, to investigate the phylogenetic relationship between genera. Terminal clades (*H. sensu stricto*, *Oceanitis*, *Remispora* and *Saagaromyces*) were well supported by bootstrap and posterior probability, but internal nodes (intergeneric relationship) were poorly supported. Genera could be separated on the basis of ascospore appendage morphology and ontogeny, while other characters were inferred to be taxonomically unimportant at the generic level, including appendage ontogeny, presence of catenophyses/periphyses and deliquescent/persistent nature of asci. Genera with unfurling ascospore appendages were polyphyletic.

Traditional ascomatal characters used in the classification of marine fungi included colour, shape, texture, neck length and size. In *Ceriosporopsis* and *Lulworthia*, colour of the ascomata was dependent on degree of maturation, and size and shape were affected by the wood type (Meyers 1957, Johnson 1958). Cavaliere and Johnson (1966) discovered that in *Lulworthia grandispora*, ascomata were membranous when grown on birch wood and balsa wood, while they were carbonaceous when grown on Whatman #1 filter paper. Nakagiri and Tubaki (1986) examined the ascomatal wall structure of *Corollospora* species and concluded a basic two-layered peridium, but they considered that this character is not taxonomically significant at the generic level.

Most taxa in the Halosphaeriaceae have deliquescing asci, in particular, oceanic species, which limit its use as a systematic character. Other taxa retain the apical structures (thickening, pore) of asci after their transition from the terrestrial habitats to the sea (Pang 2002). *Aniptodera* is one of the genera that are characterized by asci with an apical thickening, pore, retraction of plasmalemma and thick-walled ascospores (Shearer and Miller 1977). Insufficient attention has been placed on the importance of shape of asci and the length of ascus stalk in the taxonomy of the family. For instance, *Halosarpheia kandeliae* is unique in *Halosarpheia* in having an exceptionally long stalk (Abdel-Wahab et al. 1999). Phylogenetically, it is unrelated to *Halosarpheia fibrosa*, the type species of the genus (Sakayaroj et al. 2011), suggesting that the length of ascus stalk can be an important delineating character at the genus level.

Ascospore shape, colour and wall thickness have been used in the separation of taxa in the Halosphaeriaceae. *Carbosphaerella*, selected *Corollospora*, *Nereiospora* and *Phaeonectriella* are genera with coloured ascospores, but they were not monophyletic (Sakayaroj et al. 2011). *Phaeonectriella* differs from *Aniptodera* only in the dark-coloured ascospores while other characters are extremely similar (Eaton and Jones 1970). Concerning wall thickness, only selected species of *Aniptodera* possess thick ascospore walls. As a result, these characters offer little resolution in the delineation of genera. In this chapter, peridial wall layer of ascomata and the morphology of asci and ascospores are re-examined to determine if these structures offer useful information for the delineation of genera in the Halosphaeriaceae, especially those with unfurling ascospore appendages.

8.2 Material and Methods

Drift, attached and trapped wood were collected from mangrove forests and rocky shores around Taiwan. Wood samples were placed in zip-lock plastic bags and transported to the laboratory at the National Taiwan Ocean University, Keelung, Taiwan (R.O.C.). Wood pieces were incubated in sealed plastic trays lined with moist tissue paper and observed periodically for sporulating structures.

Morphology of ascomata immersed, erumpent or exposed on wood was observed under an Olympus SZ61 stereomicroscope (Tokyo, Japan) with photographs taken on an Olympus DP20 Microscope Camera (Tokyo, Japan). For sections of fruiting bodies, wood pieces with ascomata were cut out from a larger piece of collected wood and fixed by immersion in FAA solution (5% formaldehyde and 5% glacial acetic acid in 50% ethanol) overnight at 4°C. The fixed samples were rinsed three times in 50% ethanol. The samples were then dehydrated in a graduated t-butanol/ethanol/water series (10/40/50, 20/50/30, 35/50/15, 55/45/0, 75/25/0, 100/0/0, 100/0/0, in percentage), and infiltrated gradually and embedded in paraffin (Paraplast, Leica). Paraffin sections (7–10 μm) were cut on a FRM-200P rotary microtome (Tokyo, Japan), floated on 42°C water bath to relax compression and mounted on microscope slides. Dried sections were deparaffinized and rehydrated through a graded series of ethanol. The sections were then stained with 1% safranin O in 50% ethanol (10 s) and 0.5% Orange G in 95% ethanol (30 s). After washing and dehydration, each stained section was permanently mounted with a cover slip and Permount (Fisher, USA). Specimens were observed on an Olympus BX51 microscope (Tokyo, Japan) and light micrographs taken.

For asci and ascospores, ascomata were cut open with a razor under the stereomicroscope. Centrum material was removed with a fine forceps/needle and mounted in a drop of sterile sea water on a slide. Morphology of asci and ascospores was observed on the compound microscope and photographed. For the quantification of ascospore shape, ascospores were first outlined and shape parameters determined by ANALYSIS v3.2 LS image analysis software (Olympus Soft Imaging Solutions GmbH, Germany). The shape parameters included aspect ratio, convexity, elongation, shape factor, sphericity, area, perimeter, diameter max, diameter mean, diameter min and their definitions are given in Table 8.1. These data

Table 8.1 Calculation methods of shape parameters of ascospores

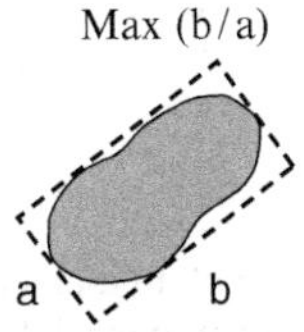	Aspect Ratio: The maximum ratio of width and height of a bounding rectangle for the particle
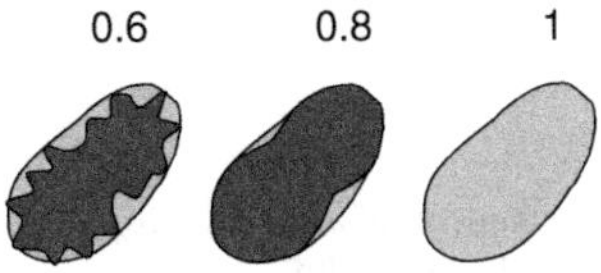	Convexity: The fraction of the particle's area and the area of its convex hull
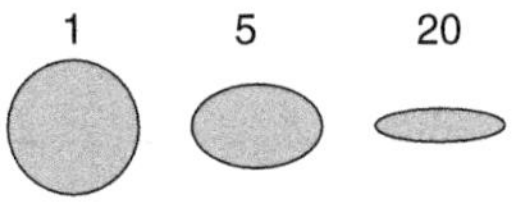	Elongation: The elongation of the particle can be considered as lack of roundness. It results from the sphericity
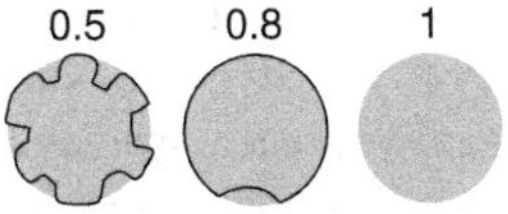	Shape Factor: The shape factor provides information about the "roundness" of the particle. For a spherical particle the shape factor is 1, for all other particles it is smaller than 1
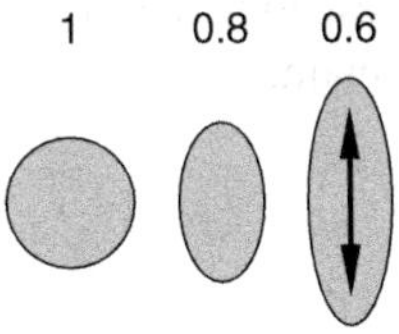	Sphericity: Describes the sphericity or "roundness" of the particle by using central moments
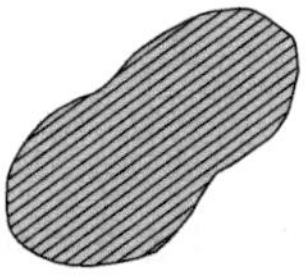	Area: The area of a particle is (number of pixels of the particles) times (calibration factors in X and Y direction)
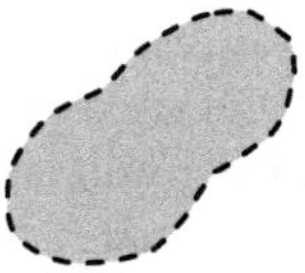	Perimeter: The sum of the pixel distance along the closed outer boundary

(continued)

Table 8.1 (continued)

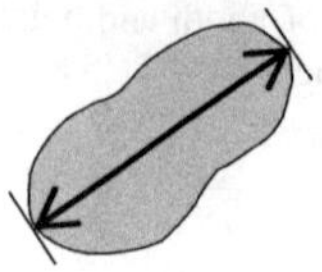	Diameter Max: To evaluate the maximum diameter of a particle, the diameter of different evaluation axes will be determined. To evaluation axis is varied in "1" steps and the maximum diameter of each angle is determined
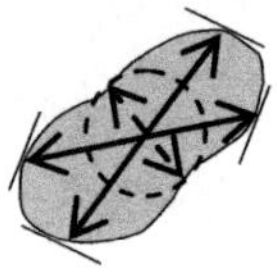	Diameter Mean: The arithmetic mean of all diameters of particle (for angles in the range 0° through 179° with step width 1°
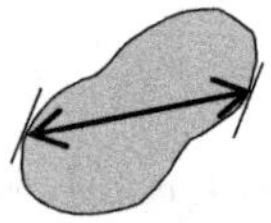	Diameter Min: The minimum diameter of a particle (for angles in range 0° through 179° with step width 1°)

were then entered into PASW Statistics 18 (SPSS Inc., Chicago, IL, USA) for a cluster analysis. In brief, the clusters are group of species with similar characteristics. To form the clusters, the procedure began with each species in a separate group. It then combined the two species that presented the closest shape parameters and formed a new group. After recomputing the distance (squared euclidean) between the groups, the two groups then closest together were combined. This procedure was iterated until only one group remained.

8.3 Results and Discussion

8.3.1 Peridial Wall Layer

Peridial wall structure of taxa with unfurling ascospore appendages and related species in the Halosphaeriaceae, based on paraffin sections, is shown in Fig. 8.1. Most species have a one-layered peridium, while in *Aniptodera chesapeakensis*, *A. salsuginosa*, *H. fibrosa*, *P. viscosus*, *Saagaromyces* spp., it is two layered. Cells are mainly elongate or forming *textura angularis*, with or without large lumina. However, the peridial wall of *Aniptodera longispora* and the outer wall layer of *H. fibrosa* are made up of *textura globulosa*, and in *Natantispora retorquens*, the wall is of flattened, rectangular cells.

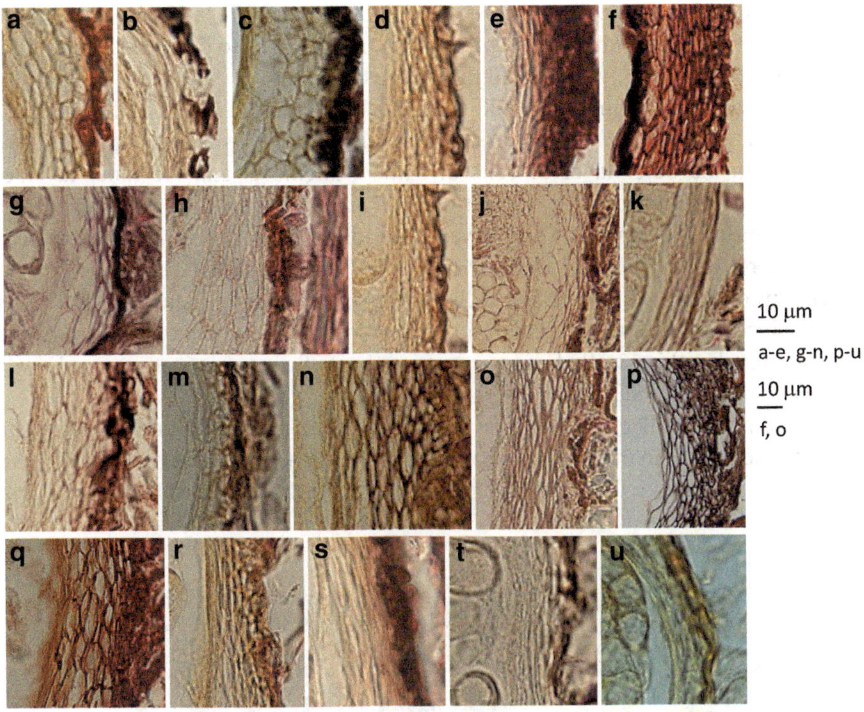

Fig. 8.1 Peridial wall layer. (**a**) *Aniptodera chesapeakensis*, (**b**) *A. lignatilis*, (**c**) *A. longispora*, (**d**) *A. salsuginosa*, (**e**) *Cucullosporella mangrovei*, (**f**) *Halosarpheia fibrosa*, (**g**) *H. marina*, (**h**) *Lignincola laevis*, (**i**) *L. tropica*, (**j**) *Moana turbinulata*, (**k**) *Natantispora retorquens*, (**l**) *Neptunella longirostris*, (**m**) *Oceanitis cincinnatula*, (**n**) *Panorbis viscosus*, (**o**) *Saagaromyces abonnis*, (**p**) *S. ratnagiriensis*, (**q**) *S. glitra*, (**r**) *S. abonnis* c.f., (**s**) *Thalassogena sphaerica*, (**t**) *Tirispora mandoviana*, (**u**) *T. unicaudata*

In *Aniptodera, A. longispora, A. lignatilis* and *A. salsuginosa* have a different wall layer pattern, when compared with *A. chesapeakensis* (an isolate with ascospore appendages). Sakayaroj et al. (2011) showed that *A. lignatilis* and *A. longispora* did not form a monophyletic group with *A. chesapeakensis* in a phylogenetic analysis of three genes. However, it is unknown if the isolate of *A. chesapeakensis* in the GenBank was with unfurling appendages. *Aniptodera longispora* grouped with *Cucullosporella mangrovei*, while *A. lignatilis* was in a group comprising *Ascosacculus*, *A. chesapeakensis*, *Nais inornata*, *Okeanomyces* and *Thalespora* (Sakayaroj et al. 2011). Peridial wall of *C. mangrovei* is composed of flattened, polyhedral cells (Fig. 8.1e), which is different from *A. longispora* (Fig. 8.1c). In *Halosarpheia*, wall of *Halosarpheia marina* (Fig. 8.1g) differs from that of *H. fibrosa* (Fig. 8.1f) in having one layer of cells of *textura angularis* with large lumina. Similarly, the peridial wall of *Lignincola laevis* (Fig. 8.1h) and *L. tropica* (Fig. 8.1i) is made up of elongate, polyhedral cells, but those cells of *L. laevis* are with large lumina. Sakayaroj et al. (2011)

also showed that *L. tropica* grouped with *H. sensu stricto* and *H. marina* with *P. viscosus*. Peridial wall of *L. tropica* and *H. marina* is also different from that of *H. sensu stricto* and *P. viscosus* (Fig. 8.1n), respectively.

Saagaromyces is a monophyletic genus (Sakayaroj et al. 2011), and *S. abonnis*, *S. glitra* and *S. ratnagiriensis* possess a similar peridial wall structure, i.e., two layers with the outer layer of cells of *textura angularis* and the inner layer of elongate, polyhedral cells. A variant of *S. abonnis* with smaller ascospores from mangroves of Taiwan has a one-layered peridium of flattened, polyhedral cells (Fig. 8.1r). Both *Tirispora mandoviana* and *T. unicaudata* have a one-layered peridium of flattened, polyhedral cells, but their monophyly requires a molecular study since they differ significantly in morphology. Ascospores of *T. mandoviana* are relatively thin walled and constricted at the septum, compared with *T. unicaudata* (Fig. 8.2u, v).

These results collectively suggest that peridial wall structure alone is insufficient to separate genera in the Halosphaeriaceae. However, the sections illustrated in this study provided only a two-dimensional view of the cells of the peridium, while cells

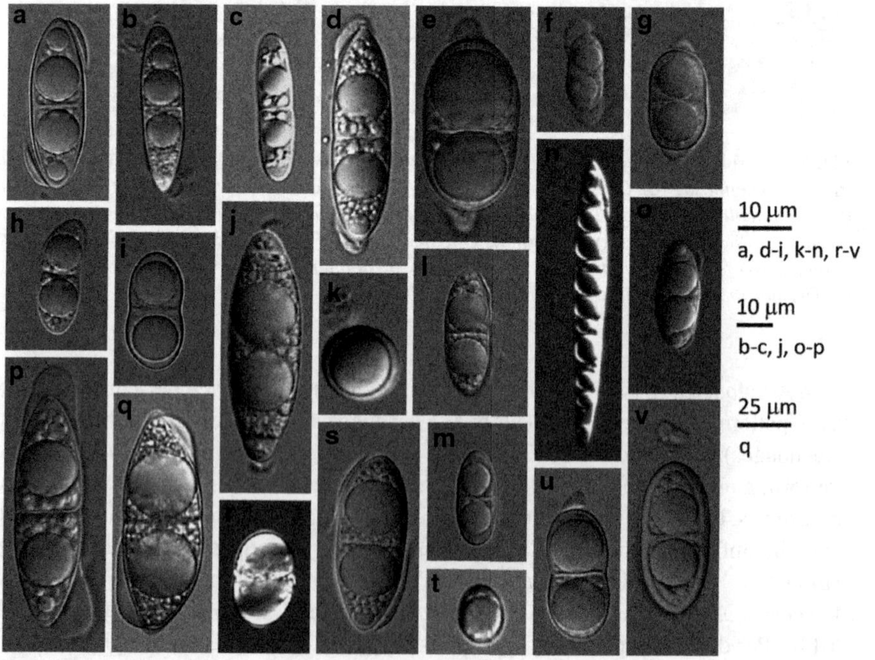

Fig. 8.2 Ascospores. (**a**) *Aniptodera chesapeakensis*, (**b**) *A. lignatilis*, (**c**) *A. longispora*, (**d**) *Cucullosporella mangrovei*, (**e**) *Halosarpheia fibrosa*, (**f**) *H. kandeliae*, (**g**) *H. marina*, (**h**) *Lignincola laevis*, (**i**) *L. tropica*, (**j**) *Magnisphaera spartinae*, (**k**) *Moana turbinulata*, (**l**) *Natantispora retorquens*, (**m**) *Neptunella longirostris*, (**n**) *Oceanitis cincinnatula*, (**o**) *Panorbis viscosus*, (**p**) *Saagaromyces abonnis*, (**q**) *S. ratnagiriensis*, (**r**) *S. glitra*, (**s**) *S. abonnis* c.f., (**t**) *Thalassogena sphaerica*, (**u**) *Tirispora mandoviana*, (**v**) *T. unicaudata*

of the peridium are three-dimensional. Sectioning from different planes is necessary to accurately reflect the shape of the cells.

8.3.2 Ascospores

Ascospores of taxa with unfurling appendages are ellipsoidal/fusiform, uni-septate and possess large oil globules (Fig. 8.2). The hamate ascospore appendages are evidently different from one another in terms of the length and width prior to unravelling, and whether the appendages readily uncoil in sea water, but there has been little attempt to interpret these characteristics to differentiate these closely related taxa. Nakagiri and Ito (1994) investigated the effect of salinity on the unfurling of ascospore appendages in *A. salsuginosa* with appendages unfurling more readily at low salinities. Very likely, these characteristics are not of evolutionary significance, since this group of fungi acquired the unfurling appendages through convergent evolution (Pang et al. 2003a). However, the length of the hamate appendages of *N. retorquens* always exceeds the mid-septum of the ascospores, an essential characteristic for the identification of this species (Shearer and Crane 1980).

Some ascospores are with pointed apices (e.g., *A. lignatilis*, *Saagaromyces* spp.) while others are with rounded ends (e.g., *H. marina*, *T. mandoviana*). Moreover, spores of *L. tropica* and *T. mandoviana* are constricted at the septum and others are not (e.g., *H. fibrosa*, *P. viscosus*). Spore shape of species within one genus is often different, e.g., (1) *A. chesapeakensis*, *A. lignatilis* and *A. longispora* (Fig. 8.2a–c), (2) *L. laevis* and *L. tropica* (Fig. 8.2h, i), (3) *H. fibrosa*, *H. kandeliae* and *H. marina* (Fig. 8.2e–g) and (4) *T. mandoviana* and *T. unicaudata* (Fig. 8.2u,v). All these species were shown to be polyphyletic (Sakayaroj et al. 2011), with the exception of *Tirispora*, since no sequence is available for *T. mandoviana*.

In order to determine if ascospore morphology can be used as a systematic character, shape parameters (aspect ratio, convexity, elongation, shape factor, sphericity, area, perimeter, diameter max, diameter mean, diameter min; Table 8.1) of the ascospores of taxa with unfurling ascospore appendages and related taxa were calculated (Table 8.2). A dendrogram from the cluster analysis is shown in Fig. 8.3. Two of the groups are concordant with the phylogenetic tree presented by Sakayaroj et al. (2011), including the group comprising *S. abonnis*, *S. glitra* and *S. ratnagiriensis* and *A. lignatilis* and *C. mangrovei*. However, other groups are inconsistent and the reason can be one of the followings: (1) ascospore shape alone cannot differentiate genera and it requires a number of morphological characters, or (2) many sequences of genera with unfurling ascospore appendages in the GenBank are early sequences of short length and a second sequence is required to confirm their identity. Nevertheless, the statistics used in this study offer an objective and accurate method to assess shape of ascospores and provide extra characters to delineate these taxa. More taxa from the Halosphaeriaceae should be included in the calculation to have a more thorough analysis of the ascospore shape.

Table 8.2 Shape parameters of ascospores of genera with unfurling appendages and related taxa ($n = 8$)

Species	Aspect ratio	Convexity	Elongation	Shape factor	Sphericity	Area (μm²)	Perimeter (μm)	Diameter max (μm)	Diameter mean (μm)	Diameter min (μm)
Aniptodera chesapeakensis	2.38	0.99	2.37	0.75	0.18	190.02	56.37	24.40	20.62	10.29
Aniptodera lignatilis	4.08	0.97	4.17	0.51	0.06	540.57	114.99	53.62	47.53	14.07
Aniptodera longispora	3.66	0.98	3.75	0.57	0.07	520.52	107.23	48.41	42.98	14.18
Cucullosporella mangrovei	3.48	0.97	3.41	0.58	0.09	776.07	130.38	60.43	51.83	18.12
Halosarpheia fibrosa	1.95	0.98	1.96	0.84	0.26	415.57	78.95	31.73	27.51	16.35
Halosarpheia marina	1.75	0.98	1.74	0.87	0.33	163.48	48.46	18.65	16.46	10.98
Lignincola laevis	3.12	0.95	3.23	0.59	0.10	124.02	51.41	23.13	20.06	7.89
Lignincola tropica	1.80	0.98	1.81	0.86	0.31	181.32	51.51	20.09	17.57	11.26
Moana turbinulata	1.08	0.98	1.08	0.99	0.87	148.51	43.39	14.58	13.93	13.24
Natantispora retorquens	3.15	0.99	3.14	0.62	0.10	162.02	57.49	26.13	22.45	8.57
Neptunella longirostris	2.01	0.98	2.04	0.82	0.25	81.59	35.39	14.47	12.43	7.37
Saagaromyces abonnis	2.52	0.98	2.54	0.71	0.17	576.64	101.40	43.61	37.17	17.83
Saagaromyces abonnis c.f.	2.18	0.98	2.19	0.77	0.21	295.23	69.27	29.63	24.97	13.43
Saagaromyces glitra	1.63	0.98	1.66	0.89	0.37	921.48	113.89	43.53	38.26	26.73
Saagaromyces ratnagiriensis	2.41	0.98	2.40	0.74	0.18	852.55	120.57	51.79	43.96	22.05
Thalassogena sphaerica	1.09	0.99	1.09	0.99	0.87	70.69	30.01	10.03	9.60	9.09
Tirispora mandoviana	1.86	0.98	1.86	0.85	0.29	162.11	48.91	19.06	16.80	10.60
Tirispora unicaudata	3.06	0.97	3.13	0.64	0.10	186.28	60.37	26.58	23.27	9.19

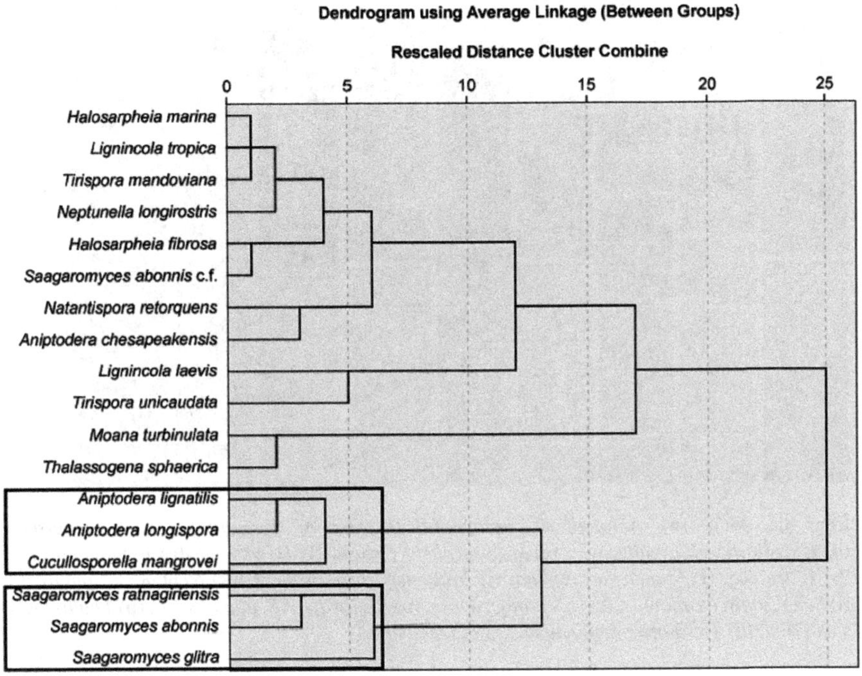

Fig. 8.3 A dendrogram from a cluster analysis of the shape parameters

8.3.3 Asci

Asci of selected taxa in the Halosphaeriaceae with unfurling ascospore appendages and related species are predominantly clavate (Fig. 8.4). Some of them have a flattened apex (e.g., *L. tropica*, *T. unicaudata*), while others have retraction of plasmalemma (e.g., *S. abonnis*, *Thalassogena sphaerica*) and a long stalk (e.g., *Saagaromyces* spp.). *Moana turbinulata* is unique in the complex in having subglobose asci with a long stalk (Fig. 8.4i). A long stalk is also a characteristic feature of *H. kandeliae*. Phylogenetically, *H. kandeliae* was unrelated to *H. sensu stricto* but clustered with *Saagaromyces* spp. but they were not monophyletic (Sakayaroj et al. 2011). A new genus should be introduced to accommodate *H. kandeliae*, but a second sequence is required to confirm its phylogenetic placement. Pang (2002) suggested the use of length/width ratio to describe the shape of the ascus to determine if it is a useful character to differentiate taxa in the Halosphaeriaceae. Shape parameters of asci should also be calculated, however, ascus wall for some of the taxa is very thin and the arrangement of ascospores inside it affects its shape. An average of many asci would give more accurate parameters, but asci of many taxa in the Halosphaeriaceae are deliquescent before maturity, especially the oceanic taxa.

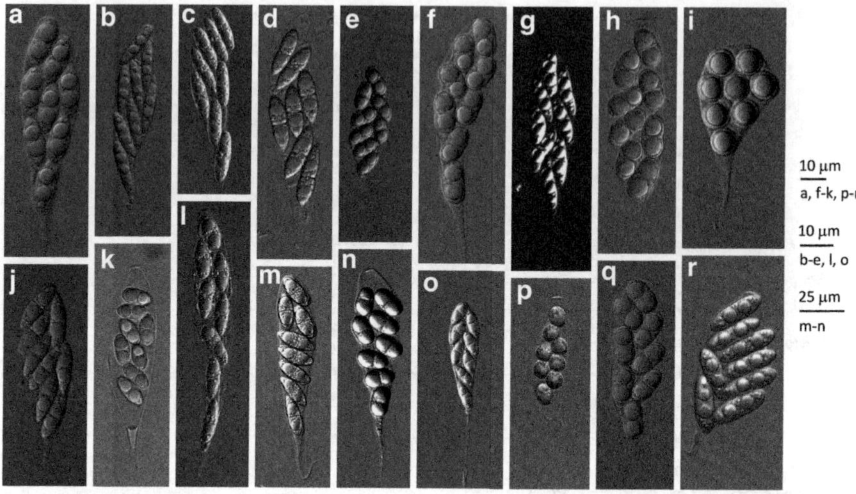

Fig. 8.4 Asci. (**a**) *Aniptodera chesapeakensis*, (**b**) *A. lignatilis*, (**c**) *A. longispora*, (**d**) *Cucullosporella mangrovei*, (**e**) *Halosarpheia fibrosa*, (**f**) *H. marina*, (**g**) *Lignincola laevis*, (**h**) *L. tropica*, (**i**) *Moana turbinulata*, (**j**) *Natantispora retorquens*, (**k**) *Neptunella longirostris*, (**l**) *Saagaromyces abonnis*, (**m**) *S. ratnagiriensis*, (**n**) *S. glitra*, (**o**) *S. abonnis* c.f., (**p**) *Thalassogena sphaerica*, (**q**) *Tirispora mandoviana*, (**r**) *T. unicaudata*

8.4 Concluding Remarks

Spatafora et al. (1998), based on phylogenetic analysis of the rRNA genes, suggested the terrestrial origin of the Halosphaeriaceae. Marine fungi probably have rapidly radiated and evolved independently a wide range of ascospore appendage morphology to increase entrapment and contact area for attachment to substrata for germination and growth (Jones 2006). About one third of taxa in the Halosphaeriaceae possess unfurling ascospore appendages and there are two possibilities: (1) it is an intermediate type of ascospore appendage between no appendage and complex appendages, or (2) it is an efficient type of appendage for attachment to substrates and is an inherited character or gained through horizontal gene transfer. Whether all these taxa are able to interbreed is also a possibility that requires further research. Due to the advancement of molecular techniques and phylogenetic analysis, many genera in the family, such as *Natantispora* and *Panorbis* were defined based on sequence data and these genera lack distinctive morphological traits from other allied taxa. In this study, characters of the ascomatal wall structure and shape of the asci and ascospores in genera with unfurling ascospore appendages in the Halosphaeriaceae are partially concordant with their phylogeny, suggesting a more thorough examination of these characters for the delineation of taxa in the family.

Acknowledgements We thank Prof. Gareth Jones for a pre-submission review. Ka-Lai Pang acknowledges the financial support by National Science Council of Taiwan (NSC 98-2321-B-019-004).

References

Abdel-Wahab MA, Jones EBG, Vrijmoed LLP (1999) *Halosarpheia kandeliae* sp. nov. on intertidal bark of the mangrove tree *Kandelia candel* in Hong Kong. Mycol Res 103:1500–1504

Campbell J, Anderson JL, Shearer CA (2003) Systematics of *Halosarpheia* based on morphological and molecular data. Mycologia 95:530–552

Cavaliere AR, Johnson TW (1966) Marine ascomycetes: ascocarp morphology and its application to taxonomy and evaluation. Nova Hedwigia 10:453–461

Dupont J, Magnin S, Rousseau F, Zbinden M, Frebourg G, Samadi S, Richer de Forges B, Jones EBG (2009) Molecular and ultrastructural characterization of two ascomycetes found on sunken wood off Vanuatu Islands in the deep Pacific Ocean. Mycol Res 113:1351–1364

Eaton RA, Jones EBG (1970) New fungi on timber from water-cooling towers. Nova Hedw 19:779–786

Farrant CA (1986) An electron microscope study of ascus and ascospore structure in *Aniptodera* and *Halosarpheia*, Halosphaeriaceae. In: Moss ST (ed) The biology of marine fungi. Cambridge University Press, Cambridge

Johnson TW (1958) Marine fungi. IV. *Lulworthia* and *Ceriosporopsis*. Mycologia 10:151–163

Jones EBG (1962) *Haligena spartinae* sp. nov. A pyrenomycete on *Spartina townsendii*. Trans Br Mycol Soc 45:246–248

Jones EBG (1995) Ultrastructure and taxonomy of the aquatic ascomycetous order Halosphaeriales. Can J Bot 73(Suppl 1):S790–S801

Jones EBG (2006) Form and function of fungal spore appendages. Mycoscience 47:167–183

Jones EBG, Johnson RG, Moss ST (1986) Taxonomic studies of the Halosphaeriaceae philosophy and rationale for the selection of characters in the delineation of genera. In: Moss ST (ed) The biology of marine fungi. Cambridge University Press, Cambridge

Jones EBG, Sakayaroj J, Suetrong S, Somrithipol S, Pang KL (2009) Classification of marine Ascomycota, anamorphic taxa and Basidiomycota. Fungal Divers 35:1–203

Kohlmeyer J (1972) A revision of Halosphaeriaceae. Can J Bot 50:1951–1963

Meyers SP (1957) Taxonomy of marine pyrenomycetes. Mycologia 49:475–528

Nakagiri A, Tubaki K (1986) Ascocarp peridial wall structure in *Corollospora* and allied genera of Halosphaeriaceae. In: Moss ST (ed) The biology of marine fungi. Cambridge University Press, Cambridge

Nakagiri A, Ito T (1994) Aniptodera salsuginosa, a new mangrove-inhabiting ascomycete, with observations on the effect of salinity on ascospore appendage morphology. Mycol Res 98:931–936

Pang KL (2002) Systematics of the Halosphaeriales: which morphological characters are important? In: Hyde KD (ed) Fungi in marine environments. Fungal Diversity Press, Hong Kong

Pang KL, Vrijmoed LLP, Kong RYC, Jones EBG (2003a) *Lignincola* and *Nais*, polyphyletic genera of the Halosphaeriales (Ascomycota). Mycol Prog 2:29–39

Pang KL, Vrijmoed LLP, Kong RYC, Jones EBG (2003b) Polyphyly of *Halosarpheia* (Halosphaeriales, Ascomycota): implications on the use of unfurling ascospore appendage as a systematic character. Nova Hedwigia 77:1–18

Sakayaroj J, Pang KL, Jones EBG (2011) Multi-gene phylogeny of the Halosphaeriaceae: its ordinal status, relationships between genera and morphological character evolution. Fungal Divers 46:87–109

Shearer CA, Crane JL (1980) Fungi of the Chesapeake Bay and its tributaries VIII. Ascomycetes with unfurling appendages. Bot Mar 23:607–615

Shearer CA, Miller M (1977) Fungi of the Chesapeake Bay and its tributaries V. *Aniptodera chesapeakensis* gen. *et* sp. nov. Mycologia 69:887–898

Spatafora JW, Volkmann-Kohlmeyer B, Kohlmeyer J (1998) Independent terrestrial origins of the Halosphaeriales (marine Ascomycota). Am J Bot 85:1569–1580

Chapter 9
Cultured and Uncultured Fungal Diversity in Deep-Sea Environments

Takahiko Nagahama and Yuriko Nagano

Contents

Abstract The importance of fungi found in deep-sea extreme environments is becoming increasingly recognized. In this chapter, current scientific findings on the fungal diversity in several deep-sea environments by conventional culture and culture-independent methods are reviewed and discussed, primarily focused on culture-independent approaches. Fungal species detected by conventional culture methods mostly belonged to Ascomycota and Basidiomycota phyla. Culture-independent approaches have revealed the presence of highly novel fungal phylotypes, including new taxonomic groups placed in deep branches within the phylum Chytridiomycota and unknown ancient fungal groups. Future attempts to culture these unknown fungal

T. Nagahama (✉)
Department of Food and Nutrition, Higashi-Chikushi Junior College, 5-1-1 Shimoitozu, Kokurakita-ku, Kitakyusyu, Fukuoka 800-0351, Japan

Institute of Biogeosciences, Japan Agency for Marine-Earth Science and Technology (JAMSTEC), 2-15, Natsushima-cho, Yokosuka 237-0061, Japan
e-mail: jamstec@gmail.com

C. Raghukumar (ed.), *Biology of Marine Fungi*,
Progress in Molecular and Subcellular Biology 53,
DOI 10.1007/978-3-642-23342-5_9,

groups may provide key insights into the early evolution of fungi and their ecological and physiological significance in deep-sea environments.

9.1 Introduction

Deep sea is recognized as an extreme environment, characterized by the absence of sunlight irradiation, predominantly low temperatures (occasionally extremely high, up to 400°C near hydrothermal vents) and high hydrostatic pressure (up to 110 MPa). Despite these extreme conditions, it is now well known that life is abundant in deep-sea environments.

Whereas marine fungi in shallow water, such as coastal and offshore areas (especially mangrove forests), have been well studied with respect to ecological roles and taxonomic positions, a knowledge of "marine fungi" in deep-sea regions of marine environments is limited. In general, it has been believed that extraordinary (occasionally large) creatures inhabit deep sea. However, fungal strains cultured from deep-sea sources have yet to illustrate any remarkable difference (in terms of taxanomic novelty and morphological characteristics) from terrestrial fungi. However, the possibility of unique fungi inhabiting the floor of deep sea remains. This theory may be answered through the application of culture-independent molecular techniques for fungal communities. In this chapter, recent progress on the biodiversity of cultured and uncultured fungi in deep-sea environment will be reviewed.

9.2 Sampling

Great depth and elevated hydrostatic pressure make deep sea a difficult environment to access. However, continuing improvements in technology have made it possible to collect samples from the bottom of the deepest ocean, at depths of 10,000 m.

There are two basic approaches to reach deep sea. One is using a human occupied vehicle (HOV), a deep-sea submarine which can carry people inside to the ocean depths. These types of research vessels have some advantages, such as allowing scientists to make direct observations of deep sea. At present, there are few HOVs which can dive to great depths. The Japanese HOV Shinkai 6500, for example, is capable of diving at depths of up to 6,500 m. An alternative method is to use a remotely operated vehicle (ROV), which is capable of diving at depths greater than 10,000 m and can collect samples from some of the ocean's deepest environments. Equipment loaded on both HOV and ROVs usually include a variety of equipment, such as cameras (television, still photograph, video) and manipulators that enable the placement of objects on the seafloor and the collection of various samples. Both methods are limited in the length of time that can be spent in deep sea to observe and collect samples, because of the time taken to reach the ocean depths.

The basic sampling methods for investigating fungal communities in deep-sea environments do not differ fundamentally from those used for shallow marine environments. However, collecting samples from deep sea involves higher costs and more complicated equipment. For example, all equipment is required to be resistant to high pressure. For fungal diversity research, sediments are the most popular and investigated areas for deep-sea environment sampling, as fungi are normally most prominently found in sediment. However, water columns, deep-sea habitants, sunken woods, and other materials are also investigated. For water sampling, some known samplers are the Nansen bottle, the Niskin sampler, and the van Dorn sampler. Several types of samplers are used to collect deep-sea sediments, such as a grab sampler, box corer, gravity corer, and piston corer. A grab sampler is useful in collecting soft, sandy, or silt sediments from the surface of the deep-sea floor. Common grab samplers used are the Smith-McIntyre and van Veen grab samplers, which have an effective sampling area of 0.1 m^2 or smaller. A box corer is used to collect the very top surface layers and the sediment underneath, and is available in various sizes. A large box corer has a sampling area of 0.25 m^2. Box corers are used to collect mud and silt but not sand. A gravity corer and piston corer are used to collect samples as a tube of sediment. These types of samplers can collect long deep-sea sediment cores and are useful for investigating fungal diversity in various depths from the seafloor. Gravity corers generally collect sediment layers >2 m in length. Piston corers are similar in appearance to gravity corers but can penetrate much deeper into the seafloor and collect much longer and older core sediment. Currently, the most advanced drilling capabilities in the world are found on the deep-sea drilling vessel, CHIKYU, which is able to bring up sediment core from 7,000 m below the seafloor. One major concern of using the above types of samplers for investigating deep-sea fungal communities is contamination from other environments, such as surface and shallow environments. In order to overcome this problem, sterile sediment samplers (Ikemoto and Kyo 1993) have been developed. These samplers can collect deep-sea sediment into sterile bottles, bringing up samples to the surface without contamination. This method is often applied in fungal diversity studies which use molecular-based methods, because even a minor contamination of samples can cause a major problem in the accuracy of results. Some disadvantages of this method are that the sterile sediment samplers are limited in collecting samples from the surface layer of sediments and can only take a small portion (up to 50 ml) of samples.

9.3 Isolation of Deep-Sea Fungi

The isolation procedure for deep-sea fungi varies depending on the type, volume, and shape of collected samples. Water samples are mostly filtered through membranes and are then used for isolation due to the low number of fungal cells in the limited volume of water samples. Solid sources, such as sunken wood or deep-sea animals, are normally applied to agar plate media or liquid media after being broken into small pieces or homogenized. Sediment samples are normally directly plated onto medium or mixed with medium agar before they get cold and solid. Mixing with an agar method can separate aggregated sediment particles and usually isolate more diverse fungi than a direct plating method. Samples can first be diluted with sterile sea water, if they contain high numbers of fungi. Particle plating techniques (Bills and Polishook 1994) are also used by some researchers (Jebaraj et al. 2010). In this method, particles that pass through 200-μm mesh but are retained on the 100-μm mesh are spread plated. When the sampling volume is too low or the content contains low numbers of fungi, the sample can be enriched with liquid media. However, it should be noted that enrichment often changes the proportion of the fungal population and also leads to a disappearance of the minority species or slower growing species.

Culture media often used for the isolation of fungi from deep-sea environments are not so different from media used for terrestrial fungi. Several media are used to investigate culturable fungal communities in deep-sea environments (Damare et al. 2006; Jebaraj et al. 2010; Burgaud et al. 2009; Le Calvez et al. 2009), and most investigators use organic media, mainly consisting of malt extract, yeast extract, peptone, glucose, and potato starch. Media are usually prepared in artificial sea water with the addition of antibiotics, such as streptomycin, penicillin, and chloramphenicol to inhibit bacterial growth. Media are also occasionally used at 1/5 strength of normal condition, in order to simulate the low nutrient conditions found in deep sea. Damare et al. (2006) compared several different media, namely, malt extract agar, malt extract broth, corn meal agar, Sabouraud's dextrose agar, Czapek Dox agar, and Czapek Dox broth for isolation of fungi from deep-sea sediments. They concluded that malt extract agar and malt extract broth were generally suitable for the isolation and growth of deep-sea fungi. It is important to note that using fresh samples dramatically increases the number of culturable fungi, compared to using old samples which may have been kept at −20°C or −80°C after sampling. Media are normally incubated between 4 and 30°C, although occasionally incubated at higher degrees, such as 45°C for samples from hydrothermal vents, where the temperature can reach very high, up to 400°C. However, to date thermophilic fungi have not yet been isolated from deep-sea environments. Culturing periods can range from 2 weeks to several months. However, a long culturing time is necessary for isolating deep-sea fungi, as some fungi grow extremely slow. It is also recommended to employ a medium that can suppress the growth of fast growing fungi, along with other fungal media for isolating slow growing deep-sea fungi. As one of the most characteristic conditions of deep sea is extremely high pressure (up to 110 MPa), pressure incubation (up to 30 MPa) has been performed in some studies (Lorenz and Molitoris 1997; Raghukumar and Raghukumar 1998; Damare et al. 2006). In these studies, some fungi showed tolerance to pressure but piezophilic fungi have not yet been reported.

The methods described earlier are what have been done in previously reported studies. However, since our knowledge on the ecological role of deep-sea fungi is very limited and many niches have not yet been explored, various new isolation media and conditions are being experimented with. Culturing with anaerobic conditions, high-pressure conditions, or in situ cultivating methods may yield interesting results in the near future.

9.4 Detecting Deep-Sea Fungal Diversity by Molecular Methods

With recent technology advances, molecular-based methods are more often used for studying fungal diversity in deep-sea environments. The advantage of using molecular methods is mainly the requirement of only a small amount of samples and the ability to detect difficult-to-culture species, including rare species. DNA-based approaches have confirmed extensive diversity of fungal communities and revealed previously undetected diversity in deep-sea environments (Bass et al. 2007; Lai et al. 2007; Le Calvez et al. 2009; Nagano et al. 2010; Nagahama et al. 2011). Typical procedures for detecting fungal diversity are (1) extracting DNA from samples, (2) amplifying fungal DNA by PCR using pan-fungal primers, (3) sequencing of PCR products, (4) identifying detected species by comparing previously published data and assigning unknown taxa to a phylogenetic clade. DNA extraction from environmental samples is normally performed with various readymade kits after homogenization or treated with six cycles of freeze-thawing in liquid nitrogen/boiling. The most popular targeted genes for fungal diversity studies in deep-sea environments are SSU rRNA and ITS regions, including 5.8 S rRNA. Targeting SSU rRNA is suitable for phylogenetic study, especially for determining the phylogenetic position of unknown or highly novel fungal taxa, which are often detected from deep-sea environments. Targeting ITS regions appears to be a sensitive means of detecting fungal DNA. However, this region is usually too short and variable for complicated phylogenetic study. It is strongly noted that detectable fungal diversity varies strongly depending on the primers used to amplify DNA (Nagano et al. 2010; Jebaraj et al. 2010). It is considered that a strong bias by the selection of PCR primers can limit detected fungal diversity (Amend et al. 2010). Therefore, it is recommended to use multiprimer sets for obtaining better resolution of true fungal diversity in deep-sea environments.

9.5 Comparison Between Culture-Independent Studies of Deep-Sea Fungi

The taxonomic distribution of fungal clones from deep-sea resources was compared according to recent studies using culture-independent approaches (Table 9.1). Of these, nine 18 S rDNA libraries were constructed and investigated in the following

Table 9.1 Appearances of fungal taxa as environmental clones in various deep-sea environments

	18S rDNA									ITS			
Author	Bass et al. (2007)	López-García et al. (2007)	Le Calvez et al. (2009)	Sauvadet et al. (2010)	Singh et al. (2010)	Eloe et al. (2011)	Jebaraj et al. (2010)	Quaiser et al. (2011)	Nagahama et al. (2011)	Lai et al. (2007)	Nagano et al. (2010)	Singh et al. (2010)	Burgaud et al. (unpublished)
Source	Various (including hydrothermal)	Hydrothermal	Hydrothermal	Water and hydrothermal	Sediment	Water	Sediment and water	Water and sediment	Sediment (cold seep)	Sediment	Sediment (including cold seep)	Sediment	Water
Depth	<4,000 m	750–900 m	1,700m, 2,630 m	500–3,000 m	3,992–5,377 m	6,000 m	200 m	1,000–1,260 m	850–1,200 m	350–3,011 m	1,174–10,131 m	3,992–5,377 m	1,750–3,750 m
Target of primers used	Fungi	Eukaryota	Fungi	Eukaryota	Fungi	Eukaryota	Both	Eukaryota	Fungi	Fungi	Fungi	Fungi	Fungi
Ascomycota													
Dothideomycetes	○	○			○	○	○		○	○		○	○
Eurotiomycetes	○	○			○	○	○	○	○	○	○	○	
Leotiomycetes									○		○		
Saccharomycetes	○			○	○		○		○	○	○	○	○
Sordariomycetes				○			○		○	○	○	○	○
Basidiomycota													
Agaricomycetes	○		○	○					○				
Cystobasidiomycetes													○
Entorrhizomycetes							○		○				
Exobasidiomycetes	○	○	○	○		○	○		○	○		○	○
Microbotryomycetes	○				○	○			○			○	○
Tremellomycetes	○		○	○		○				○	○	○	○
Ustilaginomycetes						○	○						○
Wallemiomycetes				○	○								
Chytridiomycota													
Chytridiomycetes	○		○	○									
Other basal lineage													
LKM11 clade	○								○		○		
KD14 clade	○							○	○				

○: Taxa presented by phylotypes

studies: Bass et al. (2007), López-García et al. (2007), Le Calvez et al. (2009), Jebaraj et al. (2010), Sauvadet et al. (2010), Singh et al. (2011), Eloe et al. (2011), Quaiser et al. (2011), and Nagahama et al. (2011). Four ITS libraries were constructed and investigated in the following studies: Lai et al. (2007), Singh et al. (2010), Nagano et al. (2010), and Burgaud et al. (unpublished, only in database)., In total, 16 taxa, comprising 14 fungal classes and two uncultured clone groups, have been recognized from the deep-sea resources by culture-independent methods to date.

Ten of 16 taxa were common between 18 S rDNA and ITS libraries, and five taxa, namely, Agaricomycetes, Entorrhizomycetes, Wallemiomycetes, Chytridiomycetes, and KD14 clade were only obtained from the 18 S rDNA library. Cystobasidiomycetes were only obtained from the ITS library in one study. These differences may be caused by primer biases or inhabitation to specific substrates. Geographical limitation may represent the absence of fungal taxa. However, conclusion based on such limited information is difficult.

The taxonomic distribution of deep-sea fungal clones detected by culture-independent methods will now be discussed in more detail.

9.5.1 Ascomycota

Eurotiomycetes and Saccharomycetes were the most frequently detected fungal taxa from deep-sea environments by culture-independent methods, followed by Dothideomycetes and Sordariomycetes (Table 9.1). Some operational taxonomic units (OTUs) of Leotiomycetes have occurred in limited areas (Table 9.1). The majority of OTUs belonging to Eurotiomyces were composed of members of the *Aspergillus*/*Penicillium* group, which are known to be globally distributed fungal taxa. It is suggested that these taxa were also ubiquitous in marine environments (including in deep sea), as well as terrestrial environments. It may appear doubtful that these fungi are indigenous to deep sea. However, evidence of physiological adaption of the *Aspergillus* species to deep-sea environments has been reported (Raghukumar and Raghukumar 1998; Raghukumar et al. 2004; Damare et al. 2006; Damare and Raghukumar 2008). Saccharomycetes are a class of ascomycetous yeasts which often have been isolated from oceanic regions by culturing approaches (Fell 1976). In deep-sea environment, OTUs of genera *Candida*, *Debaryomyces*, *Kodamaea*, *Metschnikowia*, *Pichia,* and their relatives have been detected. Especially, OTUs of *Metschnikowia* with very long evolutionary distances from other ascomycetes have appeared as hitherto uncultured taxa from some deep-sea environments (KD10; Bass et al. 2007; Nagano et al. 2010). Recently, closely related clones were also culture-independently detected as Daphnia parasites in lake environments (Wolinska et al. 2009). Therefore, it may be possible that the *Metschnikowia*-like yeasts in deep-sea environment would also be associated with deep-sea meiobenthos, such as small crustacean. Dothideomycetes are a class of bitunicate ascomycetes, having the order Pleosporales with many marine species.

Dothideomycetes are also reported from deep-sea environment; however, they were not the marine pleosporales but *Aureobasidium*, *Cladosporium*, or *Hortaea*. These frequent species have common characters of adaptation or resistance to/against low temperature and high osmotic pressure, which may be essential keys to survive under deep-sea conditions. Moreover, this may indicate that these fungi originated from terrestrial regions. Also, *Phoma* clones have often appeared in deep-sea studies (Lai et al. 2007; Singh et al. 2011). *Phoma* is known to associate not only with land plants but also with marine plants, such as seaweed and seagrass, as well as with marine invertebrates, such as sponge (Kohlmeyer and Volkmann-Kohlmeyer 1991). The Sordariomycetes were also detected in many studies. However, their phylotypes were few and unique to the area. From oxygen-depleted regions, a number of OTUs close to *Fusarium oxysporum*, which is known as a denitrifying fungus, have been reported (Jebaraj et al. 2010).

9.5.2 Basidiomycota

The most ubiquitous class within the phylum of Basidiomycota detected in deep-sea environments was Exobasidiomycetes (Table 9.1). The majority of the clones belonging to this class are related to the genus *Malassezia*. The genus of *Malassezia* is well known as the causative agents of skin diseases in mammals, along with marine mammals, such as seals or sea lions (Guillot et al. 1998; Nakagaki et al. 2000; Pollock et al. 2000). Whereas *Malassezia* phylotypes present ubiquitous occurrences in deep-sea environments by molecular methods, the isolates have not been cultured from seawater and sediments. It may be considered unsuccessful, because they would possess an anaerobic or hydrothermal-specific metabolism (López-García et al. 2007). In addition, because molecular detection of *Malassezia* sp. from terrestrial soil nematodes has been reported (Renker et al. 2003), the species may be living in small marine invertebrates, such as nematodes or polychaetes, which also inhabit deep-sea sediments. These may have different growth requirements from skin-inhabiting *Malassezia* spp. Moreover, this fungal group has been reported to occupy the majority of eukaryotic diversity found in deep-sea subsurface sediments (Edgcomb et al. 2011). Other classes of Ustilaginomycotina, Ustilaginomycetes, and Entorrhizomycetes were also noted. However, distribution and phylotype was limited. The ustilaginomycetous anamorphic yeasts *Pseudozyma* were detected as uncultured clones and were cultured from deep-sea bivalves (Konishi et al. 2010). The xerophilic characteristics of phylotypes of Wallemiomycetes from deep-sea environments may be a focus for discovery (Sauvadet et al. 2010; Singh et al. 2011). The parasitic smut fungus genus *Entorrhiza*, the taxon which indicated an early divergence in Basidiomycota, is also detected in deep sea. However, clones were separated by long evolutionary distances from known sequences. Two classes of the Pucciniomycotina, Cystobasidiomycetes, and Microbotryomycetes were found with the typical marine yeast orders, Sporidiobolales and Erythrobasidiales. Rare occurrences of Erythrobasidiales in deep-sea phylotypes were consistent with culturing studies (Nagahama et al. 2001). Cryptococcus-related phylotypes were a major component of Tremellomycetes, some of which were psychrotolerant and were reported from polar or alpine regions (Connell et al. 2008; Turchetti et al. 2008). It may be noted that clones of *Cryptococcus curvatus* were the major component of eukaryotic clones from deep-sea cold seep regions (Takishita et al. 2006). In marine environment, agaricomycetous fungi are scarce and mostly reported from mangrove woods (Jones et al. 2009). Unexpectedly, clones of Agaricomycetes have been detected from deep-sea environments (Bass et al. 2007; Le Calvez et al. 2009; Sauvadet et al. 2010; Nagahama et al. 2011). Because the phylotypes have high similarity (99–98%) with published sequences from the terrestrial Agaricomycetes, it is possible that these may just be contaminants from plant materials by terrestrial runoff. However, distribution was limited in the hydrothermal and cold-seep areas, which may indicate that the agaricomycetous fungi could survive in chemosynthetic communities.

9.5.3 Chytridiomycota and Other Basal Fungal Lineages

The chytrids (Phyla Chytridiomycota and Blastocladiomycota) were believed to retain ancient forms before the colonization of land by fungi. However, the majority of chytrids had not been found in marine but freshwater environments. Although reasons why they were lost are unknown, they may still remain at the bottom of deep sea. Interestingly, the culture-independent analyses have unveiled the presence of chytrids in deep-sea environments, especially environments associated with hydrothermal vents communities. The deep-sea chytridian phylotypes were almost considered novel, because their sequences differed remarkably from published species (98–93%). No phylotype of the Phylum Blastocladiomycota was found in the deep-sea chytridian clones, but all were found within the Chytridiomycetes. Some phylotypes related to the chytrids and Zygomycota with lower affinities (<90%) on the blast search belonged to the clone group known as the LKM11 clade or the KD14 clade. The LKM11 clade is an extensive clone group (except *Rozella* spp.) recognized as a sister group of the Kingdom Fungi (Lara et al. 2010). Members of the LKM11 clade have appeared from aquatic (mostly freshwater) environments, such as lake (Šlapeta et al. 2005; Lefranc et al. 2005; Takishita et al. 2007a; Lepère et al. 2006, 2007, 2008; Mangot et al. 2009), water springs (Luo et al. 2005), artificial water system (van Hannen et al. 1999; Laurin et al. 2008), and marine environment (Savin et al. 2004; Takishita et al. 2007b). Some authors consider members of the LKM11 clade to participate in anaerobic ecosystem (van Hannen et al. 1999; Takishita et al. 2005, 2007b). Also in deep-sea environments, the LKM11 clones are likely to appear only from anaerobic sediments. Interestingly, it is suggested that the LKM11 clade involved *Rozella* (Lara et al. 2010), placing the earliest divergent within fungi (James et al. 2006; Bruns 2006). Likewise, the KD14 clade (Bass et al. 2007) consists of clones associated with the

Table 9.2 Appearance and distribution of the KD14 clade

Area	Source	Depth	Clones	References
Gulf of California	Anaerobic bacterial mats	1,575	1	Bass et al. (2007)
Sagami Bay, Japan	Methane cold seep sediment	1,178	1	Takishita et al. (2007a,b)
		1,174	1	Nagano et al. (2010)
		850–1,200	>70	Nagahama et al. (2011)
Kings Bay, Svalbard, Arctic	Deep-sea sediment	20	2	Tian et al. (2008)
Sea of Marmara	Deep-sea sediment	1,260	4	Quaiser et al. (2011)

anaerobic environment. These clones have been detected (Table 9.2) only from sediments of anaerobic bacterial mat (Bass et al. 2007), methane cold seep (Takishita et al. 2007a, b; Nagano et al. 2010), Svalbard in Arctic (Tian et al. 2008), and the Sea of Marmara (Quaiser et al. 2011). Clones of this clade have not yet been detected from hydrothermal vents, in spite of occupying the majority of fungal libraries of methane cold seep (Nagahama et al. 2011). These fungi are not likely to participate with the chemosynthetic ecosystem but just anaerobic sediments in low temperature. This clade is a more coherent clone group than LKM11 and is phylogenetically isolated from known lineages.

9.6 Comparison Between Cultured and Uncultured Fungal Diversity in Deep-Sea Environments

Fungal species isolated from deep-sea environments by conventional culture methods mostly belonged to the Phylum Ascomycota and limited yeast species of Basidiomycota (Table 9.3). The majority of ascomycetous clones in deep sea were closely related to terrestrial species on SSU phylogeny and these have been cultured by conventional isolation methods as well (Burgaud et al. 2009; Singh et al. 2010).

There are no culture isolation reports of Zygomycota and Chytridiomycota from deep-sea environments to date (Table 9.3). Since Zygomycota have not been detected by culture-independent methods either, Zygomycota appears to be very rare or nonexistent in deep-sea environments. Chytridiomycota have been detected as one of the major fungal components in several deep-sea environments, such as hydrothermal vents and methane cold seep but only by culture-independent methods. Le Calvez et al. (2009) revealed the ancient evolutionary lineage within the phylum of Chytridiomycota in deep-sea hydrothermal ecosystems. Nagahama et al. (2011) also reported that deep-branching basal fungi occupied a major portion of fungal clones detected in deep-sea methane cold-seep sediments. The majority of these Chytridiomycota and basal fungi reported from deep-sea environments is highly novel. These novel fungi may have unique ecological and physiological features, so it may be extremely difficult to culture these unknown phylotypes.

Table 9.3 Appearance of fungal taxa as cultured strains in deep-sea environments

	Filamentous fungi and yeasts					Yeasts		
Author	Damare et al. (2006)	Le Calvez et al. (2009)	Burgaud et al. (2009)	Jebaraj et al. (2010)	Singh et al. (2010)	Nagahama et al. (2001)	Gadanho et al. (2005)	Burgaud et al. (2009)
Source	Sediment	Hydrothermal	Hydrothermal	Sediment and water	Sediment	Sediment	Hydrothermal	Hydrothermal
Depth	4,800–5,400 m	1,700 m, 2,630 m	700–3,700 m	<200 m	4,000–5,500 m	1,000–11,000 m	800–2,400 m	700–2,700 m
Isolation method	1–30 atm., various temp.	1 atm., 25°C	1 atm., various temp.	1 atm., room temp.	1–30 atm., various temp.	1 atm., various temp.	1 atm., various temp.	1 atm., various temp.
Ascomycota								
Dothideomycetes	○	○	○	○	○			○
Eurotiomycetes	○	○	○	○	○		○	
Leotiomycetes			○					
Saccharomycetes	○[a]					○	○	
Sordariomycetes	○	○	○	○	○			
Basidiomycota								
Agaricomycetes								
Cystobasidiomycetes	○[a]			○	○	○	○	
Entorrhizomycetes								
Exobasidiomycetes			○	○	○			
Microbotryomycetes	○[a]				○	○	○	○
Tremellomycetes					○		○	○
Ustilaginomycetes				○				
Wallemiomycetes								
Chytridiomycota								
Chytridiomycetes								
Other basal lineage								
LKM11 clade								
KD14 clade								

○: Taxa presented by phylotypes
○[a]: Taxa expected from morphology

Similar to other environmental studies, culture-dependent approaches and culture-independent approaches often provide different results on the fungal diversity. Culture-independent approaches have been able to detect diverse fungi, including many novel fungal phylotypes, which may be difficult to culture. Le Calvez et al. (2009) reported the striking difference on their results investigating deep-sea fungal diversity between culture-dependent and culture-independent methods. They isolated only Ascomycota spp. by culture-dependent methods. However, none of Ascomycota spp. were detected by culture-independent methods but Basidiomycota spp. and Chytridiomycota spp. were detected. This study suggested that the abundance of Ascomycota in the deep-sea environmental samples was extremely low, despite the frequent isolation of ascomycetous fungi from various samples investigated. Cultivation resulted in the amplification of rare organisms. Many factors can be responsible for these different results, making it very difficult to judge which methods are more accurately depicting the true fungal diversity in deep-sea environments. The absence of non-Dikarya fungi as chytrids by culture-dependent analyses can be caused as a result of culturing deep-sea fungi using an agar medium with antibiotics. Most chytrids cannot be isolated by direct plating. In addition, chloramphenicol, an antibiotic widely used in fungal isolation, is known to prevent growth of some of the chytrids (Gleason and Marano 2011). In general, culture-independent methods are thought to be able to detect wider ranges of fungi and reflect investigated environments more directly. However, culture-independent methods are also easily biased by many processes, such as PCR primer selection and DNA extraction methods. Thus, it is recommended to combine both conventional culture methods and culture-independent methods when investigating fungal diversity. Also, in order to investigate the true abundance of deep-sea fungi, it is necessary to combine with other methods, such as direct microscopic observation with appropriate staining, FISH, measuring the ergosterol (uniquely present in cell membranes of higher fungi), metagenomic approaches, and new powerful tools expected to be developed in the future.

9.7 Future Direction

Although recent studies on deep-sea fungi have been more extensively researched, our knowledge and understanding of deep-sea fungi is still limited. Unlike deep-sea bacteria, fungal taxa, which are physiologically adapted to low temperatures, high hydropressure, and play important roles in the ecosystem, have not yet been discovered. Culture-independent analyses revealed the presence of hitherto unknown fungal lineages in deep sea, which were difficult to grow via conventional culturing methods. Such uncultured taxa may be fungi which have effectively adapted to deep-sea environments. Also, these unknown fungal groups may provide key insights into the early evolution of fungi and their ecological and physiological significance in deep-sea environments. Deep-sea exploration of uncultured fungi should continue to be investigated thoroughly.

References

Amend AS, Seifert KA, Bruns TD (2010) Quantifying microbial communities with 454 pyrosequencing: does read abundance count? Mol Ecol 19:5555–5565

Bass D, Howe A, Brown N et al (2007) Yeast forms dominate fungal diversity in the deep oceans. Proc Biol Sci 274:3069–3077

Bills GF, Polishook JD (1994) Abundance and diversity of microfungi in leaf litter of a lowland rain forest in Costa Rica. Mycologia 86:187–198

Bruns T (2006) Evolutionary biology: a kingdom revised. Nature 443:758–761

Burgaud G, Calvez TL, Arzur D, Vandenkoornhuyse P, Barbier G (2009) Diversity of culturable marine filamentous fungi from deep-sea hydrothermal vents. Environ Microbiol 11:1588–1600

Connell L, Redman R, Craig S, Scorzetti G, Iszard M, Rodriguez R (2008) Diversity of soil yeasts isolated from South Victoria Land, Antarctica. Microb Ecol 56:448–459

Damare S, Raghukumar C (2008) Fungi and macroaggregation in deep-sea sediments. Microb Ecol 53:14–27

Damare S, Raghukumar C, Raghukumar S (2006) Fungi in deep-sea sediments of the Central Indian Basin. Deep-Sea Res I 53:14–27

Edgcomb VP, Beaudoin D, Gast R, Biddle JF, Teske A (2011) Marine subsurface eukaryotes: the fungal majority. Environ Microbiol 13:172–183

Eloe EA, Shulse CN, Fadrosh DW, Williamson SJ, Allen EE, Bartlett DH (2011) Compositional differences in particle-associated and free-living microbial assemblages from an extreme deep-ocean environment. Environ Microbiol Rep 3(4):449–458

Fell JW (1976) Yeasts in oceanic regions. In: Jones EBG (ed) Recent advances in aquatic mycology. Elek Science, London, UK, pp 93–124

Gadanho M, Sampaio JP (2005) Occurrence and diversity of yeasts in the Mid-Atlantic Rigde hydrothermal fields near the azores archipelago. Microb Ecol 50:408–417

Gleason FH, Marano AV (2011) The effects of antifungal substances on some zoosporic fungi (Kingdom Fungi). Hydrobiologia 659:81–92

Guillot J, Petit T, Degorce-Rubiales F, Guého E, Chermette R (1998) Dermatitis caused by *Malassezia pachydermatis* in a California sea lion (*Zalophus californianus*). Vet Rec 142:311–312

Ikemoto E, Kyo M (1993) Development of microbiological compact mud sampler. Jpn Mar Sci Technol Res 30:1–16

James TY, Kauff F, Schoch CL et al (2006) Reconstructing the early evolution of fungi using a six-gene phylogeny. Nature 443:818–822

Jebaraj CS, Raghukumar C, Behnke A, Stoeck T (2010) Fungal diversity in oxygen-depleted regions of the Arabian Sea revealed by targeted environmental sequencing combined with cultivation. FEMS Microbiol Ecol 71:399–412

Jones EBG, Sakayaroj J, Suetrong S, Somrithipol S, Pang KL (2009) Classification of marine Ascomycota, anamorphic taxa and Basidiomycota. Fungal Divers 35:1–187

Kohlmeyer J, Volkmann-Kohlmeyer B (1991) Illustrated key to the filamentous higher marine fungi. Bot Mar 34:1–61

Konishi M, Fukuoka T, Nagahama T, Morita T, Imura T, Kitamoto D, Hatada Y (2010) Biosurfactant-producing yeast isolated from *Calyptogena soyoae* (deep-sea cold-seep clam) in the deep sea. J Biosci Bioeng 110:169–175

Lai X, Cao L, Tan H, Fang S, Huang Y, Zhou S (2007) Fungal communities from methane hydrate-bearing deep-sea marine sediments in South China Sea. ISME J 1:756–762

Lara E, Moreira D, López-García P (2010) The environmental clade LKM11 and *Rozella* form the deepest branching clade of fungi. Protist 161:116–121

Laurin V, Labbé N, Parent S, Juteau P, Villemur R (2008) Microeukaryote diversity in a marine methanol-fed fluidized denitrification system. Microb Ecol 56:637–648

Le Calvez T, Gaëtan B, Mahé S, Barbier G, Vandenkoornhuyse P (2009) Fungal diversity in deep-sea hydrothermal ecosystems. Appl Environ Microbiol 75:6415–6421. doi:10.1128/AEM.00653-09

Lefranc M, Thenot A, Lepere C, Debroas D (2005) Genetic diversity of small eukaryotes in lakes differing by their trophic status. Appl Environ Microbiol 71:5935–5942

Lepère C, Boucher D, Jardillier L, Domaizon I, Debroas D (2006) Succession and regulation factors of small eukaryote community composition in a lacustrine ecosystem (Lake Pavin). Appl Environ Microbiol 72:2971–2981

Lepère C, Domaizon I, Debroas D (2007) Community composition of lacustrine small eukaryotes in hyper-eutrophic conditions in relation to top-down and bottom-up factors. FEMS Microbiol Ecol 61:483–495

Lepère C, Domaizon I, Debroas D (2008) Unexpected Importance of potential parasites in the composition of the freshwater small-eukaryote community. Appl Environ Microbiol 74:2940–2949

López-García P, Vereshchaka A, Moreira D (2007) Eukaryotic diversity associated with carbonates and fluid-seawater interface in Lost City hydrothermal field. Environ Microbiol 9:546–554

Lorenz R, Molitoris HP (1997) Cultivation of fungi under simulated deep sea conditions. Mycol Res 101(11):1355–1365

Luo Q, Krumholz LR, Najar FZ, Peacock AD, Roe BA, White DC, Elshahed MS (2005) Diversity of the microeukaryotic community in sulfide-rich Zodletone spring (Oklahoma). Appl Environ Microbiol 71:6175–6184

Mangot J-F, Lepere C, Bouvier C, Debroas D, Domaizon I (2009) Community structure and dynamics of small eukaryotes targeted by new oligonucleotide probes: new insight into the lacustrine microbial food web. Appl Environ Microbiol 75:6373–6381

Nagahama T, Hamamoto M, Nakase T, Takami H, Horikoshi K (2001) Distribution and identification of red yeasts in deep-sea environments around the northwest Pacific Ocean. Antonie Van Leeuwenhoek 80:101–110

Nagahama T, Takahashi E, Nagano Y, Abdel-Wahab MA, Miyazaki M (2011) Molecular evidence that deep-branching fungi are major fungal components in deep-sea methane cold-seep sediments. Environmental Microbiology 13:2359–2370

Nagano Y, Nagahama T, Hatada Y, Nunoura T, Takami H, Miyazaki J, Takai K, Horikoshi K (2010) Fungal diversity in deep-sea sediments – the presence of novel fungal groups. Fungal Ecol 3:316–325

Nakagaki K, Hata K, Iwata E, Takeo K (2000) *Malassezia pachydermatis* isolated from a South American sea lion (*Otaria byronia*) with dermatitis. J Vet Med Sci 62:901–903

Pollock CG, Rohrbach B, Ramsay EC (2000) Fungal dermatitis in captive pinnipeds. J Zoo Wildl Med 31:374–378

Quaiser A, Zivanovic Y, Moreira D, López-García P (2011) Comparative metagenomics of bathypelagic plankton and bottom sediment from the Sea of Marmara. ISME J 5:285–304

Raghukumar C, Raghukumar S (1998) Barotolerance of fungi isolated from deep-sea sediments of the Indian Ocean. Aquat Microb Ecol 15:153–163

Raghukumar C, Raghukumar S, Sheelu G, Gupta SM, Nagender Nath B, Rao BR (2004) Buried in time: culturable fungi in a deep-sea sediment core from the Chagos Trench, Indian Ocean. Deep Sea Res I 51:1759–1768

Renker C, Alphei J, Buscot F (2003) Soil nematodes associated with the mammal pathogenic fungal genus *Malassezia* (Basidiomycota: Ustilaginomycetes) in Central European forests. Biol Fertil Soils 37:70–72

Sauvadet A-L, Gobet A, Guillou L (2010) Comparative analysis between protist communities from the deep-sea pelagic ecosystem and specific deep hydrothermal habitats. Environ Microbiol 12:2946–2964

Savin MC, Martin JL, LeGresley M, Giewat M, Rooney-Varga J (2004) Plankton diversity in the Bay of Fundy as measured by morphological and molecular methods. Microb Ecol 48:51–65

Singh P, Raghukumar C, Verma P, Shouche Y (2010) Phylogenetic diversity of culturable fungi from the deep-sea sediments of the Central Indian Basin and their growth characteristics. Fungal Divers 40:89–102

Singh P, Raghukumar C, Verma P, Shouche Y (2011) Fungal community analysis in the deep-sea sediments of the Central Indian Basin by culture-independent approach. Microb Ecol 61:507–517

Šlapeta J, Moreira D, López-García P (2005) The extent of protist diversity: insights from molecular ecology of freshwater eukaryotes. Proc Biol Sci 272:2073–2081

Takishita K, Miyake H, Kawato M, Maruyama T (2005) Genetic diversity of microbial eukaryotes in anoxic sediment around fumaroles on a submarine caldera floor based on the small-subunit rDNA phylogeny. Extremophiles 9:185–196

Takishita K, Tsuchiya M, Reimer J, Maruyama T (2006) Molecular evidence demonstrating the basidiomycetous fungus *Cryptococcus curvatus* is the dominant microbial eukaryote in sediment at the Kuroshima Knoll methane seep. Extremophiles 10:165–169

Takishita K, Tsuchiya M, Kawato M, Oguri K, Kitazato H, Maruyama T (2007a) Genetic diversity of microbial eukaryotes in anoxic sediment of the saline meromictic lake Namako-ike (Japan): on the detection of anaerobic or anoxic-tolerant lineages of eukaryotes. Protist 158:51–64

Takishita K, Yubuki N, Kakizoe N, Inagaki Y, Maruyama T (2007b) Diversity of microbial eukaryotes in sediment at a deep-sea methane cold seep: surveys of ribosomal DNA libraries from raw sediment samples and two enrichment cultures. Extremophiles 11:563–576

Tian F, Yu Y, Chen B, Li H, Yao Y-F, Guo X-K (2008) Bacterial, archaeal and eukaryotic diversity in Arctic sediment as revealed by 16 S rRNA and 18 S rRNA gene clone libraries analysis. Polar Biol 32:93–103

Turchetti B, Buzzini P, Goretti M, Branda E, Diolaiuti G, D'Agata C, Smiraglia C, Vaughan-Martini A (2008) Psychrophilic yeasts in glacial environments of Alpine glaciers. FEMS Microbiol Ecol 63:73–83

van Hannen EJ, Mooij W, van Agterveld MP, Gons HJ, Laanbroek HJ (1999) Detritus-dependent development of the microbial community in an experimental system: qualitative analysis by denaturing gradient gel electrophoresis. Appl Environ Microbiol 65:2478–2484

Wolinska J, Giessler S, Koerner H (2009) Molecular identification and hidden diversity of novel *Daphnia* parasites from European lakes. Appl Environ Microbiol 75:7051–7059 [illegible] (2006) [illegible] fungi in deep-sea sediments of the Central Indian Basin [illegible] Fungal Divers [illegible]

[illegible] (2010) [illegible] community analysis of the [illegible] sediments of the Central Indian Basin [illegible] [illegible]

[illegible]

[illegible]

[illegible]

[illegible]

[illegible]

[illegible]

Chapter 10
Molecular Diversity of Fungi from Marine Oxygen-Deficient Environments (ODEs)

Cathrine Sumathi Jebaraj, Dominik Forster, Frank Kauff, and Thorsten Stoeck

Contents

Abstract Molecular diversity surveys of marine fungi have demonstrated that the species richness known to date is just the tip of the iceberg and that there is a large extent of unknown fungal diversity in marine habitats. Reports of novel fungal lineages at higher taxonomic levels are documented from a large number of marine habitats, including the various marine oxygen-deficient environments (ODEs). In the past few years, a strong focus of eukaryote diversity research has been on a variety of ODEs, as these environments are considered to harbor a large number of organisms, which are highly divergent to known diversity and could provide insights into the early eukaryotic evolution. ODEs that have been targeted so far include shallow water sediments, hydrothermal vent systems, deep-sea basins, intertidal habitats, and fjords. Most, if not all, molecular diversity studies in marine ODEs have shown, that contrary to previous assumptions, fungi contribute significantly to the micro-eukaryotic community in such habitats. In this chapter, we have reanalyzed the environmental fungal sequences obtained from the molecular diversity survey in 14 different sites to obtain a comprehensive picture of fungal diversity in these marine habitats. The phylogenetic analysis of the fungal environmental

C.S. Jebaraj (✉)
National Institute of Oceanography, Council of Scientific and Industrial Research, Dona Paula 403 004, Goa, India
e-mail: cathrine@nio.org

C. Raghukumar (ed.), *Biology of Marine Fungi*,
Progress in Molecular and Subcellular Biology 53,
DOI 10.1007/978-3-642-23342-5_10,

sequences from various ODEs have grouped these sequences into seven distinct clades (Clade 1–7) clustering with well-known fungal taxa. Apart from this, four environmental clades (EnvClade A, B, C, and D) with exclusive environmental sequences were also identified. This has provided information on the positioning of the environmental sequences at different taxonomic levels within the major fungal phylums. The taxonomic distribution of these environmental fungal sequences into clusters and clades has also shown that they are not restricted by geographical boundaries. The distribution pattern together with the reports on the respiratory abilities of fungi under reduced oxygen conditions shows that they are highly adaptive and may have a huge ecological role in these oxygen deficient habitats.

10.1 Marine Fungal Diversity: Historical Background

Fungi thrive in a number of habitats and display specific adaptations to these habitats in their structure and biochemistry (Maheshwari 2005). Some of these habitats colonized by fungi include the outer stratosphere (Wainwright et al. 2003; News Letter ISRO 2009), extremely cold habitats in polar regions (Robinson 2001), deep sea sediments (Raghukumar et al. 2004; Edgcomb et al. 2011), hypersaline brine with saturated salt concentrations (Edgcomb et al. 2009), extremely dry soil crusts (States and Christensen 2001), and hot springs with temperatures ranging up to 62°C (Maheshwari et al. 2000).

The presence of fungi in the marine environment was initially studied in 1936 (Sparrow 1936), followed by Barghoorn and Linder (1944). Since then, mycologists have tried to estimate the extent of fungal diversity and to understand their ecological role in the various niches of the marine ecosystem (Kohlmeyer and Kohlmeyer 1979; Gessner 1980; Raghukumar et al. 1992; Sathe-Pathak et al. 1993; Prasannari and Sridhar 2001; Raghukumar et al. 2004; Damare et al. 2006). These studies used traditional methods involving isolation and cultivation of fungal cultures. Estimates have shown that using such strategies only about 5% of the total fungal diversity is documented so far from the marine environment (Kis-Papo 2005). These diversity estimates are expected to be incomplete and down-numbered, because many of the fungal cultures show an inability to grow and sporulate under laboratory conditions (Jones and Hyde 2002). Advancements in the field of molecular ecology have revolutionized the study of fungal diversity. Molecular biological techniques, particularly the analysis of nucleotide sequences and phylogenetic approaches, have greatly influenced fungal systematics (Guarro et al. 1999). Molecular tools, such as Random Amplified Polymorphic DNA (RAPD) and Terminal Restriction Fragment Length polymorphism (T-RFLP) analysis of the amplified internal-transcribed spacer (ITS) region or the small subunit of ribosomal RNA (SSU rRNA) gene region, are widely known. The nuclear large subunit region of ribosomal RNA gene (LSU rRNA) is also used to identify the fungal cultures isolated from marine ecosystems (Roberts et al. 1996; Buchan et al. 2002), and direct PCR amplification from various environmental samples also targets these loci. Studies based on the culture-independent approach have proved to be a powerful tool and have showed that the diversity of marine fungi is more extensive than documented previously (Pang and Mitchell 2005).

A strong focus of molecular diversity research has been on micro-eukaryotes from a variety of marine environments, such as shallow water anoxic sediments (Dawson and Pace 2002; Stoeck et al. 2003), hydrothermal vent systems (Edgcomb et al. 2002; López-García et al. 2003, 2007), oxygen-depleted deep basins (Stoeck et al. 2003, 2006), anoxic fjords (Behnke et al. 2006; Stoeck et al. 2007b; Zuendorf et al. 2006), saline meromictic lakes (Takishita et al. 2007a), deep sea habitats (Takishita et al. 2005, 2007b), hypersaline systems (Alexander et al. 2009), a few oxygen-depleted freshwater systems (Dawson and Pace 2002; Luo et al. 2005; Slapeta et al. 2005), and terrestrial habitats (Schadt et al. 2003). These studies, using eukaryote specific primers, have revealed a much larger diversity in all micro-eukaryotic supergroups, including alveolates, stramenopiles, amoebozoa, excavates, discicristates, plants and ophisthokonts (Dawson and Pace 2002; Stoeck and Epstein 2003; Stoeck et al. 2003; Takishita et al. 2005; Luo et al. 2005; Behnke et al. 2006; Alexander et al. 2009). Most, if not all, environmental diversity studies conducted in oxygen-deficient systems have shown that fungal sequences are a significant component of the micro-eukaryotic community. This indicates, that molecular diversity surveys hold a great promise in the discovery of novel fungi. In order to fully exploit the potential of these taxonomically and economically novel fungi, it is inevitable to obtain cultures. Hence, an integrated approach combining molecular, cultivation-independent studies with classical isolation and cultivation techniques offers a greater chance of understanding the fungal diversity and their ecological significance in the diverse niches they occupy (Jebaraj et al. 2010).

10.2 Ecological Significance of Fungi in ODEs

A combination of high productivity and limited mixing or circulation with poorly oxygenated waters leads to the development of mid-water ODEs in the world oceans (Wyrtki 1962). These are regions where the dynamic steady state between oxygen supply and consumption is altered and the oxygen concentration is <0.2 ml l^{-1}. Such permanent anoxic zones are seen in the eastern Pacific Ocean, Indian Ocean, and in West Africa, where the fixed N is transformed in the water column to dissolved N gas through denitrification (Kamykowski and Zentara 1990). Denitrification is a major sink for fixed N and is therefore crucial to budget the loss of oceanic nitrogen through this microbial-mediated process in the ODEs.

Fungi were one of the taxonomic groups that were thought to play only a minor role in ecosystem processes of anoxic environments (Dighton 2003), but molecular analysis has shown that they occupy a greater portion in the microbial diversity of the ODEs. Studies under laboratory conditions have shown that many members of kingdom *Fungi* have physiological adaptations which allow them to thrive in environments with varying oxygen concentrations (Zhou et al. 2002). Generally, fungi are major decomposers in the marine environment and can actively degrade woody and herbaceous substances with the help of the lingo-cellulolytic, extracellular enzymes produced by them (Newell 1996; Hyde et al. 1998). Fungi play a key role in the decomposition and biochemical transformation of organic matter in the marine habitat because fungal enzymes can efficiently breakdown recalcitrant compounds like phenolics and tannins. They also play an important role in the C and N cycle and aid in the release of the inorganic nutrients back into the ecosystem (Dighton 2007). During this process, the delicate equilibrium of the oxygen consumption during the degradation of the organic matter and oxygen replenishment is maintained. A shifted balance due to increased organic load leads to oxygen depletion, and the change in oxygen concentration alters the microbial community structure (Goregues et al. 2005). In conditions where low oxygen condition prolongs, the diversity of the communities is known to reduce to a few forms which are able to tolerate anaerobic growth conditions (Levin 2003). But surprisingly, the molecular diversity from the oxygen-depleted habitats has shown that there is a wider and richer diversity than previously known (Epstein and López-García 2007).

Many fungal taxa were recently shown to possess metabolic adaptations to utilize nitrate and/or nitrite as an alternative for oxygen (Kurakov et al. 2008). Involvement of fungi in such anaerobic dissimilatory processes was largely unknown, and initial studies that reported growth of fungi in anoxic environments suggested that they required supplementation of certain inorganic salts, such as nitrate, selenite, or ferric iron in the culture media (Gunner and Alexander 1964). Bollag and Tung (1972) were the pioneers to show N_2O production by fungi when growing under anaerobic conditions. They also demonstrated growth of fungi under these conditions when supplemented with nitrate or nitrite and suggested, that nitrate respiration may be a possibility in fungi. Screening fungal cultures from terrestrial origin for their denitrification activity revealed, that indeed several different taxa were involved in this process (Shoun et al. 1992; Usuda et al. 1995). For the first time, Shoun and his group (1992) could confirm that dissimilatory nitrate reduction or microbial denitrification was also possible in fungal cultures, with fungal cultures such as *Fusarium oxysporum*, *Cylindrocarpon tonkinense,* and *Aspergillus oryzae* releasing N_2O under oxygen-depleted conditions. This was a major breakthrough, as denitrification was previously thought to be an alternative form of respiration exclusively within bacterial and archaean communities (Knowles 1982). A recent study has also demonstrated that fungi isolated from anoxic marine waters of the Arabian Sea could contribute to the denitrification process (Jebaraj and Raghukumar 2009).

Denitrification activity has been shown in terrestrial fungi belonging to *Ascomycota*, *Basidiomycota*, and *Zygomycota* (Kurakov et al. 2008). This shows that the fungal denitrification activity is prevalent among the major groups, but the molecular basis on how the members of Kingdom *Fungi* acquired the ability is still unknown. Perhaps, fungi may have a pivotal role in the nitrogen cycle in marine ODEs, but the contribution of fungi to anaerobic denitrification has not yet been quantified.

10.3 What Novel Fungal Diversity Emerges from Molecular Diversity Surveys in Marine ODEs?

In order to identify hotspots of fungal diversity and to assess the degree and distribution of fungal diversity in ODEs, we have analyzed eukaryote SSU rRNA gene meta-data from 12 environmental molecular diversity surveys, targeting 14 different sampling sites (Table 10.1). In the phylogenetic analysis of the fungal tree of life that included fungal environmental sequences ($n = 385$) from various ODEs and representative sequences of described species from all major phylogenetic fungal lineages (Hibbett et al. 2007), we identified seven major clades (clades 1–7) and four environmental clades (EnvClade A, B, C, and D) (Fig. 10.1). Clades 1–7 (Figs. 10.2–10.7) encompassed numerous environmental fungal sequences grouping within known fungal taxa, and EnvClade A, B, C, and D (Figs. 10.8a, b and 10.9a, b) contain environmental fungal sequences that are unrelated to previously described taxa.

Kingdom *Fungi* is one of the major taxonomic group where new phyla, orders, genera, and species are being added frequently with novel ecological and physiological adaptations. It includes subkingdom *Dikarya*, consisting of phylum, Ascomycota and Basidiomycota, phylum Chytridiomycota, Neocallimastigomycota, Blastocladiomycota, Microsporidia, Glomeromycota, and the basal groups (Hibbett et al. 2007). The reanalysis of the environmental fungal sequences obtained from various ODEs used in this study grouped them together with the sequences of well-described fungal taxa. The sequences clustered within members of phylum *Basidiomycota* (Clades 1, 2 and EnvClade A), *Ascomycota* (clades 3–5), *Chytridiomycota* (Clade 6 and EnvClade B, C, D), and some belonged to undescribed basal lineages (Clade 7) of kingdom *Fungi* (Fig. 10.1).

Environmental sequences of phylum *Basidiomycota* analyzed from the various ODEs grouped within subphyla *Ustilagomycotina* and *Pucciniomycotina*, and are

Table 10.1 List of oxygen-deficient environments (ODEs) analyzed

S. No.	Region	Geographical location	References
1.	Cariaco, anoxic basin	Equatorial Atlantic	Stoeck et al. (2003)
2	Lucky Strike, vent region	Mid Atlantic Ridge	Le Calvez et al. (2009)
3	Elsa, Vent region	Mid Atlantic Ridge	Le Calvez et al. (2009)
4	Menez Gwen vent region	Mid Atlantic Ridge	Le Calvez et al. (2009)
5	Greenland, tidal flat	North Atlantic	Stoeck et al. (2007a)
6	Guaymas, vent region	North Pacific	Edgcomb et al. (2002)
7	Mariager fjord	Baltic Sea	Zuendorf et al. (2006)
8	Japan, Tagiri vent site	North Pacific	Takishita et al. (2005)
9	Lost City, vent region	Mid Atlantic Ridge	López-García et al. (2007)
10	Framvaren Fjord	North Sea	Behnke et al. (2006)
11	L'Atalante saline basin	Mediterranean	Alexander et al. (2009)
12	Sippewisset salt marsh	North Atlantic	Stoeck and Epstein (2003)
13	California, sediment	North Pacific	Dawson and Pace (2002)
14	Gotland anoxic basin	Baltic Sea	Stock et al. (2009)

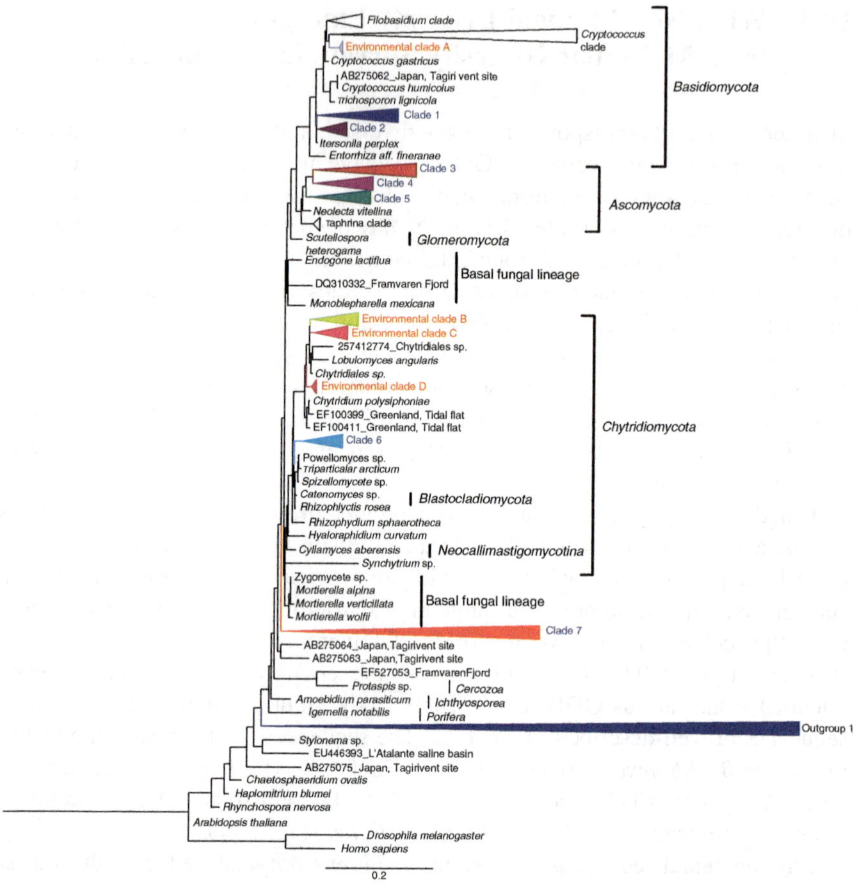

Fig. 10.1 Phylogenetic tree of the SSU rRNA sequences of Kingdom *Fungi* showing the position of environmental sequences obtained from various ODEs used for analysis. The tree shows the positioning of clades 1–7 and env clades A–D which will be explained in subsequent figures. The sequences are represented by their NCBI accession number, followed by the regional location from where each sequence was isolated

represented as clade 1 (Fig. 10.2). Clade 1 has a cluster of environmental fungal sequences obtained from two anoxic vent regions and three major ODEs of the world oceans. This cluster has been recognized as the "hydrothermal and/or anaerobic fungal group" of a similar study that surveyed the eukarotic diversity in an anoxic vent ecosystem (López-García et al. 2007). The composition of this cluster is predominated by environmental sequences, and the only named species is *Malassezia* spp. belonging to *Exobasidiomycetes* of *Ustilagomycotina*. Though *Malassezia* spp. are known to be human pathogens, environmental fungal sequences closely grouping with this genus have been identified from various habitats, such as the deep-sea regions (Lai et al. 2007), anoxic basins (Dawson and Pace 2002; Alexander et al. 2009; Jebaraj et al. 2010), and hydrothermal vent

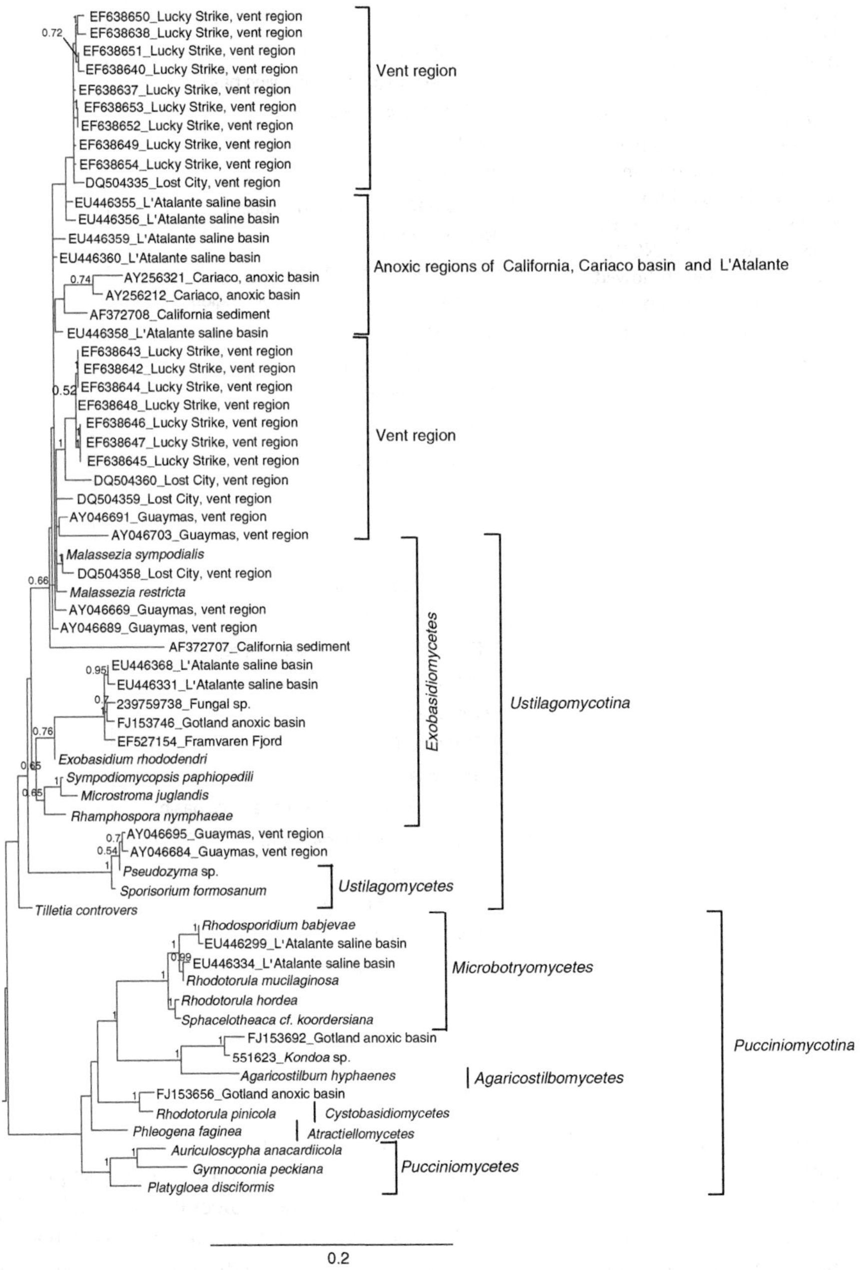

Fig. 10.2 Clade 1: Phylogenetic tree of the SSU rRNA sequences of phylum *Basidiomycota*, *Ustilagomycetes*, and *Pucciniomycetes* showing the position of environmental sequences obtained from various ODEs used for analysis. The sequences are represented by their NCBI accession number, followed by the regional location from where each sequence was isolated

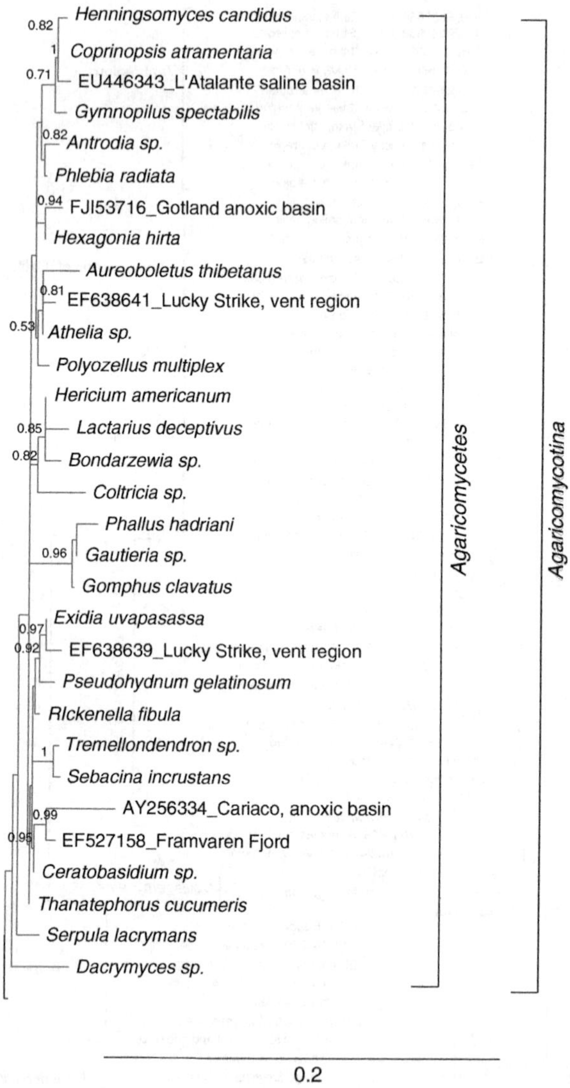

Fig. 10.3 Clade 2: Phylogenetic tree of the SSU rRNA sequences of phylum *Basidiomycota, Agaricomycetes* showing the position of environmental sequences obtained from various ODEs used for analysis. The sequences are represented by their NCBI accession number, followed by the regional location from where each sequence was isolated

regions (López-García et al. 2007). The life style of *Malassezia* does not give us any insights into the ecological significance of the fungal environmental sequences obtained from these regions. This also shows that the molecular approach using SSU rRNA as a marker to study the microbial richness and diversity has its shortcomings. A combined approach using both the cultivation based and molecular studies could help in resolving these issues. Nevertheless, this analysis has identified a promising target within the "hydrothermal and/or anaerobic fungal group" for the isolation of novel fungal lineages which have not much similarity to known fungal taxa (Jebaraj et al. 2010). The environmental sequences obtained

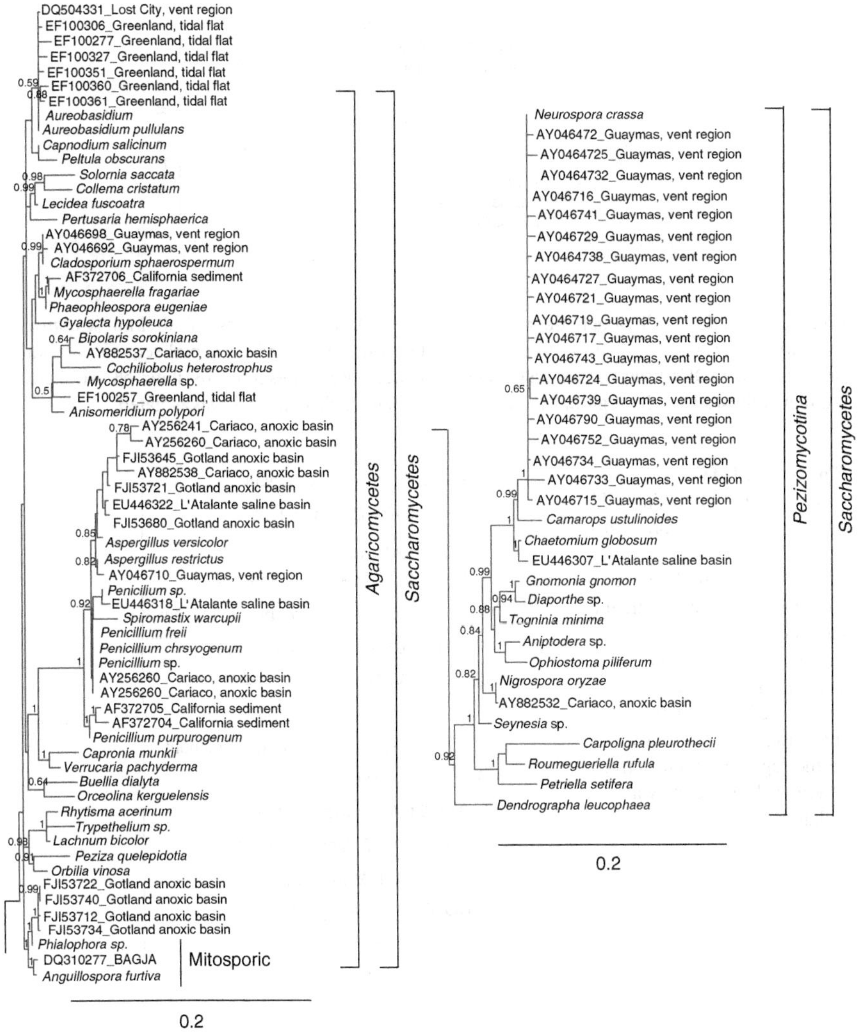

Fig. 10.4 (**a**) Clade 3 and (**b**) Clade 4: Phylogenetic tree of the SSU rRNA sequences of phylum *Ascomycota, Saccharomycetes, Pezizomycotina* showing the position of environmental sequences obtained from various ODEs used for analysis. The sequences are represented by their NCBI accession number, followed by the regional location from where each sequence was isolated

from the various ODEs also show that the microbes may have different physiological adaptations to thrive in extreme conditions, such as high sulfide conditions (Framvaren fjord), elevated temperatures (Mid-Atlantic ridge hydrothermal field), and high salt conditions (L'Atalante basin). Clade 1 (Fig. 10.2) also has another environmental cluster that groups with high divergence to *Exobasidium* sp. (Jebaraj et al. 2010) within *Ustilagomycotina*. This shows that phylum Ustilagomycotina

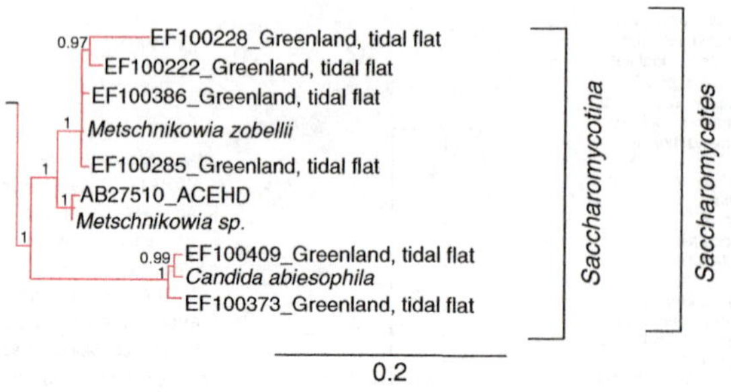

Fig. 10.5 Clade 5: Phylogenetic tree of the SSU rRNA sequences of phylum *Ascomycota, Saccharomycetes, Saccharomycotina* showing the position of environmental sequences obtained from various ODEs used for analysis. The sequences are represented by their NCBI accession number, followed by the regional location from where each sequence was isolated

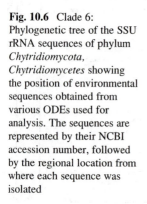

Fig. 10.6 Clade 6: Phylogenetic tree of the SSU rRNA sequences of phylum *Chytridiomycota, Chytridiomycetes* showing the position of environmental sequences obtained from various ODEs used for analysis. The sequences are represented by their NCBI accession number, followed by the regional location from where each sequence was isolated

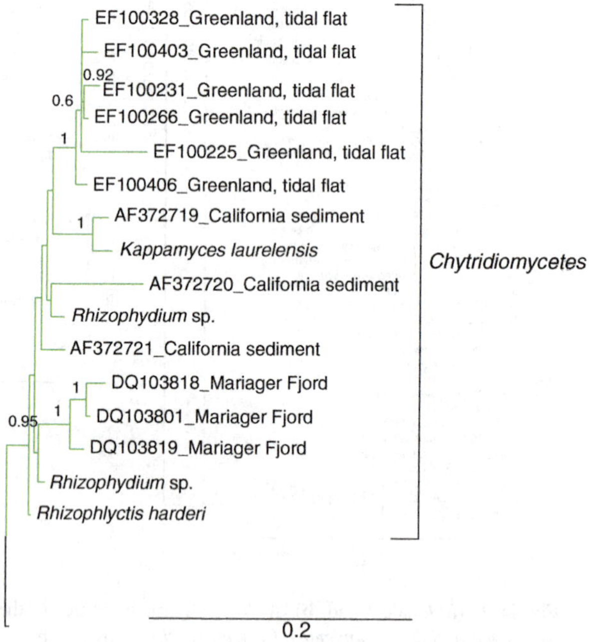

could hold a number of unknown taxa, which could be identified using group specific studies and targeted isolation of cultivable forms belonging to this group.

Clade 2 (Fig. 10.3) includes environmental sequences that group with described fungal sequences belonging to the *Agaricomycetes* of the subphylum *Agaricomycotina*. This clade has representative sequences from anoxic basins of Cariaco, L'Atalante, Gotland basins, and from the Lucky Strike vent site. *Agaricomycetes* include members of *Fungi* which have well-developed fruiting bodies, like mushrooms and wood-rot fungi. The distribution of this group is

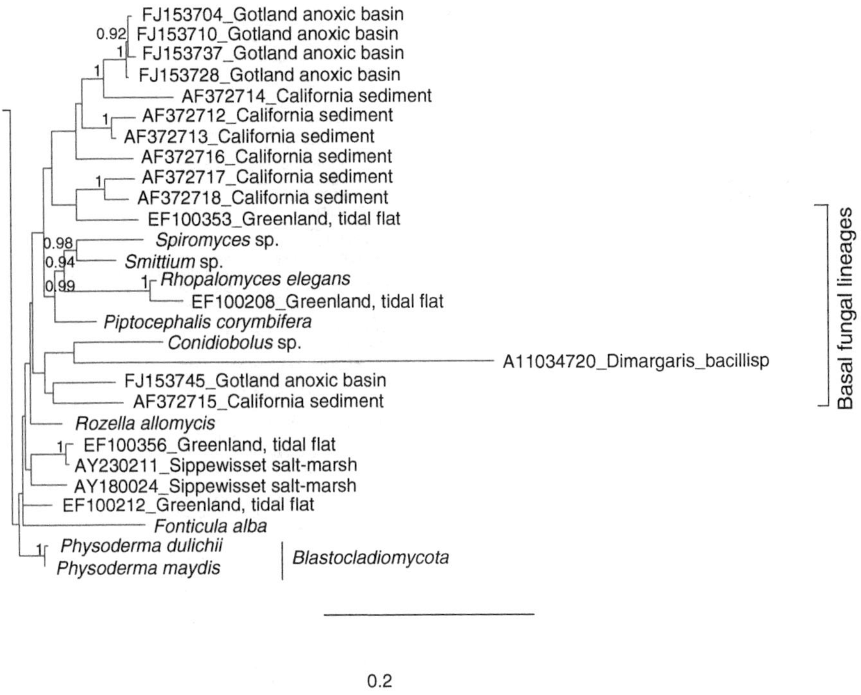

Fig. 10.7 Clade 7: Phylogenetic tree of the SSU rRNA sequences of the basal fungal lineage showing the position of environmental OTUs obtained from various ODEs used for analysis. The sequences are represented by their NCBI accession number, followed by the regional location from where each sequence was isolated

generally restricted to terrestrial habitats. Only a few of them are present in fresh water habitats, but they have rarely been isolated from the marine realm. Some species which are present in the marine environment are known only from their anamorphic stage, and the spores are released passively to the marine conditions (Jones and Choeyklin 2008). The presence of environmental sequences belonging to this group from various anoxic and vent regions shows that *Agaricomycetes* may have a broader distribution than what is documented so far. An alternative interpretation may be that the genetic signatures obtained in environmental studies come from "ancient" extracellular material or from resting stages of the fungi, which are inactive in marine waters. However, some studies amplified SSU rRNA genes from environmental RNA (cDNA) as template, which is indicative of actively growing and metabolically active cells (Stoeck et al. 2007b).

The EnvClade A (Fig. 10.8a) includes sequences exclusively from the Mid-Atlantic vent region and the Norwegian Framvaren Fjord, grouping within the *Tremellomycetes* (*Agaricomycotina*), some of which are closely related to *Cryptococcus* sp. (Fig. 10.1), also belonging to phylum *Basidiomycota*. These groups are generally yeast-like forms and are known to be widespread in the marine

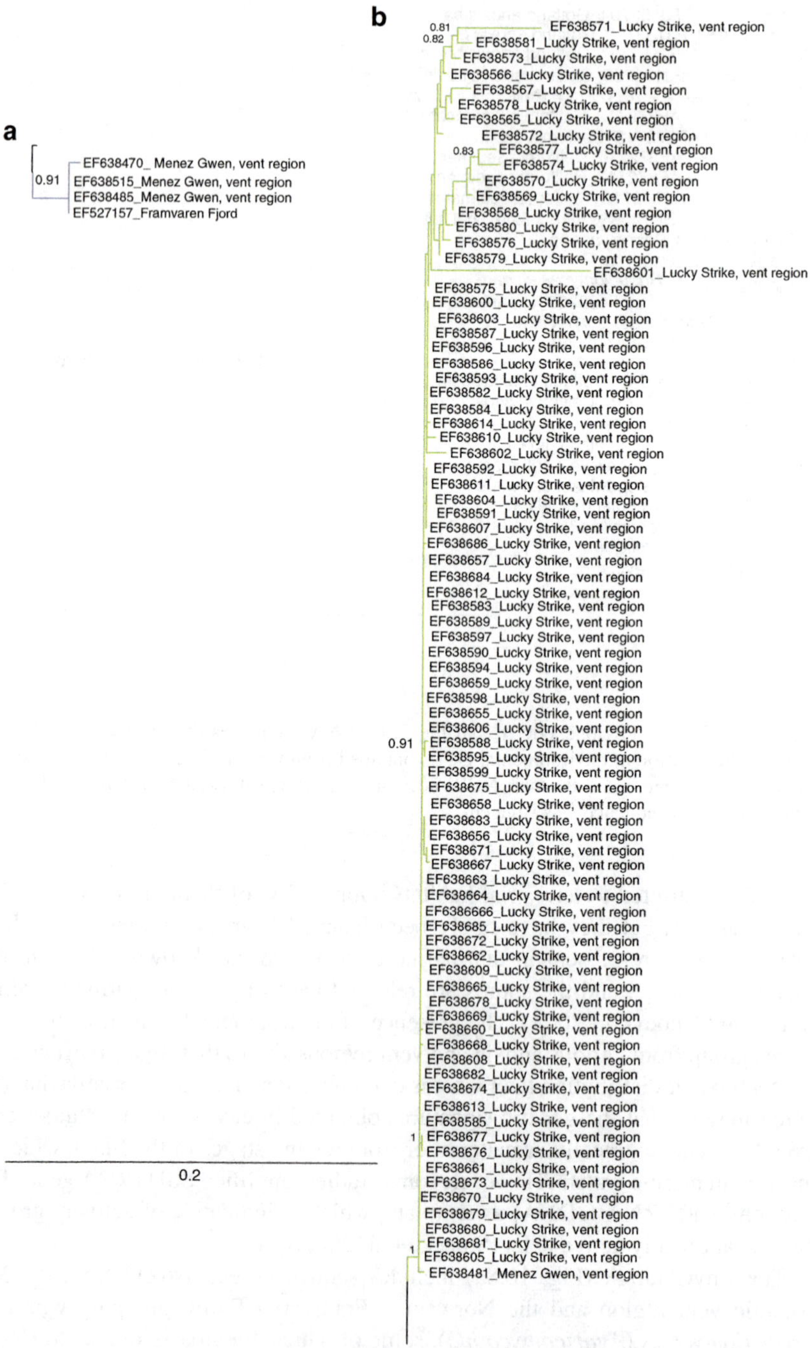

Fig. 10.8 (**a**) env clade A, (**b**) env clade B: Phylogenetic tree of the SSU rRNA sequences of environmental sequences obtained from the various ODEs used for analysis. The sequences are represented by their NCBI accession number, followed by the regional location from where each sequence was isolated

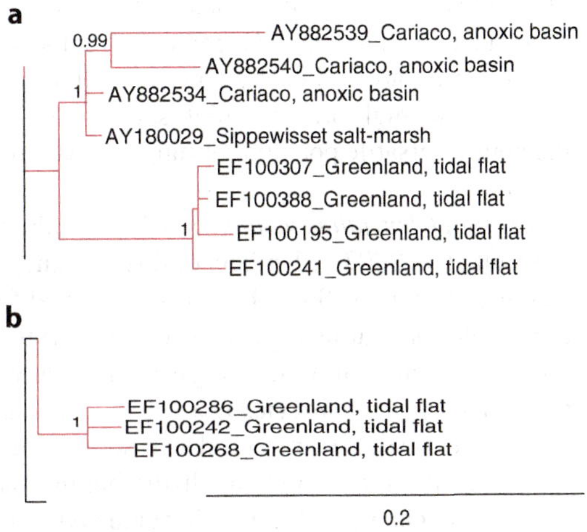

Fig. 10.9 (**a**) env clade C, (**b**) env clade D: Phylogenetic tree of the SSU rRNA sequences of environmental sequences obtained from the various ODEs used for analysis. The sequences are represented by their NCBI accession number, followed by the regional location from where each sequence was isolated

environment as pathogens of marine mammals, such as dolphins and whales (Miller et al. 2002). Other members of this clade, such as *Cystofilobasidium* sp., are also common in the marine environment (Fell et al. 2001). Other than the occurrence of these groups in marine environment, their ecological role is not well studied. The tremellomycete yeasts are rich in carotenoid pigments and are known to break down sugars by fermentative mode. Detailed studies on these groups of organisms can give us more insights into their biochemistry and physiological adaptations to marine ODEs.

Environmental sequences analyzed from the different ODEs are grouped in the classes *Pezizomycotina* (clade 3 and 4) and *Saccharomycotina* (clade 5) of the phylum *Ascomycota*, subphylum *Saccharomycetes* (Figs. 10.4 and 10.5). Environmental fungal sequences belonging to nine different ODEs are grouped within the *Ascomycota*, and these sequences are relatively closely related to cultured organisms. *Pezizomycotina* contains most of the well-known fungal species which are ubiquitous in nature. Clade 3 (Fig. 10.4a) has representative sequences belonging to oxygen-depleted regions of Greenland, intertidal sediments in California, the Cariaco, Gotland, and L'Atalante basins as well as the Framvaren fjord. Environmental sequences from the Mid-Atlantic ridge and the Guaymas vent regions were also included. Clade 4 (Fig. 10.4b) branches within the *Pezizomycotina* and includes a large number of representatives from the Guaymas vent region in the Gulf of California, Mexico. These sequences group with *Neurospora crassa*, a fungal model organism that is spread across geographical boundaries. *Neurospora* spp. are found in high numbers in predominantly tropical and subtropical terrestrial regions, such as Haiti and the Ivory Coast (Turner et al. 2001). A large number of fungal species belonging to the *Pezizomycotina* occurs in marine environments as filamentous forms, and their secondary metabolites are well studied. Clade 5 (Fig. 10.5) includes environmental sequences obtained from anoxic regions of Greenland and Japanese methane seeps. These sequences are related to known species of the *Saccharomycotina*. Our phylogenetic analyses demonstrate that none of the clusters are specific to a single anoxic habitat type. Instead, the sequences of each clade include genetic signatures from a variety of different marine ODE sites, all of which are united by oxygen deficiency, but have distinct environmental settings (like high pressure, high salinity, high or low temperature). This indicates the highly versatile potential of fungi to adapt to a variety of extreme environmental conditions.

Phylum *Chytridiomycota* was only recently described as a monophyletic phylum (Hibbett et al. 2007). Members of this phylum generally are associated with soil and freshwater habitats. Some have been reported from marine environments, generally as parasites on marine algae (Kerwin et al. 1992) and phytoplanktons (Kagami et al. 2007). In contrast to other fungal phyla, many extremophiles have been described within the *Chytridiomycetes* (Zack and Wildman 2004). They can thrive in extreme environments such as the periglacial soils at high altitudes, in the permafrost, acidic environments, mineral rich, hot hydrothermal vent regions (Gleason et al. 2010), and in ODEs (Lockhart et al. 2006). In our analyses, Clade 6 (Fig. 10.6) and Env Clades B, C, D (Figs. 10.8b and 10.9a, b) group within the *Chytridiomycota*. Where the Clade 6, and the environmental clades (Env Clade) C and D include sequences from a variety of different marine ODEs (Figs. 10.6 and 10.9a, b), Env Clade B includes exclusively hydrothermal vent sites (Fig. 10.8b). Fungal sequences from anoxic sediments in California branch in Clade 6 and group with *Kappamyces laurelensis*, a genus that has been described only few years back (Letcher and Powell 2005). Environmental fungal sequences obtained from Mariager fjord in Denmark (Baltic Sea) group with *Rhizophydium* spp., well-characterized marine parasites of phytoplanktons (Kagami et al. 2007). Interestingly, sequences obtained from intertidal anoxic sediments in western Greenland appear to be very divergent to any described fungal species and their phylogenetic position is strongly supported. The members of this cluster and the sequences from the three Env Clades B, C, and D have high bootstrap support, and most likely they represent new genera or even higher taxonomic categories. Members of phylum *Chytridiomycota* are being extensively studied for their phylogenetic affiliations as new members are being continuously added. Members of *Neocallimastigomycetes* are the only known obligate anaerobes, which have recently been upgraded to phylum level (Hibbett et al. 2007). Members of this group thrive under anaerobic conditions by a fermentative mode of respiration and can only survive in the anaerobic gut of rumens. Molecular analysis has shown that these groups are also present in natural anaerobic landfill areas (Lockhart et al. 2006).

There is a considerable number of fungal lineages, which cannot readily be assigned to any of these phyla and are therefore classified as "basal fungal lineages." The "true" phylogenetic position of many such lineages is discussed controversially. It remains to be seen whether further (environmental) sequencing efforts will strengthen the phylogenetic placement of these unassigned basal fungal lineages. In our study, environmental sequences from oxygen-depleted regions of Greenland, Gotland regions, intertidal sediments in California and Massachusetts grouped within the basal fungal group, clade 7 (Fig. 10.7). Interestingly, most of these sequences are from coastal regions (intertidal soft sediments), identifying such habitats as prime targets for the discovery of further basal fungal lineages.

The resolution of this clade needs to be clarified and further molecular diversity surveys with specific primer sets targeting basal fungal lineages might reveal more diversity within this group, and will possibly provide more sequences that help to elucidate the phylogenetic position of these taxa.

Our analyses show that, in general, the distribution of specific fungal lineages is not limited by geographical boundaries, and fungal sequences could be isolated from marine ODEs across the world oceans. This raises the question, whether one of the most discussed dogmas in microbial ecology, namely, "everything is everywhere, but the environment selects" (Baas Becking 1934) holds true for the fungi.

10.4 Methodological Considerations of Molecular Diversity Surveys and Future Directions

The number of sequences that could be assigned to *Chytridiomycota* or to basal fungal lineages is decisively lower than the number of sequences belonging to *Ascomycota* or *Basidiomycota*. This observation is likely ascribed to the following two reasons: (a) Chytridiomycota and basal fungi are indeed rare in natural (marine) habitats or have a restricted distribution in specific habitats compared to *Ascomycota* and *Basidiomycota*, for example, the basal fungal lineages are confined to anoxic intertidal soft-bottom habitats (Fig. 10.8). (b) Alternatively, methodological issues, like primer bias, could account for the low number of specific fungal taxon groups. If the latter is the case, the application of fungal-specific primer sets for the amplification of taxonomic marker genes could overcome this obstacle. Fungal-specific PCR primers generally target the ITS, SSU rRNA, or the LSU rRNA regions. Although these primer sets were designed to specifically amplify members of the fungal community from the environmental samples, cross-kingdom amplification is observed, which reduces the diversity estimates from the region of interest (Pang and Mitchell 2005). The SSU region has been the region of interest for a large number of studies because of the strong public database libraries that contain representative sequences representing all the major fungal groups. Studies targeting this region were able to amplify representatives of *Ascomycota, Basidiomycota, Chytridiomycota* and *Zygomycota* (Bass et al. 2007; Schadt et al. 2003; Malosso et al. 2006; Jebaraj et al. 2010), and also some members of the basal fungal lineages (Dawson and Pace 2002; Stoeck et al. 2007a) in various extreme habitats, such as the deep sea, frozen environments, and ODEs. Some studies suggest that the SSU rRNA gene is not the best choice for fungal diversity analysis, because of its low resolving power and because it may not always be sufficient to identify fungal species and strains (Hugenholtz and Pace 1996). A more powerful alternative might be the faster evolving ITS regions I and II which separate the ribosomal subunit genes from each other (O'Brien et al. 2005). Diversity estimates based on the ITS region may also have some disadvantages, as in some instances members belonging only to phylum *Ascomycota* and *Basidiomycota* were amplified

(Lai et al. 2007). These discrepancies may arise because not all of the fungal lineages are equally represented (Bruns 2001). The LSU rRNA region has been the most preferred region to study the distribution of some specific groups of *Fungi*, such as the mycorrhizal fungi (van Tuinen et al. 1998; Zuccaro et al. 2003). The target specificity of the primers targeting the LSU rRNA region is high, with the ability to preferentially amplify the target DNA from the fungal biomass, which is present in association with the plant material. Environmental studies using the LSU rRNA region as the target are very limited, in comparison with studies that use SSU rRNA as the marker gene. However novel fungal sequences were retrieved from the tundra soil based on LSU rRNA studies (Schadt et al. 2003). Taxon-specific studies using separate primers to target each of the fungal phyla have revealed a large diversity of *Fungi* from freshwater habitats (Nikolcheva and Bärlocher 2004). Hence, future environmental diversity studies could benefit from targeting primers specifically designed to amplify selected fungal phyla, and also following a multi-PCR approach, where all the major regions of taxonomic interest, such as the ITS, SSU rRNA, and LSU rRNA, could be amplified from the environmental DNA. This may be a versatile tool to amplify the fungal DNA across the various taxonomic communities, as the different primer sets can complement or compensate the strengths and disadvantages of each other.

The phylogenetic analyses of fungal sequences conducted in this study, which were based on publicly available SSU rRNA genes obtained from environmental diversity surveys, have revealed the extent of unknown diversity of *Fungi* in a variety of oxygen-deficient marine habitats on high taxonomic levels. The results show that *Fungi* are indeed much more diverse and abundant in anoxic habitats than previously assumed. Furthermore, most fungal lineages are widely distributed in different habitats with distinct environmental settings, indicating the versatility of *Fungi* and their capabilities to adapt to a variety of environmental conditions. This makes *Fungi* one of the most successful eukaryotic kingdoms, and with only a few exceptions (e.g. *Chytridiomycota* lineages that exhibit a more restricted distribution), they are able to colonize a large variety of natural habitats. Further studies, which specifically target fungal genes in environmental samples combined with culturable diversity studies, are needed for a deeper understanding of their distribution and their ecology in the marine ODEs.

Acknowledgments This is NIO's contribution no.4950.

References

Alexander E, Stock A, Breiner HW, Behnke A, Bunge J, Yakimov MM, Stoeck T (2009) Microbial eukaryotes in the hypersaline anoxic L'Atalante deep-sea basin. Environ Microb 11:360–381

Baas Becking LGM (1934) Geobiologie of inleiding tot de milieukunde. Van Stockum & Zoon, The Hague, The Netherlands

Barghoorn ES, Linder DH (1944) Marine fungi: their taxonomy and biology. Farlowia 1:395–467

Bass D, Howe A, Brown N, Barton H, Demidova M, Michelle H, Li L, Sanders H, Watkinson CV, Willcock S, Richards TA (2007) Yeast forms dominate fungal diversity in the deep oceans. Proc R Soc B 274:3069–3077

Behnke A, Bunge J, Barger K, Breiner HW, Alla V, Stoeck T (2006) Microeukaryote community patterns along an O_2/H_2S gradient in a supersulfidic anoxic fjord (Framvaren, Norway). Appl Environ Microbiol 72:626–3636

Bollag JM, Tung G (1972) Nitrous oxide release by soil fungi. Soil Biol Biochem 4:271–276

Bruns TD (2001) ITS reality. Inoculum 52:2–3

Buchan A, Newell SY, Moreta JIJ, Moran MA (2002) Analysis of internal transcribed spacer (ITS) regions of rRNA genes in fungal communities in a southeastern U.S. salt marsh. Microb Ecol 43:329–340

Damare S, Raghukumar C, Raghukumar S (2006) Fungi in deep-sea sediments of the central Indian basin. Deep Sea Res I 53:14–27

Dawson SC, Pace NR (2002) Novel kingdom-level eukaryotic diversity in anoxic environments. Proc Natl Acad Sci U S A 99:8324–8329

Dighton J (2003) Fungi in ecosystem processes. Marcel Dekker, New York

Dighton J (2007) Nutrient cycling by saprotrophic fungi in terrestrial habitats. In: Kubicek CP, Druzhinina IS (eds) The Mycota IV, environmental and microbial relationships, 2nd edn. Springer, Berlin

Edgcomb VP, Kysela DT, Teske A, de Vera GA, Sogin ML (2002) Benthic eukaryotic diversity in the Guaymas Basin hydrothermal vent environment. Proc Natl Acad Sci U S A 99:7658–7662

Edgcomb VP, Orsi W, Leslin C, Epstein S, Bunge J, Jeon SO et al (2009) Protistan community patterns within the brine and halocline of deep hypersaline anoxic basins in the eastern Mediterranean Sea. Extremophiles 13:151–167

Edgcomb VP, Beaudoin D, Gast R, Biddle JF, Teske A (2011) Marine subsurface eukaryotes: the fungal majority. Environ Microbiol 13:172–183

Epstein S, López-García P (2007) 'Missing' protists: a molecular prospective. Biodivers Conserv 17:261–276

Fell JW, Boekhout T, Fonseca A, Sampaio JP (2001) Basidiomycetous yeasts. In: McLaughlin DJ, McLaughlin EG, Lemke PA (eds) Mycota VII. Part B, Systematics and evolution. Springer, Berlin

Gessner RV (1980) Degradative enzyme production by salt-marsh fungi. Bot Mar 23:133–139

Gleason FH, Schmidt SK, Marano AV (2010) Can zoosporic true fungi grow or survive in extreme or stressful environments? Extremophiles 14:417–425

Goregues CM, Michotey VD, Bonin PC (2005) Molecular, biochemical, and physiological approaches for understanding the ecology of denitrification. Mol Ecol 49:198–208

Guarro J, Gene J, Stchigel AM (1999) Developments in fungal taxonomy. Clin Microbiol Rev 12:454–455

Gunner HB, Alexander M (1964) Anaerobic growth of *Fusarium oxysporum.* J Bacteriol 37:1309–1315

Hibbett DS, Binder M, Bischoff JF et al (2007) A higher-level phylogenetic classification of the Fungi. Mycol Res 111:509–547

Hugenholtz P, Pace NR (1996) Identifying microbial diversity in the natural environment: a molecular phylogenetic approach. Trends Biotechnol 14:190–197

Hyde KD, Gareth Jones EB, Leana OE, Pointing SB, Poonyth AD, Vrijmoed LLP (1998) Role of fungi in marine ecosystems. Biodivers Conserv 7:1147–1611

Jebaraj CS, Raghukumar C (2009) Anaerobic denitrification in fungi from the coastal marine sediments off Goa, India. Mycol Res 113:100–109

Jebaraj CS, Raghukumar C, Behnke A, Stoeck T (2010) Fungal diversity in oxygen-depleted regions of the Arabian Sea revealed by targeted environmental sequencing combined with cultivation. FEMS Microbiol Ecol 71:399–412

Jones EBG, Choeyklin R (2008) Ecology of marine and freshwater basidiomycetes. In: Boddy L, Franklan JC, West PV (eds) Ecology of saprotrophic basidiomycetes. Academic, New York

Jones EBG, Hyde KD (2002) Succession: where to go from here? Fungal Divers 10:241–253
Kagami M, de Bruin A, Ibelings BW, Donk EW (2007) Parasitic chytrids: their effects on phytoplankton communities and food-web dynamics. Hydrobiologia 578:113–129
Kamykowski D, Zentara SJ (1990) Hypoxia in the world ocean as recorded in the historical data set. Deep Sea Res 37:1861–1874
Kerwin JL, Johnson LM, Whisler HC, Tuiniga AR (1992) Infection and morphogenesis of *Pythium marinum* in species of Porphyra and other red algae. Can J Bot 70:1017–1024
Kis-Papo T (2005) Marine fungal communities. In: Dighton J, White JF, Oudemans P (eds) The Fungal Community: its Organization and Role in the Ecosystem. Taylor & Francis, Boca Raton, FL, pp 61–92
Knowles R (1982) Denitrification. Microbiol Rev 46:43–70
Kohlmeyer J, Kohlmeyer E (1979) Marine mycology. The higher fungi. Academic, New York
Kurakov AV, Lavrent'ev RB, Nechitailo TY, Golyshin PN, Zvyagintsev DG (2008) Diversity of facultatively anaerobic microscopic mycelial fungi in soils. Microbiology 77:90–98
Lai X, Cao L, Tan H, Fang S, Huang Y, Zhou S (2007) Fungal communities from methane hydrate-bearing deep-sea marine sediments in South China Sea. ISME J 1:756–762
Letcher PM, Powell MJ (2005) Kappamyces, a new genus in the Chytridiales (Chytridiomycota). Nova Hedwigia 80:113–133
Levin LA (2003) Oxygen minimum zone benthos: adaptation and community response to hypoxia. In: Gibson RN, Atkinson RJ (eds) A oceanography and marine biology: an annual review. Taylor and Francis, New York, pp 1–45
Le Calvez T, Burgaud G, Mahé S, Barbier G, Vandenkoornhuyse P (2009) Fungal diversity in deep-sea hydrothermal ecosystems. Appl Environ Microbiol 75:6415–6421
Lockhart RJ, van Dyke MI, Beadle IR, Humphreys P, McCarthy AJ (2006) Molecular detection of anaerobic gut fungi (Neocallimastigales) from landfill sites. Appl Environ Microbiol 72:5659–5661
López-García P, Philippe H, Gail F, Moreira D (2003) Autochthonous eukaryotic diversity in hydrothermal sediment and experimental microcolonizers at the Mid-Atlantic Ridge. Proc Natl Acad Sci U S A 100:697–702
López-García P, Vereshchaka A, Moreira D (2007) Eukaryotic diversity associated with carbonates and fluid-seawater interface in Lost City hydrothermal field. Environ Microbiol 9:546–554
Luo Q, Krumholz LR, Najar FZ, Peacock AD, Roe BA, White DC, Elshahed MS (2005) Diversity of the microeukaryotic community insulfide-rich Zodletone Spring (Oklahoma). Appl Environ Microbiol 71:6175–6184
Maheshwari R (2005) Species, their diversity and populations in fungi: experimental methods in biology. In: Maheshwari R (ed) Fungi experimental methods in biology, Mycology series 24. CRC, Boca Raton, FL, pp 191–205
Maheshwari R, Bharadwaj G, Bhat MK (2000) Thermophilic fungi: their physiology and enzymes. Microbiol Mol Biol Rev 64:461–488
Malosso E, Waite IS, English L, Hopkins DW, O'Donnell AG (2006) Fungal diversity in maritime Antarctic soils determined using a combination of culture isolation, molecular fingerprinting and cloning techniques. Polar Biol 29:552–561
Miller WG, Padhye AA, Bonn W, Jensen E, Brandt ME, Ridgway SH (2002) Cryptococcosis in a bottlenose dolphin (*Tursiops truncatus*) caused by *Cryptococcus neoformans* var. *gattii*. J Clin Microbiol 40:721–724
Newell SY (1996) Established and potential impacts of eukaryotic mycelial decomposers in marine/terrestrial ecotones. J Exp Mar Biol Ecol 200:187–206
News Letter ISRO (2009) Discovery of new microorganisms in the stratosphere. http://www.isro.org/newsletters/scripts/newslettersin.aspx?indexjan2009jun2009. Accessed 29 Nov 2010
Nikolcheva L, Bärlocher F (2004) Taxon-specific fungal primers reveal unexpectedly high diversity during leaf decomposition in a stream. Mycol Prog 3:41–49

O'Brien HE, Parrent JL, Jackson JA, Moncalvo JM, Vilgalys R (2005) Fungal community analysis by large-scale sequencing of environmental samples. Appl Environ Microbiol 71:5544–5550
Pang KL, Mitchell JI (2005) Molecular approaches for assessing fungal diversity in marine substrata. Bot Mar 48:332–347
Prasannari K, Sridhar KR (2001) Diversity and abundance of higher marine fungi on woody substrates along the west coast of India. Curr Sci 81:304–311
Raghukumar C, Raghukumar S, Sharma S, Chandramohan D (1992) Endolithic fungi from deep-sea calcareous substrata: isolation and laboratory studies. In: Desai BN (ed) Oceanography of the Indian Ocean. Oxford IBH Publication, New Delhi
Raghukumar C, Raghukumar S, Sheelu G, Gupta SM, Nath B, Rao BR (2004) Buried in time: culturable fungi in a deep-sea sediment core from the Chagos Trench, Indian Ocean. Deep Sea Res I 51:1759–1768
Roberts PL, Mitchell J, Jones EBG (1996) Morphological and taxonomical identification of marine ascomycetes: detection of races in geographical isolates of *Corollospora maritima* by RAPD analysis. In: Rossen L, Dawson MT, Frisvad J (ed.) Fungal identification techniques EU 16510 EN. European Commission, Bruxelles
Robinson CH (2001) Cold adaptation in Arctic and Antarctic fungi. New Phytol 151:341–353
Sathe-Pathak V, Raghukumar S, Raghukumar C, Sharma S (1993) Thraustochytrid and fungal component of marine detritus. 1- Field studies on decomposition of the brown alga *Sargassum cinereum*. J Agr Indian J Mar Sci 22:159–167
Schadt CW, Martin AP, Lipson DA, Schmidt SK (2003) Seasonal dynamics of previously unknown fungal lineages in tundra soils. Science 301:1359–1361
Shoun H, Kim DH, Uchiyama H, Sugiyama J (1992) Denitrification by fungi. FEMS Microbiol Lett 94:277–282
Slapeta J, Moreira D, Lopez-Garcıa P (2005) The extent of protist diversity: insights from molecular ecology of freshwater eukaryotes. Proc R Soc Lond [Biol] 272:2073–2081
Sparrow FK Jr (1936) Biological observations of the marine fungi of woods hole waters. Biol Bull 70:236–263
States JS, Christensen M (2001) fungi associated with biological soil crusts in desert grasslands of Utah and Wyoming. Mycologia 93:432–439
Stock A, Bunge J, Jurgens K, Stoeck T (2009) Protistan diversity in the suboxic and anoxic waters of the Gotland Deep (Baltic Sea) as revealed by 18S rRNA clone libraries. Aquat Microb Ecol 55:267–284
Stoeck T, Epstein S (2003) Novel eukaryotic lineages inferred from small-subunit rRNA analyses of oxygen depleted marine environments. Appl Environ Microbiol 69:2657–2663
Stoeck T, Taylor GT, Epstein SS (2003) Novel eukaryotes from the permanently anoxic Cariaco Basin (Caribbean Sea). Appl Environ Microbiol 69:5656–5663
Stoeck T, Hayward B, Taylor GT, Varela R, Epstein SS (2006) A multiple PCR-primer approach to access the microeukaryotic diversity in environmental samples. Protist 157:31–43
Stoeck T, Kasper J, Bunge J, Leslin C, Ilyin V, Epstein SS (2007a) Protistan diversity in the arctic: a case of paleoclimate shaping modern biodiversity? PLoS One 2:e728
Stoeck T, Zuendorf A, Breiner HW, Behnke A (2007b) A molecular approach to identify active microbes in environmental eukaryote clone libraries. Microb Ecol 53:328–339
Takishita K, Miyake H, Kawato M, Maruyama T (2005) Genetic diversity of microbial eukaryotes in anoxic sediment around fumaroles on a submarine caldera floor based on the small-subunit rDNA phylogeny. Extremophiles 9:185–196
Takishita K, Tsuchiyaa M, Kawatoa M, Ogurib K, Kitazatob H, Maruyamaa T (2007a) Diversity of microbial eukaryotes in anoxic sediment of the saline meromictic lake Namako-ike (Japan): on the detection of anaerobic or anoxic-tolerant lineages of eukaryotes. Protist 158:51–64
Takishita K, Yubuki N, Kakizoe N, Inagaki Y, Maruyama T (2007b) Diversity of microbial eukaryotes in sediment at a deep-sea methane cold seep: surveys of ribosomal DNA libraries from raw sediment samples and two enrichment cultures. Extremophiles 11:563–576

Turner BC, Perkins DD, Fairfield A (2001) Neurospora from natural populations: a global study. Fungal Genet Biol 32:67–92

Usuda K, Toritsuka N, Matsuo Y, Kim DH, Shoun H (1995) Denitrification by the fungus *Cylindrocarpon tonkinense*: anaerobic cell growth and two isozyme forms of cytochrome P-450nor. Appl Environ Microbiol 61:883–889

van Tuinen D, Jacquot E, Zhao B, Gollotte A, Gianinazzi-Pearson V (1998) Characterisation of root colonization profiles by a microcosm community of arbuscular mycorrhizal fungi using 25S rDNA-targeted nested PCR. Mol Ecol 7:879–887

Wainwright M, Wickramasinghe NC, Narlikar JV, Rajaratnam P (2003) Microorganisms cultured from stratospheric air samples obtained at 41 km. FEMS Microbiol Lett 218:161–165

Wyrtki K (1962) The oxygen minima in relation to ocean circulation. Deep Sea Res 9:11–23

Zack JC, Wildman HG (2004) Fungi in stressful environments. In: Mueller GM, Bills GF, Foster MS (eds) Biodiversity of fungi, inventory and monitoring methods. J Biol Chem 271:16263–16267

Zhou Z, Takaya N, Nakamura A, Yamaguchi M, Takeo K, Shoun H (2002) Ammonia fermentation, a novel anoxic metabolism of nitrate by fungi. J Biol Chem 277:1892–1896

Zuccaro A, Schulz B, Mitchell JI (2003) Molecular detection of ascomycetes associated with *Fucus serratus*. Mycol Res 107:451–466

Zuendorf A, Bunge J, Behnke A, Barger KJA, Stoeck T (2006) Diversity estimates of microeukaryotes below the chemocline of the anoxic Mariager Fjord, Denmark. Microb Ecol 58:476–491

Chapter 11
Assemblage and Diversity of Fungi on Wood and Seaweed Litter of Seven Northwest Portuguese Beaches

K.R. Sridhar, K.S. Karamchand, C. Pascoal, and F. Cássio

Contents

Abstract Three hundred and fifty woody litter and one hundred and forty seaweed litter sampled from seven beaches of Northwest Portugal were assessed for the filamentous fungal assemblage and diversity. The woody litter was screened for fungi up to 42 months using damp chamber incubation. They consisted of 36 taxa (ascomycetes, 21; basidiomycetes, 3; anamorphic taxa, 12) comprising 10 core group taxa ($\geq$10%) (ascomycetes, 8; basidiomycete, 1; anamorphic taxa, 1). The total fungal isolates ranged between 150 and 243, while the number of fungal taxa per wood ranged between 3 and 4.9. The seaweed litter was screened up to four months in damp chamber incubation. They encompassed 29 taxa (ascomycetes, 16; basidiomycetes, 2; anamorphic taxa, 11) comprising 15 core group taxa (ascomycetes, 9; basidiomycete, 1; anamorphic taxa, 5). Total fungal isolates ranged between 56 and 120, while the number of fungal taxa per seaweed segment ranged

K.R. Sridhar (✉)
Department of Biosciences, Mangalore University, Mangalore, Karnataka, India
e-mail: sirikr@yahoo.com

C. Raghukumar (ed.), *Biology of Marine Fungi*,
Progress in Molecular and Subcellular Biology 53,
DOI 10.1007/978-3-642-23342-5_11, © Springer-Verlag Berlin Heidelberg 2012

between 4.8 and 6.3. Fifteen taxa of ascomycetes, two of basidiomycetes, and four anamorphic taxa were common to wood and seaweed litter. On both the substrates, two arenicolous fungi *Arenariomyces trifurcates* and *Corollospora maritima* were the predominant fungi (72.6–85.9%). The species abundance curves showed higher frequency of occurrence of fungal taxa in seaweed than woody litter. Our study revealed rich assemblage and diversity of marine fungi on wood and seaweed litter of Northwest Portugal beaches. The fungal composition and diversity of this survey have been compared with earlier investigations on marine fungi of Portugal coast.

11.1 Introduction

The fascinating ecological group of marine fungi has worldwide geographic distribution on a variety of substrata in different habitats. Filamentous marine fungi involve in decomposition and nutrient turnover of organic matter in marine ecosystem (Fell and Master 1980). They have been extensively surveyed in temperate regions (e.g., Kohlmeyer and Kohlmeyer 1979; Koch and Petersen 1996; Petersen and Koch 1997; Shearer et al. 2007) as well as tropical mangroves (e.g., Kohlmeyer 1984; Volkmann-Kohlmeyer and Kohlmeyer 1993; Jones and Alias 1997; Schmit and Shearer 2003, 2004; Sarma and Hyde 2001; Maria and Sridhar 2002). However, survey of marine fungi has been confined to localized areas especially in Southeast Asia, Europe, and North America (Kohlmeyer 1983; Shearer et al. 2007). Therefore, extensive survey is warranted in wide geographical areas to understand the pattern of fungal distribution more precisely. The global occurrence of marine fungi based on morphological and molecular phylogeny is 530 taxa in 321 genera (Ascomycota, 424 taxa in 251 genera; anamorphic fungi, 94 taxa in 61 genera; Basidiomycota, 12 taxa in 9 genera) (Jones et al. 2009). The decadal taxonomic studies on marine fungi (1840–2009) were highest during 1990–1999 (156 taxa) followed by 1980–1989 (135 taxa) and 1960–1969 (75 taxa). The described taxa ranged between 1 and 43 in rest of the 14 decades.

The main areas surveyed in European region include continental shelf from the Canaries and Azores to Greenland and North-West Russia including Mediterranean shelf and Baltic Seas (Landy and Jones 2006). The checklist revealed occurrence of 338 taxa of marine fungi in the European marine waters (Landy and Jones 2006). In contrast, 830 km long Portuguese coast is relatively less explored for marine fungal resources. Studies on marine fungi of Portugal coast was initiated with description of new marine basidiomycete taxon (*Nia globospora*) on the baits of *Spartina maritima* in Mira River estuary by Barata et al. (1997). Subsequently, she reported the occurrence of 20 marine fungi (ascomycetes, 15; anamorphic taxa, 5) associated with the culms of salt marsh halophyte *S. maritima* in three estuaries (Tagus, Sado, and Mira rivers) (Barata 2002). In another study, sterilized stems of *S. maritima* were baited up to 13 months in the estuary of River Mira to explore marine fungi and recorded 26 taxa (ascomycetes, 17; anamorphic taxa, 7; basidiomycetes, 2) (Barata 2006). The woody litter (*Arundo donax*) and herbaceous stems (*Phragmites* sp. and *S. maritima*) immersed in two beaches (Rainha and Guincho) were incubated up to 10 months in laboratory and recorded 35 fungal taxa (ascomycetes, 27; anamorphic taxa, 6; unidentified taxa, 2) (Figueira and Barata 2007). Autoclaved wood pieces (*Fagus sylvatica* and *Pinus pinater*) were baited in two coastal locations (Cascais and Sesimbra) up to 12 months by Azevedo et al. (2010) yielded 26 fungal taxa (ascomycetes, 13; anamorphic taxa, 12; basidiomycete, 1) on laboratory incubation up to 12 months. All these studies were confined to the mid- and southwest coast of Portugal and the marine fungi of Northwest coast are not explored. Furthermore, although Portuguese coasts are endowed with rich seaweed flora (Pereira et al. 2006; Araújo et al. 2009), there seems to be no information on their association with fungi or detritus decomposing fungi. Hence, the present Chapter aims at understanding the assemblage and diversity of higher fungi on woody litter and seaweed litter of seven beaches of Northwest Portugal. Based on the present study and published reports, marine fungi on three substrata (woody litter, seaweed litter and culms of *Spartina*) of Portuguese coast are compared.

11.2 Materials and Methods

11.2.1 Study Sites and Samples

The coast of Portugal has rocky regions separated by white sandy beaches, which are often large in size and mainly used for recreation purpose. Seven beaches along the northwest coast (Moledo, Afife, Carreço, Foz do Neiva, Apúlia, Costa Nova, and Mira) were randomly selected for wood and seaweed litter collection (Fig. 11.1, Table 11.1).

About 60–70 woody litter (~1.5–2 cm diam.) preferably embedded in sand or entangled in rocky crevices from each beach were collected in clean polythene bags. They were rinsed in filter sterilized seawater in the laboratory to eliminate sediments and debris. Each wood was cut into 15 cm length and 50 wood samples per beach were incubated separately on the wet sand bed soaked in filter sterilized seawater (salinity, 34 ppt) in airtight polythene bags (23 $\pm$ 2°C) at 12-h photoperiod. They were screened once a month up to 2 years for the growth of anamorphs and teleomorphs. Moisture of sand bed was maintained by addition of sterile distilled water after each observation. From each beach, 20–30 detached seaweed thalli were sampled. They were cleaned in the laboratory, cut into segments (1–1.5 $\times$ 6 cm), and 20 pieces per beach (composite of 5–6 seaweed taxa) were incubated on sterile sand bed and observed as described below.

The fungi grown on wood and seaweed samples were mounted in seawater for morphological examination and later they were stained with cotton blue in lactophenol. Fungi isolated were identified based on monographs and primary literature (e.g. Ellis 1971, 1976; Kohlmeyer and Kohlmeyer 1979; Ellis and Ellis 1987; Hyde and Jones 1988; Kohlmeyer and Volkmann-Kohlmeyer 1991; Hyde et al. 2000; Jones et al. 2009).

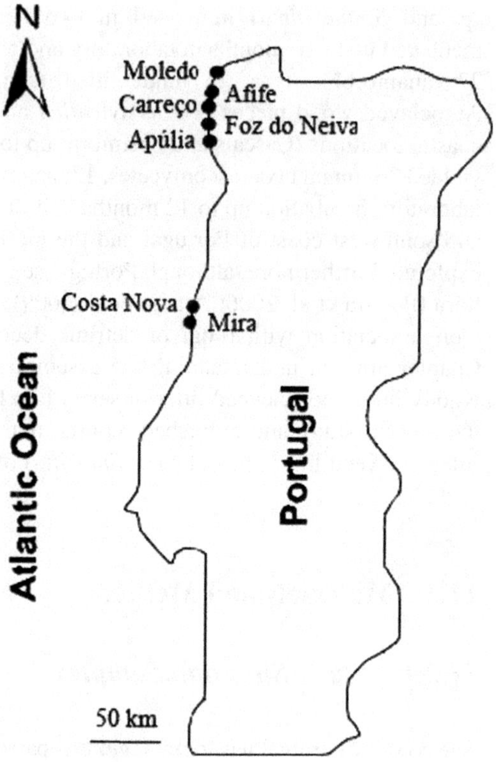

Fig. 11.1 Map showing the sampling sites of woody and seaweed litter along the Northwest Portugal coast

Table 11.1 Sampling sites of wood and seaweed litter along the Portugal coast

Beach	Location	Characteristics
Moledo	41°51′N 8°51′W	Sandy beach near Caminha Town (iodine rich)
Afife	41°46′N 8°52′W	Sandy beach near Viana do Castelo Town
Carreço	41°45′N 8°52′W	Sandy and rocky beach near Viana do Castelo Town (protected sand dunes)
Foz do Neiva	41°37′N 8°48′W	Sandy beach with adjacent endemic vegetation
Apúlia	41°29′N 8°46′W	Sandy beach (protected sand dunes)
Costa Nova	40°37′N 8°45′W	Sandy beach (fine sand grains)
Mira	40°27′N 8°48′W	Sandy beach (protected sand dunes)

11.2.2 Data Analysis

The frequency of occurrence (%) and relative abundance (%) of each fungal taxon on wood and seaweed samples were determined:

$$\text{Frequency of occurrence (\%)} = [(\text{Number of wood pieces colonized}) / (\text{Total wood pieces examined})] \times 100,$$

$$\text{Relative abundance } (\%) = [(\text{Frequency of occurrence of fungal taxon})/(\text{Total of frequency of occurrence fungal taxa})] \times 100.$$

The Simpson's (D') and Shannon's (H') diversities (Magurran 1988) and Pielou's evenness (J') (Pielou 1975) of fungal taxa on wood and seaweed samples were calculated:

$$D' = 1/\Sigma(p_i)^2,$$

$$H' = -\Sigma(p_i \ln p_i),$$

where p_i is the proportions of individual that taxon i contributes to the total number of taxa.

$$J' = (H'/H'_{\text{max}}),$$

where H'_{max} is the maximum value of diversity for the number of taxa present.

To compare the richness of fungi based on number of isolations and number of wood and seaweed samples assessed, the expected number of taxa was calculated by rarefaction indices (Ludwig and Reynolds 1988). The expected number of taxa, $E_{(t)}$, in a random sample of n isolations taken from a total population of N isolations was estimated:

$$E_{(t)} = \sum_{i=1}^{s} \left\{ 1 - \left[\binom{N - n_i}{n} \Big/ \binom{N}{n} \right] \right\},$$

where n_i is the number of fungal isolations of the ith taxon.

Sorensen's similarity coefficient (C_s) (%) was calculated based on Mueller-Dombois (1981):

$$C_s(\%) = [2c/(a + b)] \times 100,$$

where a is total number of taxa in beach 1, b is total number of taxa in beach 2, and c is number of taxa common to beach 1 and 2.

11.3 Results

11.3.1 Fungi on Woody Litter

Assessment of 50 wood samples per beach yielded isolations ranging between 150 (3 isolations/wood) (Apúlia) and 243 (4.9 isolations/wood) (Mira) (Table 11.2). Total taxa per locations ranged from 9 (Foz do Neiva, Moledo and Apúlia) to

Table 11.2 Fungal assemblage species richness and diversity on woody litter collected from seven beaches of Northwest Portugal

Beach	Assemblage[a]				Species richness[b]					Diversity		
	Sc	Sco	Fi	Fpt	Total taxa	$E_{(t140)}$	Anamorph (A)	Teleomorph (T)	A/T ratio	Simpson	Shannon	Pielou's Evenness
Moledo	50	50	235	4.7	9	48	4	5	0.80	0.842	2.787	0.897
Afife	50	50	240	4.8	15	47	3	12	0.25	0.884	3.450	0.883
Carreço	50	46	171	3.4	10	49	2	8	0.25	0.856	3.020	0.909
Foz do Neiva	50	50	194	3.9	9	49	1	8	0.13	0.821	2.698	0.851
Apúlia	50	48	150	3.0	9	45	1	8	0.13	0.856	2.953	0.931
Costa Nova	50	49	206	4.1	12	49	4	8	0.50	0.860	3.110	0.868
Mira	50	47	243	4.9	10	49	1	9	0.11	0.861	2.373	0.895

[a]Assemblage: Sc, Number of wood samples collected; Sco, Number of wood samples colonized; Fi, Total fungal isolations on wood; Fpt, Number of fungal taxa per wood

[b]Species richness: $E_{(t140)}$, Expected number of fungal taxa out of 140 random isolations)

15 (Afife); however, the range of expected number of taxa out of 140 random isolations in seven beaches was narrow (45–49 taxa). The range of teleomorphs was higher than anamorphic taxa per beach (5–12 vs. 1–4) with A/T ratio ranging from 0.11 (Mira) to 0.80 (Moledo). The Simpson and Shannon diversities were highest in Afife (0.884 and 3.450), while they were least in Foz do Neiva (0.821) and Mira (2.373), respectively. The Pielou's evenness was highest in Apúlia (0.931) and least in Foz do Neiva (0.851).

Three hundred and fifty woody litter showed 36 taxa (ascomycetes, 21; basidiomycetes, 3; anamorphic taxa, 12). Eight ascomycetes and one each of basidiomycete and anamorphic taxon belonged to core group (≥10%) (ascomycetes: *Arenariomyces parvulus, A. trifurcates, Chaetomium cochlioides, Corollospora angusta, C. maritima, C. pulchella, Dryosphaera navigans, Halosphaeriopsis* sp.; basidiomycete: *Nia vibrissa*; anamorphic taxon: *Bactrodesmium linderi*) (Table 11.3). Among them, *Arenariomyces trifurcatus, B. linderi,* and *Carollospora maritima* were the most frequent taxa (42.1–82.1%) with high relative abundance (10.1–19.7%). Five taxa out of 21 ascomycetes have been reported from the Portugal and up to 11 typical marine ascomycetes are new records to Portugal coast. Interestingly, among three taxa of *Chaetomium* on woody litter (*C. cochlioides, C. gobosum,* and *C. reflexum*), *C. cochlioides* belonged to core group (16.2%). Out of three basidiomycetes, an unknown taxon possessing sigmoid basidiospores was recorded from Costa Nova. *N. vibrissa* and *N. globospora* have been reported from Portugal. Typical marine anamorphic taxa, *B. linderi* and *Astreomyces cruciatus* are new record to Portugal coast. Up to 10 anamorphic taxa seems to be terrestrial fungi and unlike marine fungi their frequency of occurrence was very low (0.9–1.8%). Sorensen's similarity of fungi between 21 beach pairs was least between Afife and Costa Nova (14.8%) and highest between Foz do Neiva and Mira (73.7%) (Table 11.4). Only seven beach pairs showed similarity 50% and above, while it was ranged between 14.8 and 44.4% in rest of the pairs.

11.3.2 Fungi on Seaweed Litter

Assessment of 20 seaweed samples per beach yielded isolations ranging between 56 (2.8 isolations/wood) (Costa Nova) and 125 (6.3 isolations/wood) (Carreço) (Table 11.5). Total taxa per location ranged from 4 (Costa Nova) to 13 (Apúlia), however, the expected number of taxa out of 45 random isolations in all beaches was uniform (19 taxa). The range of teleomorphs was higher than anamorphic taxa per beach (3–8 vs. 1–6) with A/T ratio ranging from 0.33 (Costa Nova) to 0.86 (Apúlia). The Simpson diversity was highest in Apúlia (0.902) and the Shannon diversity in Moledo (3.923), while they were least in Costa Nova (0.705 and 1.876). The Pielou's evenness was highest in Foz do Neiva (0.956) and least in Moledo (0.924).

Table 11.3 Occurrence of fungal taxa on woody litter sampled from seven beaches of Northwest Portugal (out of 50 samples in each beach)

Taxon	Occurrence in beaches							FO (%)	RA (%)	[a]Report
	Mo	Af	Ca	FN	Ap	CN	Mi			
Ascomycetes										
[b]*Corollospora maritima* Werderm.	49	45	45	45	20	39	35	82.1	19.7	2, 3, 5
[b]*Arenariomyces trifurcates* Höhnk	42	50	40	38	35	31	39	80.9	19.4	3
[b,c]*Dryosphaera navigans* Jørg. Koch & E.B.G. Jones	–	15	–	40	–	–	35	26.5	6.4	–
[b,c]*Corollospora pulchella* Kohlm., I. Schmidt & N.B. Nair	–	–	09	35	–	–	27	20.9	5.0	–
[b]*Halosphaeriopsis* sp.	–	29	15	04	15	–	03	19.1	4.6	–
[b,c]*Arenariomyces parvulus* Jørg. Koch	–	05	10	09	15	–	19	17.1	4.1	–
[b]*Chaetomium cochlioides* Palliser	24	10	–	–	21	–	–	16.2	3.9	–
[b,c]*Corollospora angusta* Nakagiri & Tokura	–	–	35	–	–	05	–	11.8	2.8	–
[b]*Nereiospora* sp.	–	20	11	–	–	–	–	9.1	2.2	–
[b,c]*Carbosphaerella leptosphaerioides* I. Schmidt	–	–	–	05	–	25	–	8.8	2.1	–
[c]*Lignincola laevis* Höhnk	30	–	–	–	–	–	–	8.8	2.1	2, 4
Chaetomium globosum Kunze	–	15	–	10	–	–	–	7.4	1.8	–
[c]*Dryosphaera tropicalis* Kohlm. & Volkm.-Kohlm.	–	05	–	–	–	–	20	7.4	1.8	–
Phaeosphaeria orae-maris (Linder) Khashn. & Shearer	–	–	17	–	–	–	–	5.0	1.2	3, 5
[b,c]*Verruculina enalia* (Kohlm.) Kohlm. & Volkm.-Kohlm.	–	11	–	–	–	–	04	4.4	1.1	–
Lulworthia fucicola G.K. Sutherl.	–	–	–	–	–	11	–	3.2	0.8	3, 4
[b,c]*Dactylospora haliotrepha* (Kohlm. & E. Kohlm.) Hafellner	–	–	–	–	06	–	03	2.7	0.7	–
[b,c]*Lignincola tropica* Kohlm.	–	–	–	–	09	–	–	2.7	0.7	–
[c]*Leptosphaeria orae-maris* Linder	–	–	–	–	–	06	–	1.8	0.4	–
Chaetomium reflexum Skolko & J.W. Groves	–	05	–	–		–	–	1.5	0.4	–
[b,c]*Hydronectria tethys* Kohlm. & E. Kohlm.	–	–	–	–	05	–	–	1.5	0.4	–
Basidiomycetes										
[b]*Nia vibrissa* R.T. Moore & Meyers	42	–	–	–	–	24	–	19.4	4.7	2, 5
[b]*Nia globospora* Barata & Basilio	–	05	–	–	–	–	–	1.5	0.4	1, 2
[c]Unknown sp. (sigmoid spores)	–	–	–	–	–	06	–	0.3	0.1	–
Anamorphic taxa										
[b,c]*Bactrodesmium linderi* (J.L. Crane & Shearer) M.E. Palm & E.L. Stewart	38	–	–	05	25	45	35	42.1	10.1	–
[b]*Aspergillus flavus* Link	–	20	–	–	–	–	–	5.9	1.4	–
[b,c]*Astreomyces cruciatus* Moreau & R. Moreau	–	–	16	–	–	–	–	4.7	1.1	–
	–	06	–	–	–		–	1.8	0.4	–

(continued)

Table 11.3 (continued)

Taxon	Occurrence in beaches							FO (%)	RA (%)	[a]Report
	Mo	Af	Ca	FN	Ap	CN	Mi			
Camposporium pellucidum (Grove) S. Hughes										
Curvularia brachyspora Boedijn	–	–	–	–	–	06	–	1.8	0.4	–
[b]*Aspergillus puniceus* Kwon-Chung & Fennell	–	–	05	–	–	–	–	1.5	0.4	–
Curvularia crepinii (Westend.) Boedijn	05	–	–	–	–	–	–	1.5	0.4	–
Dortomyces stemonitis (Pers.) F.J. Morton & G. Sm.	–	05	–	–	–	–	–	1.5	0.4	–
Drechslera fugax (Wallr.) Shoemaker	04	–	–	–	–		–	1.2	0.4	–
Tetracoccosporium paxianum Szabó	–	–	–	–	–	05	–	1.5	0.4	–
Wardomyces anomalus F.T. Brooks & Hansf.		–	–	–	–	04	–	1.2	0.4	–
Diplocladiella scalaroides G. Arnaud	03	–	–	–	–	–	–	0.9	0.1	–

FO%, Percent frequency of occurrence; RA%, Percent relative abundance; arranged in decreasing order

Beaches: *Mo* Moledo; *Af* Afife; *Ca* Carreço; *FN* Foz do Neiva; *Ap* Apúlia; *CN* Costa Nova; *Mi* Mira

[a]Reported from Portugal: 1, Dead culms of *Spartina maritima*: Barata et al. 1997; 2, Baited sterilized young stem of *S. maritima*: Barata 2006; 3, Stems of *Phragmites* sp. and *Spartina maritima*: Figueira and Barata 2007; 4, Beach woody litter: Figueira and Barata 2007; 5, Baited wood pieces of *Fagus sylvatica* and *Pinus pinaster*:Azevedo et al. 2010

[b]Also found in seaweed litter

[c]New to Portugal marine habitats

A total of 140 seaweed samples consist of 29 taxa (ascomycetes, 16; basidiomycetes, 2; anamorphic taxa, 11). Nine ascomycetes, one basidomycete, and five anamorphic taxa belonged to core group (≥10%) (ascomycetes: *A. trifurcatus*, *C. cochlioides*, *C. angusta*, *C. maritima*, *C. pulchella*, *Halosphaeriopsis* sp., *Dactylospora haliotrepha*, *D. navigans*, and *Verruculina enalia*; basidiomycete: *N. vibrissa*; anamorphic taxa: *Aspergillus flavus*, *A. puniceus*, *A. oxysporum*, *A. tamari*, and *B. linderi*) (Table 11.6). Among core group taxa, *A. trifurcatus*, *Aspergillus tamari*, *B. linderi*, and *C. maritima* were the most frequent taxa (40.7–85.9%) with relative abundance ranging from 6.2 to 16.3%. Three taxa out of 16 ascomycetes have been reported from the Portugal and up to 12 typical marine ascomycetes are new records. Among ascomycetes, *C. cochlioides* belonged to core group (13.3%). Among basidiomycetes, *N. vibrissa* and *N. globospora* have been reported from Portugal coast. Out of 11 anamorphic taxa, *Periconia prolifica* is known from Portugal, while three taxa *A. cruciatus*, *B. linderi*, and *Dendryphiella arenariae* are new records. Up to seven anamorphic taxa seems to be typical terrestrial fungi, and four of them were more frequent (14.8–40.7%). Soresen's similarity of fungi between 21 pairs of beaches was least between Costa Nova and Mira (26.7%) and highest between Afife and Carreço and Foz do Neiva and Mira (60%) (Table 11.4). Only five pairs of beaches showed similarity up to 50% and above, while it was ranged between 26.7 and 47.6% in rest of the beach pairs.

Table 11.4 Sorensen's similarity coefficient (%) of fungal taxa on woody litter and seaweed litter (in parenthesis) from seven beaches of Northwest Portugal

	Afife	Carreço	Foz do Neiva	Apúlia	Costa Nova	Mira
Moledo	25 (52.6)	21.1 (31.6)	33.3 (55.6)	44.4 (36.4)	38.1 (46.2)	31.6 (30)
	Afife	40 (60)	50 (31.6)	41.7 (34.8)	14.8 (28.6)	56 (28.6)
		Carreço	52.6 (42.1)	42.1 (43.5)	27.3 (28.6)	50 (47.6)
			Foz do Neiva	55.6 (45.5)	38.1 (46.2)	73.7 (60)
				Apúlia	28.6 (35.3)	63.2 (50)
					Costa Nova	27.3 (26.7)

11.3.3 Fungal Composition in Wood and Seaweed Litter

Altogether wood and seaweed litter yielded 44 taxa (ascomycetes, 22; basidiomycetes, 3; anamorphic taxa, 19). Woody litter represented by 81% of total fungi, while seaweed litter consists of 70% (Tables 11.3 and 11.6). Twenty one taxa were common to both substrates (ascomycetes, 15; basidiomycetes 2; anamorphic taxa, 4). Teleomorphs of wood and seaweed litter were more similar than anamorphic taxa. Anamorph/teleomorph ratio (0.61 vs. 0.50), Simpson diversity (0.923 vs. 0.900), Shannon diversity (4.116 vs. 3.994), and Pielou's equitability (0.867 vs. 0.773) were higher in seaweed than woody litter.

Seven teleomorphs were confined to woody litter (*Chaetomium globosum, C. reflexum, Dryosphaera tropicalis, Leptosphaeria orae-maris, Lulworthia fucicola, Phaeosphaeria orae-maris,* and unknown basidiomycete), while only one ascomycete was restricted to seaweed litter (*Zopfiella latipes*) (Table 11.3). Among the anamorphic taxa, eight were confined to woody litter (*Camposporium pellucidum, Curvularia brachyspora, C. crepinii, Diplocladiella scalaroides, Dortomyces stemonitis, Drechslera fugax, Tetracoccosporium paxianum,* and *Wardomyces anomalus*), while seven were confined to seaweed litter (*Aspergillus niger, A. tamari, D. arenariae, Fusarium oxysporum, P. prolifica, Sporidesmium afrormosiae,* and *Trichoderma harzianum*) (Table 11.6).

Eight core group teleomorphs were common to both substrates (*A. trifurcatus, C. cochlioides, C. angusta, C. maritima, C. pulchella, Halosphaeriopsis* sp., *D. navigans,* and *N. vibrissa*) (Tables 11.3 and 11.6). Only one core group anamorphic taxon was common to wood and seaweed (*B. linderi*). Interestingly, on both substrates two arenicolous fungi *Arenariomyces trifurcates* and *Corollorpora maritima* were dominated in all beaches (72.6–85.9%).

Beach-wise comparison of fungal assemblage, richness, and diversity between wood and seaweed litter drastically varied (Tables 11.2 and 11.5). Although overall occurrence of fungi was higher in woody litter, on plotting frequency of occurrence against species sequence, the frequency of occurrence of fungi in seaweed litter was higher than woody litter (Fig. 11.2). The overall rarefaction curve of woody litter resembles that of all beaches except for the beach Apúlia, so also the curve of seaweed litter except for the beach Costa Nova (Fig. 11.3). The range of expected number of taxa, $E_{(t)}$, out of 140 random isolations of woody litter per beach was

Table 11.5 Fungal assemblage species richness and diversity on seaweed litter collected from seven beaches of Northwest Portugal

Beach	Assemblage[a]				Species richness[b]					Diversity		
	Sc	Sco	Fi	Fpt	Total taxa	$E_{(t45)}$	Anamorph (A)	Teleomorph (T)	A/T ratio	Simpson	Shannon	Pielou's Evenness
Moledo	20	20	104	5.2	9	19	3	6	0.50	0.856	3.928	0.924
Afife	20	20	95	4.8	10	19	4	6	0.67	0.880	3.173	0.955
Carreço	20	17	125	6.3	10	19	4	6	0.67	0.878	3.115	0.950
Foz do Neiva	20	20	110	5.5	9	19	3	6	0.50	0.868	3.030	0.956
Apúlia	20	20	121	6.1	13	19	6	7	0.86	0.902	3.505	0.947
Costa Nova	20	19	56	2.8	4	19	1	3	0.33	0.705	1.876	0.938
Mira	20	19	120	6.0	11	19	3	8	0.38	0.888	3.208	0.948

[a]Assemblage: Sc, Number of seaweed samples collected; Sco, Number of seaweed samples colonized; Fi, Total fungal isolations on seaweed samples; Fpt, Number of fungal taxa per seaweed sample

[b]Species richness: $E_{(t45)}$, Expected number of fungal taxa out of 45 random isolations)

Table 11.6 Occurrence of fungal taxa on seaweed litter sampled from seven beaches of Northwest Portugal (out of 50 samples in each beach)

Taxon	Occurrence in beaches							FO (%)	RA (%)	[a]Report
	Mo	Af	Ca	FN	Ap	CN	Mi			
Ascomycetes										
[b]*Arenariomyces trifurcatus* Höhnk	15	14	9	20	15	29	14	85.9	16.3	3
[b]*Corollospora maritima* Werderm.	19	15	17	10	05	16	16	72.6	13.8	2, 3, 5
[b, c]*Verruculina enalia* (Kohlm.) Kohlm. & Volkm.-Kohlm.	–	–	–	10	15	–	11	26.7	5.1	–
[b]*Halosphaeriopsis* sp.	–	10	05	–	14	–	–	21.5	4.1	–
[b, c]*Corollospora pulchella* Kohlm., I. Schmidt & N. B. Nair	–	–	04	19	–	–	05	20.7	3.9	–
[b, c]*Dryosphaera navigans* Jørg. Koch & E.B.G. Jones	–	–	–	10	–	–	15	18.5	3.5	–
[b]*Chaetomium cochlioides* Palliser	14	04	–	–	–	–	–	13.3	2.5	–
[b, c]*Corollospora angusta* Nakagiri & Tokura	–	–	17	–	–	–	–	12.6	2.4	–
[b, c]*Dactylospora haliotrepha* (Kohlm. & E. Kohlm.) Hafellner	–	–	–	–	09	–	05	10.4	1.2	–
[b, c]*Hydronectria tethys* Kohlm. & E. Kohlm.	–	–	–	–	10	–	–	7.4	1.4	–
[b, c]*Lignincola tropica* Kohlm.	–	–	–	–	10	–	–	7.4	1.4	–
[b, c]*Arenariomyces parvulus* Jørg. Koch	–	05	04	–	–	–	–	6.7	1.3	–
[b, c]*Carbosphaerella leptosphaerioides* I. Schmidt	–	–	–	–	–	09	–	6.7	1.3	–
[c]*Zopfiella latipes* (N. Lundq.) Malloch & Cain	04	–	–	05	–	–	–	6.7	1.3	–
[b]*Nereiospora* sp.	–	05	–	–	–	–	–	3.7	0.7	–
[b]*Lignincola laevis* Höhnk	03	–	–	–	–	–	–	2.2	0.4	2, 4
Basidiomycetes										
[b]*Nia vibrissa* R.T. Moore & Meyers	16	–	–	–	–	–	19	25.9	4.9	2, 5
[b]*Nia globospora* Barata & Basilio	–	–	–	–	–	–	05	3.7	0.7	1, 2
Anamorphic taxa										
[b, c]*Bactrodesmium linderi* (J.L. Crane & Shearer) M.E. Palm & E.L. Stewart	20	–	–	15	09	15	–	43.7	8.3	–
Aspergillus tamari Kita	09	10	15	–	05	–	16	40.7	7.7	–
[b]*Aspergillus flavus* Link	–	–	15	16	04	–	09	32.6	6.2	–
[b]*Aspergillus puniceus* Kwon-Chung & Fennell	–	15	10	–	–	–	–	18.5	3.5	–
Fusarium oxysporum E.F. Sm. & Swingle	04	11	–	05	–	–	–	14.8	2.8	–
[b, c]*Astreomyces cruciatus* Moreau & R. Moreau	–	–	10	–	–	–	–	7.4	1.4	–
Aspergillus niger Tiegh.	–	05	–	–	–	–	–	3.7	0.7	–
[c]*Dendryphiella arenariae* Nicot	–	–	–	–	05	–	–	3.7	0.7	–
Periconia prolifica Anastasiou	–	–	–	–	–	–	04	3.0	0.6	5
Sporidesmium afrormosiae M.B. Ellis	–	–	–	–	03	–	–	2.2	0.4	–
Trichoderma harzianum Rifai	–	–	–	–	03	–	–	2.2	0.4	–

FO%, Percent frequency of occurrence; RA%, Percent relative abundance; arranged in decreasing order

Beaches: *Mo* Moledo; *Af* Afife; *Ca* Carreço; *FN* Foz do Neiva; *Ap* Apúlia; *CN* Costa Nova; *Mi* Mira

[a]Reported from Portugal: 1, Dead culms of *Spartina maritima*: Barata et al. 1997; 2, Baited sterilized young stem of *S. maritima*: Barata 2006; 3, Stems of *Phragmites* sp. and *Spartina maritima*: Figueira and Barata 2007; 4, Beach woody litter: Figueira and Barata 2007; 5, Baited wood pieces of *Fagus sylvatica* and *Pinus pinaster*: Azevedo et al. 2010

[b]Also found in woody litter

[c]New to Portugal marine habitats

narrow (45–49), but on pooling isolates of all beaches the $E_{(t)}$ raised to 117 taxa (Table 11.2 and Fig. 11.3). Similarly, the $E_{(t)}$ out of 45 random isolations of seaweed litter was 19 which raised to 40 on pooling isolates of all beaches (Table 11.5 and Fig. 11.3). The $E_{(t)}$ out of 700 random isolations of pooled data was 308 in woody litter, while it was 140 in seaweed litter. On comparison of Soresen's similarity of fungi between wood and seaweed litter, it was least (wood) or near to least (seaweed) between Afife and Costa Nova (14.8 and 28.6%), while it was highest between Foz do Neiva and Mira (73.7 and 60%) (Table 11.4). The similarity of fungi on seaweed litter of Afife vs. Carreço was also highest (60%)

11.4 Discussion

Driftwood in marine habitats has been considered as the most important ecological niche for marine fungi (Hyde et al. 2000). Although marine fungi of European region are relatively well known (Kohlmeyer 1983; Landy and Jones 2006; Shearer et al. 2007), studies on marine fungi of Portugal coast are scanty. Available literature deals with marine fungi mainly on salt marsh halophyte (*Spartima maritima*) in estuaries of Portugal (Barata et al. 1997; Barata 2002, 2006). Stems of monocots (*Phragmites* sp. and *S. maritima*) collected from two beaches were studied for marine fungi by Figueira and Barata (2007). Two studies dealt with marine fungi associated with naturally accumulated woody litter (*A. donax*) (Figueira and Barata 2007) and baited woody litter (*F. sylvatica* and *P. pinater*)

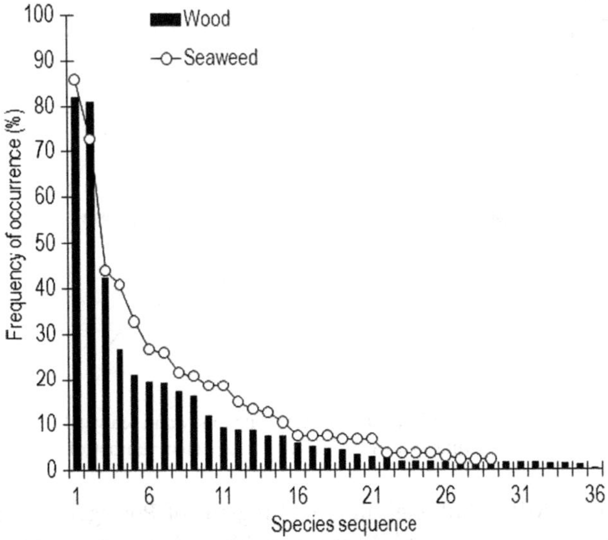

Fig. 11.2 Species abundance curves of fungi recorded on woody and seaweed litter (Fungal taxa on each substrate were arranged in descending order)

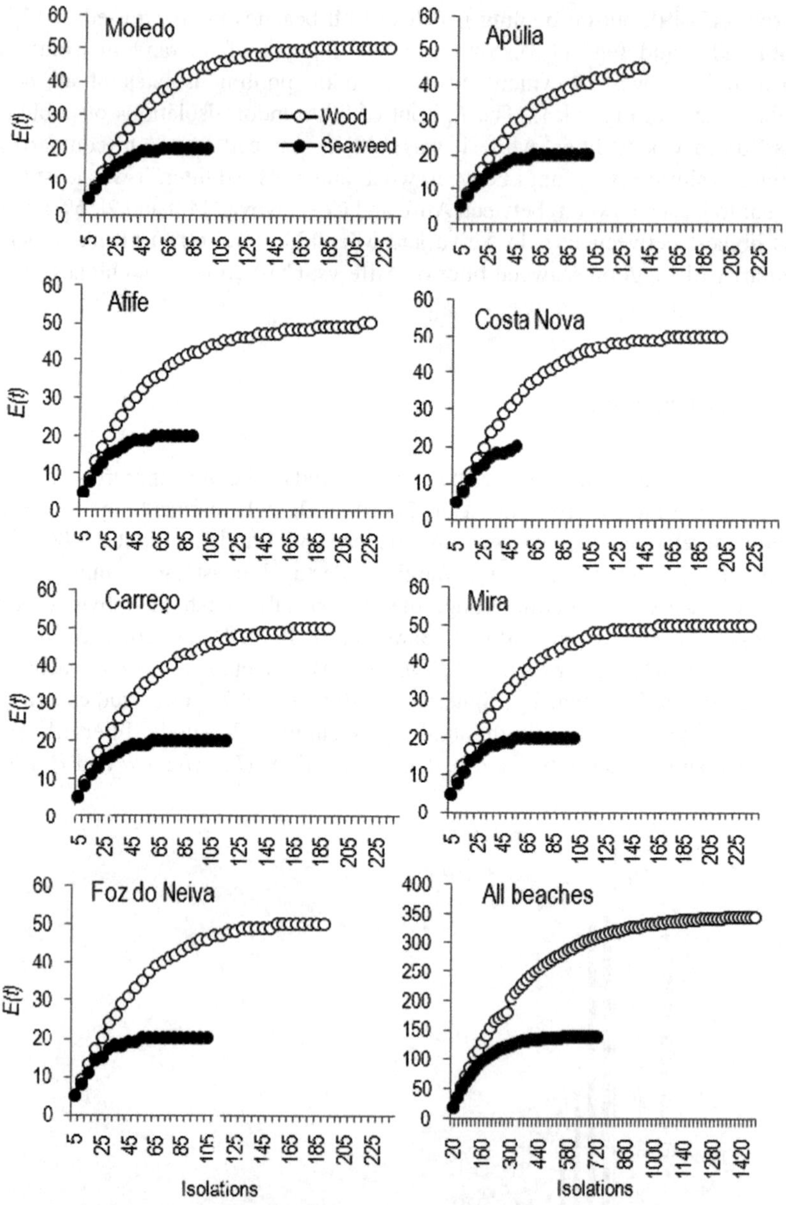

Fig. 11.3 Comparison of rarefaction curve of fungi on woody and seaweed litter of seven beaches of Northwest Portugal coast (number of isolations vs. expected number of taxa)

(Azevedo et al. 2010). Although intertidal regions of Portugal coast are endowed with a variety of seaweeds (e.g., Pereira et al. 2006; Araújo et al. 2009) there seems to be no studies on their association or decomposition by fungi.

11.4.1 Fungi on Woody Litter

Incubation of woody litter sampled from seven beaches of Northwest Portugal up to 24 months in our study yielded 24 typical marine fungi (teleomorphs, 22; anamorphic taxa, 2) (Table 11.3). Among them, only seven teleomorphs are known from the Portugal coast, but none of anamorphic taxa are common to earlier studies. Overall 14 marine fungi (teleomorphs, 12; anamorphic taxa, 2) are new records to Portugal coast.

Our study also yielded many arenicolous fungi on woody litter. Interestingly, the top-ranked two taxa belonged to arenicolous group (*Areneriomyces trufurcatus* and *Corollospora maritima*) with cosmopolitan distribution (Kohlmeyer 1983; Kohlmeyer and Volkmann-Kohlmeyer 1993; Nakagiri and Tokura 1994). Fruit bodies of arenicolous fungi usually develop on wood or animal substrates on long-term incubation (e.g., Ananda et al. 1998; Prasannarai and Sridhar 2003; Ananda and Sridhar 2004). The dominance of arenicolous or nonarenicolous fungi on woody litter may support or suppress other fungi (Tan and Leong 1992; Tan et al. 1995; Jones 2000). For example, *Aigialus parvus* and *V. enalia* suppressed *Lignincola laevis* in co-culture, while *L. laevis* enhances formation of ascomata in association with *V. enalia*. Similarly, *C. maritima* and *V. enalia* have been considered as "aggressive" taxa (Miller et al. 1985; Tan et al. 1995). As diverse fungal flora was seen in our study, the dominant arenicolous fungi might have positive rather than negative interaction with co-occurring fungi.

Short-term incubation (2–4 months) of woody litter yields primarily terrestrial fungi, but long-term incubation (6–12 months) leads to dominance of marine fungi (e.g., Ravikumar et al. 2009). The present study revealed some commonly occurring terrestrial fungi on woody litter (e.g., Ascomycetes: *Chaetomium* spp.; Anamorphic taxa: *Aspergillus* spp., *Camposporium* sp., *Curvularia* spp., *Diplocladiella* sp., *Dortomyces* sp., *Drechslera* sp., *Tetracoccosporium* sp., and *Wardomyces* sp.). However, except for *A. flavus, C. cochlioides,* and *C. globosum* (5.9–16.2%), the rest were found in low frequency (0.9–1.8%). Contribution toward decomposition of organic matter by so-called terrestrial fungi in marine habitats cannot be ruled out (Sridhar et al. 2010). It is interesting to note that seawater, sea foam, and beach sand in Saudi Arabia yielded many terrestrial fungi along with typical marine fungi (Bokhary et al. 1992). Similarly, about 36 taxa of terrestrial fungi were isolated from the hypersaline salterns in Puerto Rico (Cantrell et al. 2006). Among them, *A. flavus, A. niger,* and *C. globosum* were common to our study. Fungal taxa such as *A. niger* and *C. globosum* were also inhabitants of hypersaline waters in temperate and tropical habitats (see Cantrell et al. 2006). Cantrell et al. (2006) classified *A. flavus* as moderately halotolerant taxon, while *A. niger* and *C. globosum* as highly halotolerant taxa. Diverse taxa of *Aspergillus* and *Penicillium* have been reported from the marine sediments from different region (e.g., Sponga et al. 1999; Takami 1999; Damare et al. 2006; Ramesh et al. 2006; Jebraj and Raghukumar 2009; Das et al. 2009).

The data on the mean number of fungi per sample in a substrate help comparing different hosts or regions. The mean number of fungi per wood sample in the present study was ranged between 3 and 4.9. It is higher than mangrove woody litter in Indian ocean (1–1.7), Hong Kong, Macau (1.2–1.7), Southeast Asia (0.9–2.6) (see Maria and Sridhar 2003), Red Sea mangroves of Egypt (1.1–1.3) (Abdel-Wahab 2005), Mediterranean Sea (1–2.8) (Abdel-Aziz 2010), Bahamas Islands (0.9–1.9) (Jones and Abdel-Wahab 2005), and partially resembles the reports on mangrove wood in Malaysia (3.3) and west coast of India (1.5–4.6) (Maria and Sridhar 2004). However, the mean fungi per wood in our study is higher than the report on woody litter (3–4.9 vs. 0.96) (Figueira and Barata 2007) and immersed wood segments (1.11–1.72) (Azevedo et al. 2010) from Portugal. But mean fungi per wood partially resemble the reports on stems of *S. maritima* (1.2–3.7) from Portugal coast (Barata 2006).

11.4.2 Fungi on Seaweed Litter

Incubation of seaweed litter collected from seven beaches of Northwest Portugal up to 4 months in our study yielded 21 typical marine fungi (teleomorphs, 17; anamorphic taxa, 4 taxa) (Table 11.6). Among them, five teleomorphs and an anamorphic taxon are known from the Portugal coast (Barata et al. 1997; Barata 2006; Figueira and Barata 2007; Azevedo et al. 2010). Overall 13 marine fungi (teleomorphs, 10; anamorphic taxa, 3) are new records to the Portugal coast. The mean number of fungi per seaweed litter ranged between 2.8 and 6.3, which is equivalent or higher than woody litter (Tables 11.2 and 11.5) and other substrata assessed in marine and mangrove habitats (see Sect. 11.4.1). The mean fungi per seaweed litter is higher than woody litter and immersed wood segments (0.96–1.72) (Figueira and Barata 2007; Azevedo et al. 2010) and partially resembles with that of stems of *S. maritima* (1.2–3.7) (Barata 2006).

A variety of arenicolous fungi were inhabitants of seaweed litter in our study and top-ranked two species belonged to cosmopolitan arenicolous fungi (*A. trifurcatus* and *C. maritima*). Unlike woody litter, in seaweed detritus in our study the arenicolous fungi grew within 2–4 months of incubation. Beyond this period, they cannot be handled as they become more amorphous. Seaweed detritus were host for many saprophytic marine fungi including arenicolous fungi (see Zuccaro and Mitchell 2005). Molecular studies on healthy thalli of brown seaweed *Fucus serratus* from temperate region yielded several arenicolous fungi along with nonarenicolous fungi (Zuccaro and Mitchell 2005). As seen in woody litter, besides typical marine fungi, a variety of terrestrial fungi were inhabitants of seaweed detritus in our study (*Aspergillus* spp., *Chaetomium* sp., *Fusarium* sp., *Sporidesmium* sp., and *Trichoderma* sp.). Some of these fungi were also endophytic in seaweeds of the southeast coast of India (Suryanarayanan et al. 2010). Molecular studies of live and dead thalli of *Fucus serratus* and other seaweeds (of Antarctica) also yielded several terrestrial fungi and arenicolous fungi (Zuccaro and Mitchell 2008; Loque et al. 2009).

11.5 Conclusions

Our study revealed rich fungal population because of assessment of mixture of wood and seaweed litter from many beaches of Northwest Portugal. Most of the earlier studies were confined to assess fungal assemblage mainly from the saltmarsh plant species *S. maritima* (Barata et al. 1997; Barata 2002, 2006; Figueira and Barata 2007) except for two studies on beach woody litter (Figueira and Barata 2007) and immersed wood segments (Azevedo et al. 2010). As seen in the present study, the frequency of occurrence or percentage occurrence of some fungi were very high on immersed wood (e.g., *Cirrenalia macrocephala*, 48.6%; *Lulworthia* sp., 88.9%) (Azevedo et al. 2010) and monocot stems (*Dictysporium pelagicum*, 66.7%; *Halosarpheia retorquens*, 70%; *Lulworthia* sp., 88.6%; *Phialophorophoma litoralis*, 59.2%; *Phoma* sp., 56.7%; *Sphaerulina oraemaris*, 57.5%) (Barata 2002, 2006) in the Portugal coast. The basidiomycete, *N. globospora* isolated from *S. maritima* of Mira river estuary (Barata et al. 1997), was also reported to colonize on the baited stem of *S. maritima* in Mira saltmarsh (Barata 2002, 2006). In our study, this fungus was an inhabitant of seaweed litter of Mira beach and the woody litter of the beach Afife. None of the fungi in our study was comparable to fungi found on *S. maritima* segments submerged in two salt marshes (Barata 2006). This may be due to assessment of young and senescent stems of *S. maritima* or possibly due to substrate specificity. There seems to be considerable difference in fungal population on organic matter between Northwest and Southwest Portugal as these coasts are influenced by the major currents Canary and Azores, respectively. Our study provided a baseline data on the fungal composition and diversity on wood and seaweed litter of Northwest Portugal. Future explorations may consider sampling and assessment of organic matter from Northwest and Southwest locations simultaneously to provide a clear picture on the fungal assemblage and diversity to strengthen the mycogeography of marine fungi.

Acknowledgments KRS is thankful to Mangalore University for permission to visit the University of Braga during May–June 2006. KSK acknowledges the award of research fellowship by the Mangalore University under Rajeev Gandhi Fellowship, University Grants Commission, New Delhi, India.

References

Abdel-Aziz FA (2010) Marine fungi from two sandy Mediterranean beaches on the Egyptian north coast. Bot Mar 53:283–289

Abdel-Wahab MA (2005) Diversity of marine fungi from Egyptian Red Sea mangroves. Bot Mar 48:348–355

Ananda K, Sridhar KR (2004) Diversity of filamentous fungi on decomposing leaf and woody litter of mangrove forests of southwest coast of India. Curr Sci 87:1431–1437

Ananda K, Prasannarai K, Sridhar KR (1998) Occurrence of higher marine fungi on animal substrates along the west coast of India. Indian J Mar Sci 27:233–236

Araújo R, Bárbara I, Tibaldo M, Berecibar E, Tapia PD, Pereira R, Santos R, Pinto IS (2009) Checklist of benthic marine algae and cyanobacteria of northern Portugal. Bot Mar 52:24–46

Azevedo E, Rebelo R, Caerio MF, Barata M (2010) Diversity and richness of marine fungi on two Portuguese marinas. Nova Hedwigia 90:521–531

Barata M (2002) Fungi on the halophyte *Spartina maritima* in salt marshes. In: Hyde KD (ed) Fungi in marine environments, Fungal diversity research series # 7. Hong Kong University Press, Hong Kong, pp 179–193

Barata M (2006) Marine fungi from Mira river salt marsh in Portugal. Rev Iberoam Micol 23:179–184

Barata M, Basilo MC, Baptista-Ferreira JL (1997) *Nia globospora,* a new marine gasteromycete on baits of *Spartina maritima* in Portugal. Mycol Res 101:687–690

Bokhary HA, Moslem MA, Parvez S (1992) Marine fungi of the Arabian Gulf Coast and Saudi Arabia. Microbiologica 15:281–290

Cantrell SA, Casills-Martinez L, Molina M (2006) Characterization of fungi from hypersaline environments of solar salterns using morphological and molecular techniques. Mycol Res 110:962–970

Damare S, Raghukumar C, Raghukumar S (2006) Fungi in deep-sea sediments of the Central Indian Basin. Deep Sea Res I 53:14–27

Das S, Lyla PS, Khan SA (2009) Filamentous fungal population and species diversity from the continental slope of Bay of Bengal, India. Acta Oecol 35:269–279

Ellis MB (1971) Dematiaceous hyphomycetes. Commonwealth Mycological Institute, Kew

Ellis MB (1976) More dematiaceous hyphomycetes. Commonwealth Mycological Institute, Kew

Ellis MB, Ellis JP (1987) Microfungi on land plants: an identification handbook. Croom Helm, London

Fell JW, Master IM (1980) The association and potential role of fungi in mangrove detrital systems. Bot Mar 23:257–263

Figueira D, Barata M (2007) Marine fungi of two sandy beaches. Mycologia 99:20–23

Hyde KD, Jones EBG (1988) Marine mangrove fungi. PSZNI Mar Ecol 9:15–33

Hyde KD, Sarma VV, Jones EBG (2000) Morphology and taxonomy of higher marine fungi. In: Hyde KD, Pointing SB (eds) Marine mycology – a practical approach. Fungal Diversity Press, Hong Kong, pp 172–204

Jebraj CS, Raghukumar C (2009) Anaerobic denitrification in fungi from coastal marine sediments off Goa, India. Mycol Res 113:100–109

Jones EBG (2000) Marine fungi: some factors influencing biodiversity. Fungal Divers 4:53–73

Jones EBG, Abdel-Wahab MA (2005) Marine fungi from the Bahamas islands. Bot Mar 48:356–364

Jones EBG, Alias SA (1997) Biodiversity of mangrove fungi. In: Hyde KD (ed) Biodiversity of tropical microfungi. Hong Kong University Press, Hong Kong, pp 71–92

Jones EBG, Sakayaroj J, Suetrong S, Somrithipol S, Pang KL (2009) Classification of marine Ascomycota, anamorphic taxa and Basidiomycota. Fungal Divers 35:1–203

Koch J, Petersen KRL (1996) A check-list of higher marine fungi on wood from Danish coasts. Mycotaxon 60:397–414

Kohlmeyer J (1983) Geography of marine fungi. Aust J Bot Suppl Ser 10:67–76

Kohlmeyer J (1984) Tropical marine fungi. PSZNI Mar Ecol 5:329–378

Kohlmeyer J, Kohlmeyer E (1979) Marine mycology: the higher fungi. Academic Press, New York

Kohlmeyer J, Volkmann-Kohlmeyer B (1991) Illustrated key to the filamentous higher marine fungi. Bot Mar 34:1–61

Kohlmeyer J, Volkmann-Kohlmeyer B (1993) Biogeographic observations on Pacific marine fungi. Mycologia 85:337–346

Landy ET, Jones GM (2006) What is the fungal diversity of marine ecosystems in Europe? Mycologist 20:15–21

Loque CP, Medeiros AO, Pellizzari EM, Oliveira EC, Rosa CA, Rosa LH (2009) Fungal community associated with marine macroalgae from Antarctica. Polar Biol 33:641–648

Ludwig JA, Reynolds JF (1988) Statistical ecology – a primer on methods and computing. Wiley, New York

Magurran AE (1988) Ecological diversity and its measurement. Princeton University Press, New Jersey

Maria GL, Sridhar KR (2002) Richness and diversity of filamentous fungi on woody litter of mangroves along the west coast of India. Curr Sci 83:1573–1580

Maria GL, Sridhar KR (2003) Diversity of filamentous fungi on woody litter of five mangrove plant species from the southwest coast of India. Fungal Divers 14:109–126

Maria GL, Sridhar KR (2004) Fungal colonization of immersed wood in mangroves of the southwest coast of India. Can J Bot 82:1409–1418

Miller JD, Jones EBG, Moharir YE, Findaly JA (1985) Colonization of wood blocks by marine fungi in Langstone harbour. Bot Mar 28:251–257

Mueller-Dombois D (1981) Ecological measurements and microbial populations. In: Wicklow DT, Caroll GC (eds) The fungal community: organization and role in the ecosystems. Marcel Dekker, New York, pp 123–184

Nakagiri A, Tokura R (1994) Taxonomic studies of the genus *Corollospora* (Halosphaeriales, Ascomycotina) with descriptions of seven new species. Trans Mycol Soc Jpn 28:413–436

Pereira SG, Lima FP, Queiroz NC, Ribeiro PA, Santos AM (2006) Biogeographic patterns of intertidal macroinvertebrates and their association with macroalgae distribution along the Portuguese coast. Hydrobiologia 555:185–192

Petersen KRL, Koch K (1997) Substrate preference and vertical zonation of lignicolous marine fungi on mooring posts of oak (*Quercus* sp.) and Larch (*Larix* sp.) in Svanemollen Harbour, Denmark. Bot Mar 40:451–463

Pielou FD (1975) Ecological diversity. Wiley InterScience, New York

Prasannarai K, Sridhar KR (2003) Abundance and diversity of marine fungi on Intertidal woody litter of the west coast of India on prolonged incubation. Fungal Divers 14:127–141

Ramesh S, Jayaprakashvel M, Mathivannan N (2006) Microbial status in seawater and coastal sediments during pre- and post-tsunami periods in the Bay of Bengal, India. Mar Ecol 27:198–203

Ravikumar M, Sridhar KR, Sivakumar T, Karamchand KS, Sivakumar N, Vellaiyan R (2009) Diversity of filamentous fungi on coastal woody debris after tsunami on the southeast coast of India. Czech Mycol 61:107–115

Sarma VV, Hyde KD (2001) A review on frequently occurring fungi in mangroves. Fungal Divers 8:1–34

Schmit JP, Shearer CA (2003) A checklist of mangrove associated fungi, their geographical distribution and known host plants. Mycotaxon 85:423–477

Schmit JP, Shearer CA (2004) Geographic and host distribution of mangrove-associated fungi. Bot Mar 47:496–500

Shearer CA, Descals E, Volkmann-Kohlmeyer B, Kohlmeyer J, Marvanová L, Padgett D, Porter D, Thorton HA, Voglymayr H, Raja HA, Schmit JP (2007) Fungal biodiversity in aquatic habitats. Biodivers Conserv 19:49–67

Sponga F, Cavaletti L, Lazzarini A, Borghi A, Ciciliato I, Losi D, Marinelli F (1999) Biodiversity and potentials of marine-derived microorganisms. J Biotechnol 70:65–69

Sridhar KR, Karamchand KS, Sumathi P (2010) Fungal colonization and breakdown of sedge (*Cyperus malaccensis* Lam.) in a southwest mangrove, India. Bot Mar 53:525–533

Suryanarayanan TS, Venkatachalam A, Thirunavukkarasu N, Ravishankar JP, Doble M, Geetha V (2010) Internal mycobiota of marine macroalgae from the Tamilnadu coast: distribution, diversity and biotechnological potential. Bot Mar 53:457–468

Takami H (1999) Isolation and characterization of microorganisms from deep-sea mud. In: Horikoshi K, Tsujii K (eds) Extremophiles in deep-sea environments. Springer, Tokyo, pp 3–26

Tan TK, Leong WF (1992) Lignicolous fungi of tropical mangrove wood. Mycol Res 96:413–414
Tan TK, Teng CL, Jones EBG (1995) Substrate and microbial interactions as factors affecting ascocarp formation by mangrove fungi. Hydrobiologia 295:127–134
Volkmann-Kohlmeyer B, Kohlmeyer J (1993) Biogeographic observations on Pacific marine fungi. Mycologia 85:337–346
Zuccaro A, Mitchell JI (2005) Fungal communities of seaweeds. In: Dighton J, White JF, Oudemans P (eds) The fungal community its organization and role in the ecosystem. CRC Press, New York, pp 533–579
Zuccaro A, Mitchell JI (2008) Detection and identification of fungi intimately associated with the brown seaweed *Fucus serratus*. Appl Environ Microbiol 74:931–941

Chapter 12
Xylariaceae on the Fringe

Sukanyanee Chareprasert, Mohamed T. Abdelghany, Hussain H. El-sheikh, Ayman Farrag Ahmed, Ahmed M.A. Khalil, George P. Sharples, Prakitsin Sihanonth, Hamdy G. Soliman, Nuttika Suwannasai, Anthony J.S. Whalley, and Margaret A. Whalley

Contents

Abstract The Xylariaceae is one of the best-known pyrenomycete families (Ascomycota) and is distributed throughout the world. The majority are wood inhabitants and are prevalent in tropical and subtropical regions. *Halorosellinia oceanicum* is the most widely distributed in mangroves and can be regarded as truly manglicolous being frequently recorded as the dominant member of the family in such environments in S.E. Asia. In Malaysian mangroves, members of the Xylariaceae have been found to be numerically important with up to 9% present in one mangrove ecosystem. A further twelve xylariaceous genera are reported as occurring as their teleomorphs in mangrove forest and their immediate surroundings including *Anthostomella*, *Astrocystis*, *Biscogniauxia*, *Camillea*, *Daldinia*, *Fasciatispora*, *Hypoxylon*, *Kretzschmaria*, *Nemania*, *Nipicola*, *Rosellinia* and *Xylaria*. Furthermore, the presence of species from a number of these taxa, especially species of *Anthostomella* and *Xylaria*, are regularly isolated as endophytes from a variety of mangrove plant species. Mangrove Xylariaceae are also well known for their ability to produce novel and often bioactive metabolites.

We dedicate this chapter to the late Dr Steve Moss in memory of many years of collaboration and friendship.

A.J.S. Whalley (✉)
School of Pharmacy and Biomolecular Sciences, Liverpool John Moores University, Byrom Street, Liverpool L3 3AF, UK
e-mail: A.J.Whalley@ljmu.ac.uk

C. Raghukumar (ed.), *Biology of Marine Fungi*,
Progress in Molecular and Subcellular Biology 53,
DOI 10.1007/978-3-642-23342-5_12,

12.1 Introduction

What constitutes a marine fungus is open to debate but a generally accepted definition of marine fungi is that provided by Kohlmeyer and Kohlmeyer (1979) "obligate marine fungi are those that grow and sporulate exclusively in a marine or estuarine habitat; facultative marine fungi are those from fresh water or terrestrial milieus able to grow (and possibly sporulate) in the marine environment". Jones et al. (2009) adopted a broad interpretation of what we consider to be marine and included species saprotrophic on decaying culms of maritime grasses such as *Spartina* species, *Juncus roemerianus*, *Phragmites communis*, mangrove fungi, and especially those on the palm, *Nypa fruticans*. We review the recorded Xylariaceae, including endophytes, although many of these belong to *Xylaria* Hill ex Schrank and are only identified to generic level. Also included are descriptions of *Halorosellinia oceanica* (Schatz) Whalley, E.B.G. Jones, K.D. Hyde and Laessoe *Camillea selangorensis* M.A. Whalley, Whalley and E.B.G. Jones and *C. malaysianensis* M.A. Whalley. *Halorosellinia oceanica* because it is the most widely distributed and is similar to, and may be confused with *Nemania maritima* Y.-M. Ju and J.D. Rogers, which occurs in Taiwan, Malaysia and appears to be common in Hong Kong (Ju and Rogers 2002; Jones et al. 2009), *C. selangorensis*, which is only the second species of *Camillea* to be recorded in the old world and which may prove to occur more extensively on *Ficus* bordering mangroves in S.E. Asia and elsewhere and *C. malaysianensis* because of possible confusion with the widely distributed *C. tinctor* (Berk.) Læssøe, J.D. Rogers and Whalley (Whalley et al. 1999). The Xylariaceae is one of the best known and widely distributed families of the Ascomycota. Ju and Rogers (1996) recognized 38 and Whalley (1996) 40 genera but with subsequent additions, the number is now 65 or more depending on individual opinion (Lumbsch and Huhndorf 2007). Although the genera *Hypocopra* (Fr.) J. Kickx fil. *Poronia* Willd., *Podosordaria* Ellis & Holw. and *Wawelia* Namyslowski occur on dung and a number of *Xylaria* species occur on leaves, seeds or are associated with insects, the family is mainly one of wood inhabitants. The vast majority of these wood inhabitants are considered to be saprobes but several species are well-known phytopathogens. *Rosellinia necatrix* Prill. causes a world-wide serious white root rot of many commercial plants, *Entoleuca mammata* (Wahlenberg:Fr.) J.D. Rogers & Y.-M. Ju is a major pathogen of aspen trees in North America and species of *Biscogniauxia* Kuntze induce canker formation in trees, which are water-stressed (Edwards et al. 2003; Nugent et al. 2004). Currently there is considerable interest in the Xylariaceae as they are now found to be almost ubiquitous endophytic inhabitants with a high presence in tropical plants (Whalley 1996). They are also known to produce an impressive array of novel natural compounds (Whalley and Edwards 1995; Stadler and Hellwig 2005). It is this ability to produce a wide range of chemical structures coupled with their common endophytic occurrence, which has stimulated considerable research into the family.

12.1.1 The Xylariaceae and the Marine Environment

Not many members of the Xylariaceae are recognized by researchers in the family as belonging to the marine environment, but there are scattered references to species of endophytic *Xylaria* isolates from mangroves (Lin et al. 2001; Okane et al. 2001; Liu et al. 2006; Xu et al. 2008) and "marine-derived" isolates (Rukachaisirkul et al. 2009), in relation to their ability to produce novel metabolites. *Hypoxylon croceum* J.H. Miller isolated from driftwood in a mangrove in Florida, USA, was found to produce a new sordarin derivative, hypoxysordarin and a novel hypoxylactone in addition to the previously known sordarin with both hypoxysordarin and sordarin exhibiting antifungal properties (Daferner et al. 1999). It appears that *H. croceum* is a frequent inhabitant of mangroves occurring in the Bahamas and East Coast USA (Stadler pers. comm.). Furthermore, Jones et al. (2008) provide a recent review of the potential of marine fungi for the discovery of new chemicals whilst Pan et al. (2008) reviewed the bioactive compounds from fungi in the South China Sea, which included compounds from *Xylaria* species.

There are representatives of twelve xylariaceous genera occurring as teleomorphs, *Anthostomella* Sacc., *Astrocystis* Berk. & Broome, *Biscogniauxia* Kuntze, *Camillea* Fr. *Daldinia* Ces. & De Not., *Fasciatispora* K.D. Hyde, *Halorosellinia* Whalley, E.B.G. Jones, K.D Hyde & Læssøe, *Hypoxylon* Bull., *Kretschmaria* Fr., *Nemania* S.F. Gray *Nipicola* K.D. Hyde, *Rosellinia* De Not. and *Xylaria* Hill ex Schrank which have to date been found in mangroves or in vegetation immediately adjacent to mangroves (Hyde 1992; Whalley et al. 1996, 1999, 2000; Smith and Hyde 2001; Jones and Abdel-Wahib 2005; Jones et al. 2009). It is mainly through the studies of E.B. Gareth Jones and Kevin D. Hyde, their co-workers and students that a substantial number of these taxa have been described from the marine environment. There are at least four species of *Anthostomella* recorded from *Nypa fruticans* or growing on *Spartina* or *Juncus* (Lu and Hyde 2000; Jones et al. 2009) and there are considerably more species occurring on mangrove wood from the Bahamas (Abdel-Wahab and Jones unpublished data). Species of *Anthostomella* together with *H. oceanica* and possibly *N. maritima* Y.-M. Ju & J.D. Rogers are members of the Xylariaceae which are widely distributed in the marine environment. *Nipicola* also occurs on the intertidal palm, *N. fruticans* and is clearly very closely related to *Anthostomella* but lacks an iodine-positive subapical apparatus and has black ascospores in addition to other microscopical differences (Hyde 1992; Jones et al. 2009). We are not convinced that the two species of *Nipicola, N. carbospora* K.D. Hyde and *N. selangorensis* K. D. Hyde can be logically separated from *Anthostomella*. Smith and Hyde (2001) in describing *Astrocystis nypae* and *A. selangorensis* from Malaysia maintained the separation of the genus from *Rosellinia*. We agree with Smith and Hyde (2001) in keeping separate status for *Astrocystis* and although we recognize that these genera are closely related, we do not follow Ju and Rogers (1990) in their merger of the genera and consider in particular the *Acanthodochium* anamorph of *Asterocystis* to provide strong evidence for their separation (Whalley 1996; Smith and Hyde 2001).

Two species of *Fasciatispora*, *F. nypae* K.D. Hyde and *F. lignicolla* Alias, E.B.G. Jones & Kuthub. are reported as marine inhabitants, although other members of the genus are non-marine in their distribution (Jones et al. 2009). Species of *Biscogniauxia*, *Daldinia*, *Hypoxylon*, *Kretschmaria*, *Rosellinia* and *Xylaria* were not included by Jones et al. (2009) in their classification of marine fungi and their inclusion here is open to debate. Certainly a wide range of these taxa were recorded for Kuala Selangor Nature Park in woodland immediately adjacent to the mangrove (Whalley et al. 1999) and several species of *Hypoxylon* have been found in a mangrove environment in Thailand (Whalley et al. 1994). It is noteworthy that representatives of these genera are reported to occur commonly as endophytes, especially in tropical plants (Whalley 1996).

Halorosellina oceanica was originally described from Florida, USA, by Schatz (1988) as *Hypoxylon oceanicum* Schatz for an ascomycete fungus growing on *Rhizophora* mangle in a Floridian mangrove, although later considered to be a member of the Xylariaceae, it was not accepted as a *Hypoxylon* species and the genus *Halorosellinia* was, therefore, erected to accommodate it (Whalley et al. 2000). *Halorosellina oceanica* is a widespread mangrove inhabitant originally recorded as a *Rosellinia* sp from Sierra Leone by Aleem (1980). It has since been reported from the Seychelles (Hyde and Jones 1988; Jones and Hyde 1990), Brunei (Hyde 1988), N. Sumatra (Hyde 1989a), Thailand (Hyde 1989b; Hyde et al. 1990), Malaysia (Jones and Tan 1987; Jones and Kuthubutheen 1989; Alias et al. 1995), Philippines (Jones et al. 1988), Singapore (Tan et al. 1989), South Africa (Steinke and Jones 1993), Taiwan (Ju 1990), Hong Kong (Vrijmoed et al. 1994) and Australia (Hyde 1990). It has recently been reported as one of the most frequently collected marine fungi in Malaysian mangroves with an occurrence of 6.2% of all species recorded (Alias et al. 2010).

Camillea selangorensis, described from the Kuala Selangor Nature Park, Selangor, Malaysia, was only the second species of the genus to be recorded from outside of the Americas and West Africa. It has also been recorded from the edge of a mangrove in Phuket Island, Thailand (Whalley et al. 1999). The Kuala Selangor Nature Park was also the type locality for a second species of *Camillea*, *C. malaysianensis* M.A. Whalley (Whalley et al. 1999). Kuala Selangor Nature Park contains a wide range of habitats including mangrove forest, secondary forest, estuary, mud flats and a brackish water lake system. From the late 1980s, the water table in the park has been dropping and mangrove species are declining and being replaced by more dry-adapted species. The two species of *Rhizophora*, four species of *Bruguiera*, three species of *Avicennia* and two species of *Sonneratia*, which occur in the mangrove, have decreasing numbers of representatives in the secondary forest. As these taxa decreased, the presence of *Ficus microcarpa* L. has increased and this is the host for *C. selangorensis*. The increase in number of specimens of *C. selangorensis* from surveys in 1993 and 1997 may well be the result of the increase in the numbers of the *Ficus*. Whalley et al. (2002) recorded two species of *Biscogniauxia*, two species of *Camillea*, Daldinia *eschscholzii* (Ehrenb.:Fr.) Rehm, fourteen species of *Hypoxylon*, including one considered likely to be previously undescribed, three species of *Nemania*, a small-spored

Kretschmaria deusta (Hoffm.:Fr.) P.M.D. Martin, two species of *Rosellinia* and ten species of *Xylaria*. In 1993, *H. haematostroma* Mont. and *H. subgilvum* Berk. & Broome were the most frequently recorded species but by 1997, *C. selangorensis*, *H. lenormandii* Berk. & M.A. Curtis apud Berk. and *X. cubensis* (Mont.) Fr. had become more dominant, which is probably a result of the drier conditions. *Halorosellinia oceanica* was confined to the true mangrove forest. It is probable that one of the *Nemania* species found will, with re-examination, prove to be *N. maritima*. In a preliminary study of Xylariaceae in mangrove forest or bordering mangroves in Malaysia and Thailand, Whalley et al. (1994) recorded *Halorosellinia oceanica* (as *Hypoxylon oceanicum*), *Hypoxylon bovei* var. *microspora* Speg.var.*microspora* J.H. Miller, *H. subgilvum* and *H. lenormandii* (as *H. nectriodeum* Sacc. & Trott.). Thus H. *lenormandii* and *H. subgilvum* would appear to be regular inhabitants of the mangrove fringe in S.E. Asia.

12.1.2 The Xylariaceae as Endophytes of Mangrove Plants

In their examination of endophytes from *Bruguiera gymnorrhiza* in the Shira River Basin, Iriomote Is. (Okane et al. 2001) reported on xylariaceaeous fungi and Pang et al. (2008) isolated an endophytic *Xylaria* species from *Kandelia candel* (Rhizophoraceae) in Mai Po Nature Reserve, Hong Kong. Although members of the Xylariaceae are common endophytic inhabitants of tropical plants (Whalley 1996; Rodrigues and Petrini 1997), they do not appear to be common or universally present in mangrove plants unlike the situation in other tropical plants. In our studies on Thai plants and their endophytes (Mekkamol 1998; Ruchikackorn 2005; Chareprasert et al. 2006) and those of Lumyong et al. (2004) and in Hong Kong by Photita et al. (2001), the Xylariaceae are indeed common and well represented. Promputtha et al. (2007) in their study of endophytic fungi from *Magnolia liliifera* recorded a range of species of *Xylaria* and other members of the Xylariaceae and indicated that *Xylaria* species from leaves were endophytic, although teleomorphic material for these species was not present even in the immediate vicinity. We also believe that many of the leaf-inhabiting *Xylaria* species are endophytic and under suitable environmental conditions develop their teleomorph on fallen leaves. *Xylaria aristata* Mont. and *X. amphithele* F.S.M. Gonzalez & J.D. Rogers are species found on leaves in Thai forests but their occurrence is rare (Thienhirun and Whalley 2004). We suggest that this is a direct reflection of the nutritional status of the leaf and its immediate environment, which either allows or prevents continued development of the fungus. It is also significant that these and other leaf-inhabiting teleomorphic *Xylaria* species are small and fragile and usually, if found, only contain a few perithecia or more frequently are immature (Thienhirun 1997). The seed-inhabiting species of *Xylaria* are different and in our view are a result of the seed or fruit acting as a "bait" once the fruit or seed is in the litter/soil layer (Rogers 1979a, b; Whalley 1996). Certainly *X. carpophila* Pers ex Fr., which occurs on fallen beech mast in temperate countries, there is clear evidence to show that this is the case. It is also noteworthy that *Xylaria magnolia* var. *microspora* J.D. Rogers, Ju and Whalley, which is common on fallen fruits of *Magnolia* in Northern Thailand (Rogers and Ju 2002), was not isolated in the Hong Kong study. This perhaps provides additional support that the fruits act as baits and that *X. magnolia* var. *microspora* is not endophytic. Interestingly Sakayaroj et al. (2010), when reporting on endophytic assemblages from the tropical seagrass *Enhalus acoroides* in Thailand, failed to isolate *Xylaria* or other xylariaceaous fungi and noted that this was in agreement with studies on seagrasses in Bermuda, USA, Hong Kong, the Philippines and Puerto Rico. Kumaresan and Suryanarayanan (2001) investigated seven dominant mangrove species of an estuarine forest in south India and did not isolate any *Xylaria* species or other xylariaceaous taxa. They reported that many endophytes were common to more than one host and that the endophyte assemblage of each mangrove species was dominated by different endophytic species. In an earlier study of endophytes of halophytic plants from the same mangrove, no Xylariaceae were isolated with species of *Colletotrichum*, *Phomopsis*, *Phyllosticta* and *Sporormiella* being ubiquitous (Suryanarayanan and Kumaresan 2001).

In extensive surveys of three mangrove sites in Thailand, selected for their differences in dominant mangrove tree species and different environmental parameters, it was found that the presence of endophytic xylariaceous taxa also differed (Chareprasert et al. 2010). Thus *Daldinia eschscholtzii* was isolated from *Avicennia alba* Bl., *Rhizophora apiculata* Bl. and *R. mucronata* Poir. from the Chanthaburi site and from *R. apiculata*, *R. muccronata*, *Xylocarpus granatum* Koen. and *X. moluccensis* Roem from the Ranong mangrove. The percentage colonization frequency ranged from 0.67–3.33% at the Chanthaburi site to 6.00–7.33% at Ranong. *Daldinia eschscholtzii* is a common endophyte in Thailand and S.E. Asia and has been isolated from a wide range of plant species. It proved to be a dominant isolate from teak trees in Chiang Mai Province, Northern Thailand (Mekkamol 1998) and was a frequent isolate from teak in the vicinity of Bangkok (Chareprasert et al. 2006). Endophytic species of *Xylaria* were generally less widespread, with only *Xylaria* sp. 1 occurring on more than a single mangrove species, being isolated from nine out of thirteen mangrove species investigated. The remaining *Xylaria* species were restricted to one or two hosts (Chareprasert et al. 2010). The species of Xylariaceae recorded, although regularly present, were always in the minority, with *Phyllosticta* as the dominant genus in all mangrove plants from the Chanthaburi site and also in two plant species from Prachuap Khiri Khan. The dominant fungi in almost all mangrove species at the Ranong location were different for each host plant (Chareprasert et al. 2010), which is in agreement with the findings of studies on Indian mangrove endophytes where the dominant endophytes were also different for each host plant in the mangrove community (Kumaresan and Suryanarayanan 2001). In common with other reports on endophytic fungi as a source of bioactive compounds (e.g. Huang et al. 2007), many of the culture extracts exhibited a range of antibacterial activities, with *Xylaria* sp. 1, which was isolated from *A. ilicifolius* leaves, showing strong inhibition of both Gram- positive and Gram-negative bacteria. Additionally, crude extracts of many of the endophytic fungal isolates exhibited anticancer activity by the MTT assay with *Xylaria* species 1, 3 and 5 exhibiting cytotoxicity against the Acute T cell leukaemia Jurkat cell line and *D. eschscholzii* proving to be strongly inhibitory to the colorectal adenocarcinoma SW620 cell line.

In their recent review of marine fungi from Malaysian mangroves, Alias et al. (2010) reported several Xylariaceae as common species of fungi occurring in this environment at between 5 and 9% occurrences. In addition to *H. oceanica*, other Xylariaceae included *N. carbospora* and *F. selangorensis*, *Fasciatispora lignicola* and *F. nypae*, *N. maritima* and *A. nypae*, *A. nypensis* and *A. nypicola*. *Nipicola carbospora* was reported to be cosmopolitan. Thus, from our own observations and those referred to here, members of the Xylariaceae are regular inhabitants of mangroves and their fringes.

12.2 Taxonomic Aspects

Halorosellinia oceanica (Schatz) Whalley, E.B.G. Jones, K.D Hyde & Læssøe. Mycological Research 104: 370 (2000).

Pseudostromata seated on decorticated wood, occasionally embedded at the base, pulvinate to hemispherical, 0.4–0.8 mm diam., single, in clusters of up to 30 uniperitheciate pseudostromata, linear to suborbicular, surface leathery in fresh material, when young covered with a whitish hyphal layer bearing the anamorph (Ju 1990), at maturity black, generally with conspicuous ascomatal projections. In section pseudostromata comprising host cells filled with light-brown fungal cells in the form of *textura globulosa* or amorphous black fungal material. *Ascomata* immersed in pseudostroma, subglobose to hemispherical, soft to leathery, black, *ostioles* papillate. *Peridium* 25–35 μm wide, comprising tissue of *textura porrecta*, fusing at the outside with the pseudostromata. *Paraphyses* 2–2.5 μm wide at the base, abundant, persistent, remotely septate. *Asci* eight-spored, 177–219 μm, spore-bearing part (112)135–140 μm, stipe 37–79 μm, cylindrical, unitunicate, subapical apparatus dark blue in Melzer's iodine reagent, tapering with a distinct apical rim, (4.7–)5.6–6.6 × 4.2–4.7 μm. *Ascospores* uniseriate to obliquely uniseriate or partially biseriate at the upper end of the ascus, dark grey-olive to opaque brown, more or less inequilaterally ellipsoid, ventral side varying in degree of convex curvature, upper end broadly rounded, lower end slightly pointed, one-celled throughout ascospore development, (17.9–)18.7–26(–28) × 7.5–13(–13.5) μm, biguttulate, wall smooth and relatively thick, without appendages or loosening perispore, germination slit usually clearly seen on the ventral side, straight, conspicuous, ½ to total length of spore (Fig. 12.1a–f).

Ascospores germinated freely to form strongly growing white to whitish grey colonies similar to those produced by other xylariaceous taxa such as *Xylaria*, *Kretzschmaria*, *Nemania* but no anamorphs detected. According to Ju (1990), there is an anamorphic form with strongly geniculate conidiophores, which is typical for a number of xylariaceous genera.

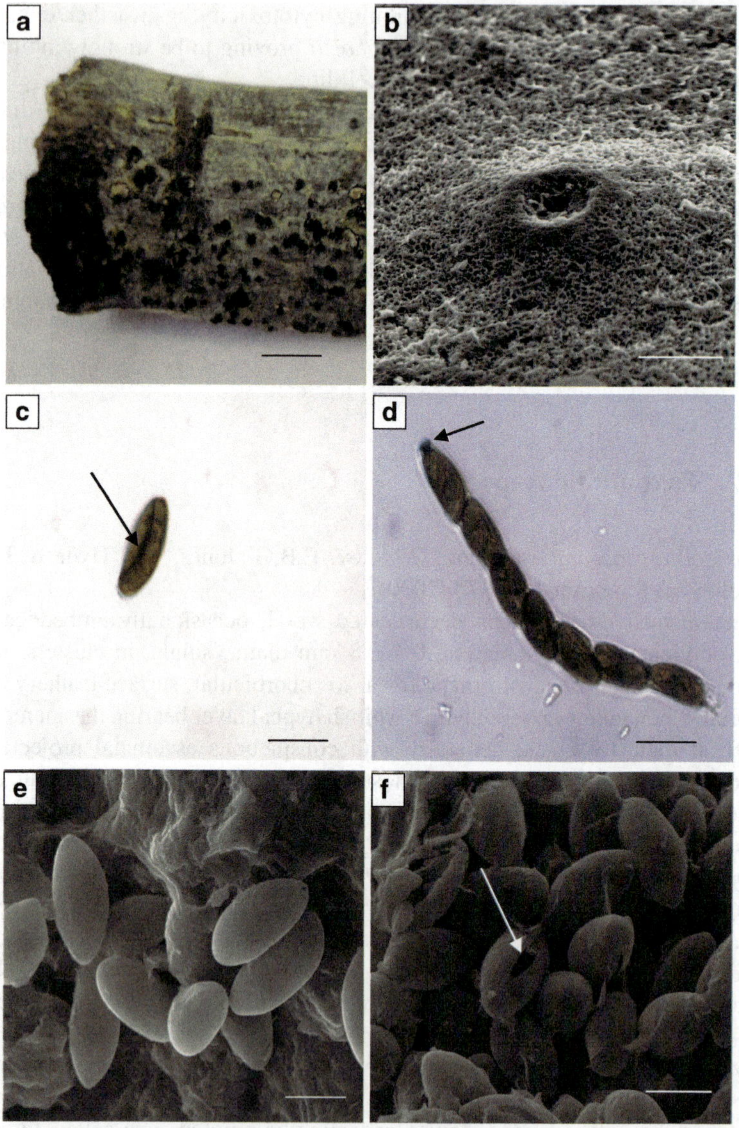

Fig.12.1 *Halorosellinia oceanica.* (**a**) Stromata. (**b**) Ostiolum. (**c**) Ascospore with germ slit (*arrow*). (**d**) Ascus with apical apparatus (*arrow*). (**e**) Ascospores. (**f**) Ascospores with germ slit (*arrow*). Bar markers: (**a**) 1 cm; (**b**) 100 μm; (**c–f**) 15 μm

Camillea selangorensis M.A. Whalley, Whalley & E.B.G. Jones. Sydowia 48:246 (1996).

Stromata erumpent through bark, circular to orbicular, 7–16 × 6–10 mm or occasionally elongated, 11–29 × 6–10 mm c.2–3 mm thick, with applanate to convex apex, surrounded by a slightly raised rim, black. *Ostioles* finely papillate becoming punctate with age. *Perithecia* packed in a brownish-black, brittle, entostroma, basally seated. *Asci* cylindrical, 106–146 × 3.5–5.8 μm, sp. p. 85–103 μm, stipe 17–53 μm. *Apical* apparatus prominent, dark blue in Melzer's iodine reagent, more or less rhomboid, 2.5–3.8 μm tall and 3.5–5.8 μm wide. *Paraphyses* long, tread-like, unbranched, 152–182 μm long and 7.5 μm at their widest point, persistent. *Ascospores* obliquely uniseriate, hyaline or dilute yellow, one-celled, ellipsoid inequilateral. Minutely warted by light microscopy, strongly verrucose by scanning electron microscopy, 10.0–13.8 × 3.8–6.3 μm, germ slit absent. Anamorph not known (Fig. 12.2a–d).

Camillea malaysianensis M.A. Whalley Kew Bulletin 54: 716 (1999).

Stromata erumpent through bark, orbicular to elongate, 4–5 × 2–3 cm, c. 1.5–2 mm thick, with applanate to convex apex, black. *Ostioles* papillate. *Perithecia* basally seated in black, brittle entostroma, 1 mm high and 0.2–0.4 mm in diam. *Asci* cylindrical, 8-spored, 129–179 × 7.5–11.25 μm, spp. 88–126 μm, stipe 41–53 μm. Apical apparatus prominent, dark blue in Melzer's iodine reagent, more or less rhomboid, 5 μm tall and 3.75 μm wide. Ascospores obliquely uniseriate, hyaline or dilute yellow, one-celled, ellipsoidal, smooth by light microscopy, distinct poroid ornamentation by scanning electron microscopy, 8.75–15 × 5–6.25 μm, germ slit absent. Anamorph unknown.

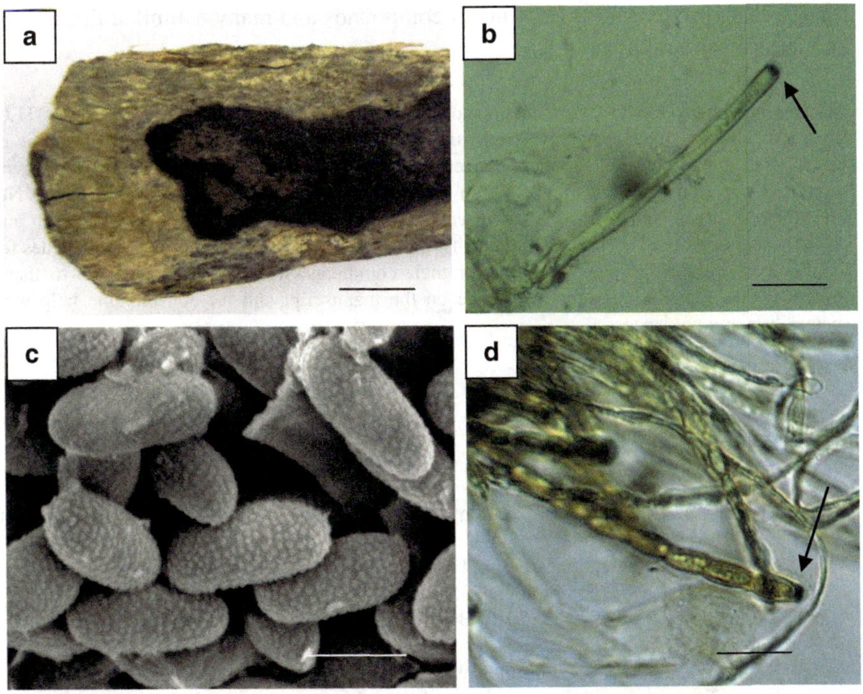

Fig.12.2 *Camillea selangorensis*. (**a**) Stromata. (**b**) Immature ascus with apical apparatus (*arrow*). (**c**) Ascospores. (**d**) Ascus with apical apparatus (*arrow*). Bar markers: (**a**) 2 cm; (**b, d**) 20 μm; (**c**) 10 μm

Camillea malaysianensis is superficially similar to *C. tinctor* (Berk.) Læssøe, J.D. Rogers & Whalley but differs in its considerably smaller ascospores and also does not produce the yellow staining of the wood which is characteristic for that species and which usually appears to be present in collections from Malaysia and Thailand. Unlike *C. malaysiensis*, which so far has only been recorded from the type locality, *C. tinctor* is widespread occurring in the Americas, West Africa, Thailand, Malaysia, Singapore and Papua New Guinea (Læssøe et al. 1989).

12.3 Conclusions

Halorosellinia oceanica is a manglicolous member of the Xylariaceae, which appears to be restricted to this habitat. Other taxa which might also be included here are several species of *Anthostomella*, *Nemania maritima* and *Nipicola carbospora* and *N. lignicola*. *Camillea selangorensis* and *C. malaysianensis* have to date only been reported from forest immediately adjacent to mangroves. *Hypoxylon subgilvum* and *H. lenormandii* are regular inhabitants in the mangrove fringe in S.E. Asia. Species of *Xylaria* and *D. eschscholzii* occur as endophytes in a variety of mangrove species. Many endophytic isolates from mangroves, including Xylariaceae, produce a range of novel compounds and many exhibit antimicrobial and anticancer activities.

Acknowledgements We thank Chulalongkorn University, the University of Malaya, BIOTEC, Liverpool John Moores University, the British Council and the government of the Republic of Egypt for financial and logistic support over many years. We gratefully acknowledge financial support of the Thai Research Fund through the Royal Golden Jubilee PhD Program (Grant No. PHD/0206/2548) to Sukanyanee Chaeprasert and Associate Professor Prakitsin Sihanonth and partially supported by CU Graduate School Thesis Grant. We also thank our many colleagues for their help in collecting specimens and for their comments. It is a special pleasure to thank Professor E.B. Gareth Jones for his advice on the manuscript and for considerable help with species lists and guidance.

References

Aleem AA (1980) Distribution and ecology of marine fungi in Sierra Leone (Tropical West Africa). Bot Mar 22:679–688

Alias SA, Kuthubutheen AJ, Jones EBG (1995) Frequency of occurrence of fungi on wood in Malaysian mangroves. Hydrobiologia 295:97–106

Alias SA, Zainuddin N, Jones EBG (2010) Biodiversity of marine fungi in Malaysian mangroves. Bot Mar 53:545–554

Chareprasert S, Piapukiew J, Thienhirun S, Whalley AJS, Sihanonth P (2006) Endophytic fungi of teak leaves *Tectona grandis* L. and rain tree leaves *Samanea saman*, Merr. World J Microbiol Biotechnol 22:481–486

Chareprasert S, Piapukiew J, Whalley AJS, Sihanonth P (2010) Endophytic fungi from mangrove plant species of Thailand: their antimicrobial and anticancer potentials. Bot Mar 53:555–564

Daferner M, Mensch S, Anke T, Sterner O (1999) Hypoxysordarin, a new sordarin derivative from *Hypoxylon croceum*. Z Naturforsch C 54:474–480

Edwards RL, Jonglaekha N, Kshirsagar A, Maitland DJ, Mekkamol S, Nugent LK, Phosri C, Rodtong S, Ruchichachorn N, Sangvichien E, Sharples GP, Sihanonth P, Suwannasai N, Thienhirun S, Whalley AJS, Whalley MA (2003) The Xylariaceae as phytopathogens. Recent Res Dev Plant Sci 1:1–19

Huang H, She Z, Lin Y, Vrijmoed LLP, Lin W (2007) Cyclic peptides from an endophytic fungus obtained from a mangrove leaf (*Kandelia candel*). J Nat Prod 70:1696–1699

Hyde KD (1988) Studies on the tropical marine fungi of Brunei. Bot J Linn Soc 98:135–151

Hyde KD (1989a) Intertidal mangrove fungi from north Sumatra. Can J Bot 67:3078–3082

Hyde KD (1989b) *Caryospora mangrovei* sp. nov. and notes on marine fungi from Thailand. Trans Mycol Soc Jpn 30:333–341

Hyde KD (1990) Intertidal fungi from warm temperate mangroves of Australia, including *Tunicalispora australiensis*, gen. et sp. nov. Aust Syst Bot 3:711–718

Hyde KD (1992) Fungi from *Nypa fruticans*, *Nipicola carbospora* gen. et sp.- nov. (Ascomycotina). Cryptogam Bot 21:330–332

Hyde KD, Jones EBG (1988) Marine mangrove fungi. PSZNI Mar Ecol 9:15–23

Hyde KD, Chalermpongse A, Boonthavikoon T (1990) Ecology of intertidal fungi at Ranong mangrove, Thailand. Trans Mycol Soc Jpn 31:17–27

Jones EBG, Abdel-Wahib MA (2005) Marine fungi from the Bahamas Islands. Bot Mar 48:356–364

Jones EBG, Hyde KD (1990) Observations on poorly known mangrove fungi and a nomenclatural correction. Mycotaxon 37:197–201

Jones EBG, Kuthubutheen AJ (1989) Malaysian mangrove fungi. Sydowia 42:160–169

Jones EBG, Tan TK (1987) Observations on manglicolous fungi from Malaysia. Trans Br Mycol Soc 89:390–392

Jones EBG, Uyenco FR, Follosco MP (1988) Fungi on driftwood collected in the intertidal zone from the Philippines. Asian Mar Biol 5:103–106

Jones EBG, Stanley SJ, Pinruan U (2008) Marine endophytes sources of new chemical natural products: a review. Bot Mar 51:163–170

Jones EBG, Sakayaroj J, Suetrong S, Somrithipol A, Pang KL (2009) Classification of marine Ascomycota, anamorphic taxa and Basidiomycota. Fungal Divers 35:1–187

Ju Y-M (1990) Studies of Xylariaceae from Taiwan. M.S. Thesis, Washington State University, Pullman, Washington State, USA

Ju Y-M, Rogers JD (1990) Astrocystis reconsidered. Mycologia 82:342–349

Ju Y-M, Rogers JD (1996) A revision of the genus *Hypoxylon*. APS Press, Minnesota

Ju Y-M, Rogers JD (2002) The genus *Nemania* (Xylariaceae). Nova Hedwigia 74:75–120

Kohlmeyer J, Kohlmeyer E (1979) Marine mycology: the higher fungi. Academic Press, New York

Kumaresan V, Suryanarayanan TS (2001) Occurrence and distribution of endophytic fungi in a mangrove community. Mycol Res 105:1388–1391

Læssøe T, Rogers JD, Whalley AJS (1989) *Camillea*, *Jongiella* and light-spored species of *Hypoxylon*. Mycol Res 93:121–155

Lin Y, Wu X, Feng S, Jiang G, Luo J, Zhou S, Vrijmoed LLP, Jones EBG, Krohn K, Steingrover K, Zsila F (2001) Five unique compounds: xyloketals from mangrove fungus *Xylaria* sp. from the South China Sea. J Org Chem 66:6252–6256

Liu X, Xu F, Zhang Y, Liu L, Huang H, Cai X, Lin Y, Chan W (2006) Xyloketal from the mangrove endophytic fungus *Xylaria* sp. 2508. Russ Chem Bull 55:1091–1092

Lu B, Hyde KD (2000) A world monograph of *Anthostomella*. Fungal Divers 35:1–187

Lumyong, S, Lumyong P, Hyde KD (2004) Endophytes . In: Jones EBG, Tantichareon M, Hyde KD (eds) Thai fungal diversity. BIOTEC, Thailand, pp197–205

Lumbsch HT, Huhndorf SM (2007) Outline of Ascomycota – 2007. Myconet 13:1–58

Mekkamol S (1998) Endophytic fungi of *Tectona grandis* L. (teak). Ph.D. Thesis, Liverpool John Moores University, Liverpool, UK

Nugent LK, Sihanonth P, Thienhirun S, Whalley AJS (2004) *Biscogniauxia*: a genus of latent invaders. Mycologist 20:111–114

Okane I, Nakagiri A (2001) Assemblages of endophytic fungi on *Bruguiera gymnorrhiza* in the Shiira River basin, Iriomote Is. J Inst Ferment Osaka Res Commun 20:41–49

Pang KL, Vrijmoed LLP, Goh TK, Plaingam N, Jones EBG (2008) Fungal endophytes associated with Kandelia candel (Rhizophoraceae) in Mai Po Nature Reserve, Hong Kong. Bot Mar 51:171–178

Pan JY, Jones EBG, She ZY, Ling YC (2008) Review of bioactive compounds from fungi in the South China Sea. Bot Mar 51: 179–190

Photita W, Lumyong S, Lumyong P, Ho WH, McKenzie EHC, Hyde KD (2001) Fungi on *Musa acuminata* in Hong Kong. Fungal Divers 6:99–106

Promputtha I, Lumyong S, Dhanasekaran V, McKenzie EHC, Hyde KD, Jeewan R (2007) A phylogenetic evaluation of whether endophytes become saprotrophs at host senescence. Microbial Ecol 55:579–590

Rodrigues KE, Petrini O (1997) Biodiversity of endophytic fungi in tropical regions. In: Hyde KD (ed) Biodiversity of tropical microfungi. Hong Kong University Press, Hong Kong, pp 57–69

Rogers JD (1979a) The Xylariaceae: systematic, biological and evolutionary aspects. Mycologia 71:1–42

Rogers JD (1979b) *Xylaria magnolia* sp. nov. and comments on several other fruit inhabiting species. Can J Bot 57:941–9445

Rogers JD, Ju Y-J (2002) A reassessment of the *Xylaria* on Magnolia fruits. Sydowia 54:91–97

Ruchikackorn N (2005) Endophytic fungi of Cassia fistula L. Ph.D Thesis, Liverpool John Moores University, Liverpool, UK

Rukachaisirkul V, Khamthong N, Sukpondma Y, Pakawatchai C, Phongpaichit S, Sakayaroj J, Kirtikara K (2009) An [11]cytochalasin derivative from the marine-derived fungus *Xylaria* sp. PSU-F100. Chem Pharm Bull (Tokyo) 57(12):1409–1411

Sakayaroj J, Preedanon S, Supaphon O, Jones EBG, Phongpaichit S (2010) Phylogenetic diversity of endophytic assemblages associated with the tropical seagrass *Enhalus acoroides* in Thailand. Fungal Divers 42:27–45

Schatz S (1988) *Hypoxylon oceanicum* sp. nov. from mangroves. Mycotaxon 33:413–418

Smith GJD, Hyde KD (2001) Fungi from palms. XLIX. *Astrocystis*, *Biscogniauxia*, *Cyanopulvis*, *Hypoxylon*, *Nemania*, *Guestia*, *Rosellinia* and *Stillbohypoxylon*. Fungal Divers 7:89–127

Stadler M, Hellwig V (2005) Chemotaxonomy of the Xylariaceae and remarkable bioactive compounds from Xylariales and their associated asexual stages. Recent Res Dev Phytochem 9:41–93

Steinke TD, Jones EBG (1993) Marine and mangrove fungi from the Indian ocean coast of South Africa. S Afr J Bot 67:2686–2691

Suryanarayanan TS, Kumaresan V (2001) Endophytic fungi of some halophytes from an estuarine mangrove forest. Mycol Res 104:1465–1467

Tan TK, Leong WF, Jones EBG (1989) Succession of fungi on wood of *Avicennia alba* and *A. lanata* in Singapore. Can J Bot 67:2680–2691

Thienhirun S (1997) A preliminary account of the Xylariaceae of Thailand. Ph.D Thesis, Liverpool John Moores University, Liverpool, UK

Thienhirun S, Whalley AJS (2004) Xylariaceae. In: Jones EBG, Tantichareon M, Hyde KD (eds) Thai fungal diversity. BIOTEC, Thailand, pp 71–77

Vrijmoed LLP, Jones EBG, Hyde KD (1994) Observations on subtropical mangrove fungi in the Pearl River estuary. Acta Scientiarum Naturalium Universitalis Sunyatseni 33:78–85

Whalley AJS (1996) The xylariaceous way of life. Mycol Res 100:897–922

Whalley AJS, Edwards RL (1995) Secondary metabolites and systematic arrangement within the Xylariaceae. Can J Bot 73(suppl 1):S802–S810

Whalley AJS, Jones EBG, Alias SA (1994) The Xylariaceae (ascomycetes) of mangroves in Malaysia and South East Asia. Nova Hedwigia 59:207–218
Whalley MA, Whalley AJS, Jones EBG (1996) *Camillea selangorensis* sp. nov. from Malaysia. Sydowia 48:145–151
Whalley MA, Whalley AJS, Thienhirun S, Sihanonth P (1999) *Camillea malaysiensis* sp. nov. and the distribution of *Camillea* in Southeast Asia. Kew Bull 54:715–722
Whalley AJS, Jones EBG, Hyde KD, Læssøe T (2000) *Halorosellinia* gen. nov. to accommodate *Hypoxylon oceanicum*, a common mangrove species. Mycol Res 104:368–374
Whalley MA, Whalley AJS, Abdulla F (2002) The Xylariaceae of Kuala Selangor Nature Park. Malay Nat J 56:5–13
Xu F, Zhang Y, Wang J, Pang J, Huang C, Wu X, She Z, Vrijmoed LLP, Jones EBG, Lin Y (2008) Benzofuran derivatives from the mangrove endophytic fungus *Xylaria* sp. (#2508). J Nat Prod 71:1251–1253

Chapter 13
Diversity and Distribution of Marine Fungi on *Rhizophora* spp. in Mangroves

Vemuri Venkateswara Sarma

Contents

Abstract *Rhizophora* spp. occurring in mangrove habitats are excellent hosts for marine fungi. The morphological adaptations of this host plant provide suitable niche for the marine fungi. This review deals with diversity and ecology of marine fungi occurring on *Rhizophora* spp. Two hundred and one fungal species have been recorded on *Rhizophora* plant species consisting of more than ten fungal species specific to this host. In mangroves, no other genus other than *Rhizophora* accommodates as many as 201 marine fungi. *Rhizophora* plant species are highly suitable hosts as they show a variety of niches with vertical zonation, specificity, preferential colonization and succession. Thus this host could be considered as one of the important hosts to investigate marine fungi. The current review provides updated

---

This article is dedicated to Dr. K.D. Hyde for his extensive contribution to the studies and our understanding on marine fungi in mangroves.

V.V. Sarma (✉)
Department of Biotechnology, School of Life Sciences, Pondicherry University, Kalapet, Pondicherry 605 014, India
e-mail: sarmavv@yahoo.com

C. Raghukumar (ed.), *Biology of Marine Fungi*,
Progress in Molecular and Subcellular Biology 53,
DOI 10.1007/978-3-642-23342-5_13,

information based on the available literature with checklists on diversity and distribution of fungi on *Rhizophora* spp.

13.1 Introduction

Mangroves are the important "land builders" in tropical and subtropical coastal regions. They are considered "interface" habitats of upland terrestrial and coastal estuarine ecosystems (Lugo and Snedaker 1974). They reduce tidal currents, cause extensive deposition of mud and silt, and provide surfaces for attachment of marine organisms (Odum 1971). Their importance lies in generation of leaf and other detritus, which supply nutrients for primary production and thus major energy source for fisheries (Heald and Odum 1970). Mangroves with their rich load of organic detritus support a large number of animal communities and are breeding and nursery grounds for shell fisheries (Jones and Alias 1997). Marine Fungi play an important role in the production of organic detritus in mangroves (Kohlmeyer and Kohlmeyer 1979). Their importance in nutrient regeneration cycles as decomposers of dead and decaying organic matter in marine ecosystems has been well documented (Fell and Master 1980; Raghukumar et al. 1994; Hyde et al. 1998).

Marine fungi are not restricted to any particular taxonomic group but are rather an ecologically and physiologically defined group. They comprise an estimated 1,500 species, excluding those that form lichens (Hyde et al. 1998), although 444 formerly described higher marine fungi have been reported (Hyde et al. 2000). Recently Jones et al. (2009) reported 530 marine fungi.

13.1.1 Substrata for Marine Fungi

Marine fungi grow on diverse substrata including wood, sediments, algae, dead corals, calcareous tubes of mollusks, decaying leaves, seedlings, prop roots and pneumatophores of mangroves, seagrasses and living animals, and in the guts of crustaceans (Kohlmeyer and Kohlmeyer 1979; Hyde 1996). Wood in the open ocean (drift wood) and decaying/decomposing substrata in mangrove habitats are the two predominant substrata, which support marine fungi. Mangrove plants are categorized as "typical mangrove plants" and "mangrove associated plants." There are around 30 "typical mangrove plants" and more or less equal number of "mangrove associated plants" that occur in mangroves forests around the world.

13.1.2 Marine Fungi on Mangroves

Kohlmeyer and Kohlmeyer (1979) listed 42 fungi from mangroves. After a lapse of 24 years roughly around 600 fungi have been reported from mangroves, which include marine, estuarine and terrestrial fungi (Schmidt and Shearer 2003). Terrestrial fungi and lichens occupy the aerial parts of mangrove plants, whereas the marine fungi occur at lower parts such as trunks and roots, which are permanently or intermittently inundated with water (Kohlmeyer 1969; Kohlmeyer and Kohlmeyer 1979). There will be an interface and an overlap of marine and terrestrial fungi occurring at the high tide mark (Kohlmeyer 1969; Kohlmeyer and Kohlmeyer 1979). The interest of mycologists, however, has been more intense on marine fungi occurring in mangroves with special reference to hard substrata such as wood, roots and prop roots.

13.1.3 Marine Fungi on Rhizophora

Among the "typical mangrove plants" species belonging to genera *Aegiceras*, *Avicennia*, *Bruguiera*, *Ceriops*, *Excoecaria*, *Kandelia*, *Lumnitzera*, *Nypa*, *Rhizophora*, *Sonneratia* and *Xylocarpus* are the most prominent. Within these species *Avicennia* and *Rhizophora* are not only found most commonly but are also abundant throughout the world. Between these two genera several interesting fungi are known from *Rhizophora* spp. Further, the species belonging to this genus have been excellent experimental hosts for several biodiversity and ecological investigations including nutrient recycling, frequency of occurrence, vertical zonation, substrate and seasonal occurrence. Several studies have been conducted on biodiversity and ecology of fungi on *Rhizophora*. Hence it is worthwhile to review the diversity and geographical distribution of fungi reported from *Rhizophora* spp. worldwide.

13.2 Diversity

Kuthubutheen (1981) investigated the phylloplane fungi of *Rhizophora* spp. and reported: *Cephalosporium*, *Cladosporium oxysporum*, *Codinaea simplex*, *Corynespora*, *Fusarium*, *Pestalotiopsis guepini*, *Pestalotiopsis versicolor*, *Zygosporium masoni*. Similarly, in an earlier investigation on the leaves of *Rhizophora*, Fell et al. (1975) found that the majority of the genera were ubiquitous saprophytes (e.g., *Aspergillus*, *Aureobasidium*, *Cladosporium*, *Curvularia*, *Cylindrocephalum*, *Drechslera*, *Fusarium*, *Myrothecium*, *Nigrospora*, *Penicillium*, *Pestalotia*, *Phyllosticta*, *Trichoderma* and *Verticilium*).

More than 201 lignicolous filamentous fungi belonging to 139 genera are known from *Rhizophora* spp. (Table 13.1). Thus, approximately one-third of total number of marine fungi are recorded on *Rhizophora* spp. These comprise 126 species of ascomycetes in 85 genera; 5 species of basidiomycetes in 3 genera; 70 species of anamorphic fungi in 51 genera of which 10 species in 9 genera are coelomycetes; 60 species in 42 genera are hyphomycetes. A total of 51 bitunicate ascomycetes and

Table 13.1 Lignincolous filamentous marine fungi recorded on *Rhizophora* spp. from Neotropics, Indian Coasts, South East Asia and other sites

S. No.	Name of the species	Neotropics (N & S America)[1]	Belize, Fiji, Hawai[2]	Mandai, Siingapore[3]	Ranong, Thailand[4]	Kuala Selangor, Malaysia[5]	Morib, Malaysia[6]	Irimoto Island, Japan[7]	North Sumatra[8]	Queensland, Austrlaia[9]	Pichavaram, TN, India[10]
1	*Acrocordiopsis patilii* Borse & K.D. Hyde	–	–	–	+	–	–	–	–	–	–
2	*Aigialus grandis* Kohlm. & S. Schatz	–	+	+	+	–	+	+	+	+	+
3	*A. mangrovei* Borse	–	+	+	+	–	–	–	–	–	+
4	*A. parvus* S. Schatz & Kohlm.	–	–	+	+	+	–	–	–	–	+
5	*A. striatispora* K.D. Hyde	–	–	–	+	–	–	–	–	–	–
6	*Aigialus* sp.	–	+	–	–	–	–	–	–	–	–
7	*Aniptodera chesapeakensis* Shearer & M.A. Mill.	–	–	+	+	–	–	–	+	–	+
8	*A. longispora* K.D. Hyde	–	+	–	+	–	–	+	–	+	–
9	*A. haispora* Vrijmoed, K.D. Hyde & E.B.G. Jones	–	–	–	–	–	–	–	–	–	–
10	*A. mangrovei* K.D. Hyde	–	–	–	+	+	–	–	–	–	–
11	*Aniptodera* sp.	–	–	–	–	+	–	–	–	–	–
12	*Antennospora quadricornuta* (Cribb & J.W. Cribb) T.W. Johnson	+	+	–	+	+	+	–	–	–	+
13	*Anthostomella* sp.	–	–	–	–	–	–	–	–	–	–
14	*Ascocratera manglicola* Kohlm.	–	+	–	+	–	+	–	–	–	+
15	*Ascocratera* cf. *manglicola* Kohlm.	–	–	–	+	–	–	–	–	–	–
16	*Astrosphaeriella striataspora* (K.D. Hyde) K.D. Hyde ≡*Trematosphaeria striatispora* K.D. Hyde	–	–	–	–	–	–	–	–	–	–
17	*Bathyascus grandisporus* K.D. Hyde & E.B.G. Jones	–	–	–	+	–	–	–	–	–	–
18	*Bathyascus mangrovei* Ravikumar & Vittal	–	–	–	–	–	–	–	–	–	+
19	*Bathyascus* sp.	–	–	–	+	–	–	–	–	–	–
20	*Belizeana tuberculata* Kohlm. & Volkm.–Kohlm.	–	+	–	+	–	–	–	–	–	+
21	*Biatriospora marina* K.D. Hyde & Borse	–	+	–	–	–	–	–	–	–	–
22	*Byssothecium obiones* (P. Crouan & H. Crouan) M.E. Barr	–	–	–	–	–	–	–	–	–	–

23	*Capillataspora corticola* K.D. Hyde	–	–	–	–	–	–	–	–	–	–
24	*Caryosporella rhizophorae* Kohlm.	–	+	–	+	–	–	+	–	+	–
25	*Coccostroma* sp.	–	–	–	–	–	–	–	–	–	+
26	*Corollospora pulchella* Kohlm., I. Schmidt & N.B. Nair	–	–	–	–	–	–	–	–	–	–
27	*Coronopapilla mangrovei* (K.D. Hyde) Kohlm. & Volkm.–Kohlm.	–	+	–	–	–	–	–	–	–	–
28	*Cryptosphaeria mangrovei* K.D. Hyde	–	–	–	+	–	–	–	–	–	–
29	*Cucullosporella mangrovei* K.D. Hyde & E.B.G. Jones	–	–	–	+	–	–	+	–	–	–
30	*Dactylospora haliotrepha* (Kohlm . & E. Kohlm.) Hafellner	–	+	+	+	+	+	+	+	+	+
31[a]	[a]*Diatrypasimilis australiensis* J. Zhou & Kohlm	–	–	–	–	–	–	–	–	–	–
32	[b]*Didymella avicenniae* Patil & Borse	–	–	–	–	–	–	–	–	–	+
33	*Durella* sp.	–	–	–	–	–	–	–	–	–	–
34	*Etheirophora bleapharospora* (Kohlm. & E. Kohlm.) Kohlm. & Volkm.–Kohlm.	+	+	–	–	–	–	–	–	+	–
35	[b]*Eutypa bathurstensis* K.D. Hyde & Rappaz	–	–	–	–	–	–	–	–	–	–
36	*Gnomonia* sp.-like	–	–	–	–	–	–	–	–	–	+
37	*Haiyanga salina* (Meyers) K.L. Pang & E. B. G. Jones, Yusoff & S. T.Moss ≡*Antennospora salina* (Meyers) Yusoff, E.B.G. Jones & S.T. Moss,	–	–	–	–	–	+	–	–	–	–
38	*Haligena* sp.	–	+	–	–	–	–	–	–	–	–
39	*Halomassarina thalassiae* (Kohlm. & Volkm.–Kohlm.) Suetrong, Sakayaroj, E.B.G. Jones, Kohlm., Volkm.–Kohlm. & C.L. Schoch, ≡*M. thalassiae* Kohlm. & Volkm.–Kohlm.	–	+	–	+	–	+	–	–	+	+
40	*Halorosellinia oceanica* Whalley, E.B.G. Jones, K.D. Hyde & Laessøe	–	+	+	+	+	+	–	+	–	+
41	*Halosarpheia fibrosa* Kohlm. & E. Kohlm.	+	–	–	–	–	+	+	–	+	–
42	*H. marina* (Cribb & J.W. Cribb) Kohlm.	–	–	+	+	+	+	–	+	+	+

(continued)

Table 13.1 (continued)

S. No.	Name of the species	Neotropics (N & S America)[1]	Belize, Fiji, Hawai[2]	Mandai, Siingapore[3]	Ranong, Thailand[4]	Kuala Selangor, Malaysia[5]	Morib, Malaysia[6]	Irimoto Island, Japan[7]	North Sumatra[8]	Queensland, Austrlaia[9]	Pichavaram, TN, India[10]
43	*H. minuta* Leong	–	–	+	+	–	–	–	–	–	–
44	*Halosarpheia* sp.	–	+	–	–	–	–	–	–	–	+
45	*Hapsidascus hadrus* Kohlm. & Volkm.–Kohlm.	–	+	–	–	–	–	–	–	–	–
46	*Heleococcum japonense* Tubaki	–	–	–	–	–	–	–	–	–	+
47	*Helicascus kanaloanus* Kohlm.	+	–	–	–	–	–	–	–	–	–
48	*Hypocrea* sp.	–	–	–	–	–	–	–	–	–	–
49	*Hypophloeda rhizospora* K.D. Hyde & E.B.G. Jones	–	+	–	+	–	+	–	–	–	–
50	*Hypxylon hypomiltum* Montagne	–	–	–	+	–	–	–	–	–	–
51	*Hypoxylon* sp.	–	–	–	–	–	+	–	–	–	–
52	[c]*Hysterium* sp.[b]	–	–	–	–	–	–	–	–	–	–
53	*Kallichroma tethys* (Kohlm. & E. Kohlm.) Kohlm. & Volkm.–Kohlm.	+	+	–	+	+	+	–	–	–	–
54	*Kirschsteiniothelia maritima* -like	–	–	–	–	–	–	–	–	–	–
55	[b]*Julella avicenniae* (Borse) K.D. Hyde	–	–	–	–	–	+	–	–	–	–
56	*Leptosphaeria australiensis* (Cribb & J.W. Cribb) G.C. Hughes	+	+	–	+	–	+	–	–	+	+
57	[b]*Leptosphaeria* cf. *avicenniae* Kohlm. & E. Kohlm.	–	–	–	–	–	–	–	–	–	–
58	[b]*Leptosphaeria*.cf. *peruviana* Speg	–	–	–	–	–	–	–	–	–	–
59	[b]*Leptosphaeria salvinii* Catt.	–	–	–	–	–	–	–	–	–	–
60	*Leptosphaeria* sp.	–	–	–	–	–	–	–	+	–	–
61	*Lignincola laevis* Höhnk	+	+	+	+	+	+	–	+	+	–
62	*L. tropica* Kohlm.	–	+	–	–	–	–	–	–	–	–
63	*Limacospora sundica* (Jørg. Koch & E.B.G. Jones) Jørg. ≡*Ceriosporopsis sundica* Jørg Koch & E.B.G. Jones	–	–	–	–	–	–	–	–	–	–

64	*Lineolata rhizophorae* (Kohlm.& E. Kohlm.) Kohlm.& Volkm.–Kohlm	+	+	–	+	–	–	+	–	+	+
65	*Lophiostoma armatisporum* (K.D. Hyde, Vrijmoed, Chinnaraj & E.B.G. Jones) Liew, Aptroot & K.D. Hyde, *Massarina armatispora* K.D. Hyde, Vrijmoed, Chinnaraj & E.B.G. Jones.	–	–	–	–	–	–	–	–	–	–
66[d]	[d]*Lophiostoma. rhizophorae* (Poonyth, K.D. Hyde, Aptroot & Peerally) Aptroot & K.D. Hyde	–	–	–	–	–	–	–	–	–	–
67	*Lulworthia grandispora* Meyers	–	–	–	+	–	+	–	+	–	+
68	*Lulworthia sp.*	+	+	+	+	+	+	–	+	+	+
69	*Manglicola guatemalensis* Kohlm. & E. Kohlm.	+	–	–	–	–	–	–	–	–	–
70	*Marinosphaera mangrovei* K.D. Hyde	–	+	–	+	+	+	–	+	–	–
71	*Massarina lacertensis* Kohlm. & Volkm.–Kohlm.	–	–	–	–	–	–	–	–	+	–
72	*Massarina* sp.	–	–	–	–	–	–	–	–	–	+
73	*Massariosphaeria typhicola* (P. Karst.) Barr	–	–	–	–	–		–	–	–	–
74	*Melaspilea mangrovei* Vrijmoed, K.D. Hyde & E.B.G. Jones	–	–	–	–	–	+	–	–	–	–
75	*Morosphaeria ramunculicola* (K.D. Hyde) Suetrong, Sakayaroj, E.B.G. Jones & C.L. Schoch ≡*Massarina ramunculicola* K.D. Hyde	–	–	–	+	–	+	+	–	–	–
76	*Morosphaeria velataspora* (K.D. Hyde & Borse) Suetrong, Sakayaroj, E.B.G. Jones & C.L. Schoch ≡*Massarina velataspora* K.D. Hyde & Borse	–	+	–	+	–	+	–	–	–	+
77	[b]*Mycosphaerella pneumatophorae* Kohlm.	–	–	+	+	–	–	–	–	–	–
78	*Nais inornata* Kohlm.	–	–	+	+	–	–	–	–	–	–
79	*Nais* sp.	–	–	–	–	–	–	–	–	–	+

(continued)

Table 13.1 (continued)

S. No.	Name of the species	Neotropics (N & S America)[1]	Belize, Fiji, Hawai[2]	Mandai, Siingapore[3]	Ranong, Thailand[4]	Kuala Selangor, Malaysia[5]	Morib, Malaysia[6]	Irimoto Island, Japan[7]	North Sumatra[8]	Queensland, Austrlaia[9]	Pichavaram, TN, India[10]
80	*Natantispora retorquens* (Shearer & J.L. Crane) J. Campb., J.L. Anderson & Shearer ≡*Halosarpheia retorquens* Shearer & J.L. Crane	–	–	+	–	–	–	–	–	–	–
81	*Neptunella longirostris* (Cribb & J.W. Cribb) K.L. Pang & E.B.G. Jones ≡*Lignincola longirostris* (Cribb & J.W. Cribb) Kohlm	–	–	–	+	+	–	–	–	–	+
82	*Okeanomyces cucullatus* (Kohlm.) K.L. Pang & E.B.G. Jones ≡*Halosphaeria cucullata* (Kohlm.) Kohlm	–	+	–	–	–	+	–	–	–	+
83	*Ophiodeira monosemeia* Kohlm. & Volkm.–Kohlm.-like	–	+	–	+	–	–	–	–	–	–
84	*Oceanitis cincinnatula* (Shearer & J.L. Crane) J. Dupont & E.B.G. Jones ≡*Halosarpheia cincinnatula* Shearer & J.L. Crane	–	–	+	–	–	–	–	–	–	+
85	*O. viscidula* (Kohlm. & E. Kohlm.) J. Dupont & E. B.G. Jones ≡*Haligena viscidula* Kohlm. & E. Kohlm., Nova Hedw. **9**: 92, 1965	+	–	–	–	–	–	–	–	–	–
86	*Panorbis viscosus* (I. Schmidt) J. Campb., J.L. Anderson & Shearer ≡*Halosarpheia viscosa* (I. Schmidt) Shearer & J.L. Crane	–	–	+	+	–	–	–	–	–	–
87	*Paraliomyces lentifer* Kohlm.	+	–	–	–	–	–	–	–	–	–
88	*Patellaria atra* (Hedw.) Fr., ≡*Lecanidion atratum* (Hedw.) Endl.	–	–	–	–	–	–	–	–	–	–
89	*Passeriniella savoryellopsis* K.D. Hyde & Mouzouras	–	+	+	+	–	–	–	–	–	–

90	*Passereniella mangrovei* G.L. Maria & K.R. Sridhar	–	–	–	–	–	–	–	–	–	–
91	*Payosphaeria minuta* W.F. Leong	–	–	+	–	–	–	–	–	–	–
92	*Payosphaeria* sp.	–	–	–	–	–	–	–	–	–	+
93	*Pedumispora rhizophorae* K.D. Hyde & E.B.G. Jones	–	–	–	–	–	–	–	–	–	–
94	*Phaeosphaeria neomaritima* (R.V. Gessner & Kohlm.) Shoemaker & C.E. Babc	+	–	–	–	–	–	–	–	–	–
95	*Phaeosphaeria oraemaris* (Linder) Khashn. & Shearer	–	–	+	–	–	–	–	–	–	–
96	*Polystigmina* sp.	–	–	–	–	–	–	–	–	–	–
97	*Pyrenographa xylographoides* Aptroot	–	–	–	–	–	+	–	–	–	–
98	*Quintaria lignatilis* (Kohlm.) Kohlm. & Volkm.–Kohlm.	–	+	+	+	+	+	+	–	+	+
99	*Rhizophila marina* K.D. Hyde & E.B.G. Jones	–	–	–	+	–	+	–	+	+	+
100	*Rimora mangrovei* (Kohlm. & Vittal) Kohlm., Volkm–Kohlm., Suetrong, Sakayaroj & E.B.G. Jones ≡*Lophiostoma mangrovei* Kohlm. & Vital	–	+	+	+	–	–	–	–	–	+
101	*Saccardoella mangrovei* K.D. Hyde	–	–	–	+	–	–	–	–	–	–
102	*S. rhizophorae* K.D. Hyde	–	+	–	–	–	+	–	–	–	+
103	*Saagaromyces abonnis* (Kohlm.) K.L. Pang & E.B.G. Jones ≡*Halosarpheia abonnis* Kohlm.	–	+	–	+	–	+	+	–	–	–
104	*S. glitra* (J.L. Crane & Shearer) K.L. Pang & E.B.G. Jones, ≡*Nais glitra* J.L. Crane & Shearer	–	–	–	+	+	–	–	–	–	+
105	*S. ratnagiriensis* (S.D. Patil & Borse) K.L. Pang & E.B.G. Jones ≡*Halosarpheia ratnagiriensis* S.D. Patil & Borse	–	–	–	+	–	+	–	–	–	–
106	*Savoryella lignicola* E.B.G. Jones & R.A. Eaton	–	–	–	–	+	+	–	–	–	–

(continued)

Table 13.1 (continued)

S. No.	Name of the species	Neotropics (N & S America)[1]	Belize, Fiji, Hawai[2]	Mandai, Siingapore[3]	Ranong, Thailand[4]	Kuala Selangor, Malaysia[5]	Morib, Malaysia[6]	Irimoto Island, Japan[7]	North Sumatra[8]	Queensland, Austrlaia[9]	Pichavaram, TN, India[10]
107	*S. longispora* E.B.G. Jones & K.D. Hyde	–	–	–	+	–	–	–	–	–	–
108	*S. paucispora* (Cribb & J.W. Cribb) Jørg. Koch	–	–	+	–	–	–	–	–	–	–
109	*Splanchnonema brtizelmayriana*-like	–	–	–	–	–	–	–	–	–	–
110	*Splanchnonema* sp.	–	–	–	–	–	–	–	–	–	–
111	*Sporormiella grandispora* (Speg.) Ahmed & Cain	–	–	–	–	–	–	–	–	–	+
112	*S. minima* (Auersw.)S.I. Ahmed & Cain	–	–	–	–	–	–	–	–	–	+
113	*Swampomyces armeniacus* Kohlm. & Volkm.–Kohlm.	–	+	–	–	–	–	–	–	–	+
114	*S.wampomyces* cf. *armeniacus* Kohlm. & Volkm.–Kohlm.	–	–	–	+	–	–	–	–	–	+
115	*S. triseptatus* K.D. Hyde & Nakagiri	–	–	–	+	–	–	+	–	–	–
116	*Thalassogena sphaerica* Kolhm. & Volkm.–Kohlm.	–	+	–	+	–	–	–	–	–	–
117	*Torpedospora radiata* Meyers	+	+	–	–	–	+	–	–	–	–
118	*Trematosphaeria mangrovei* Kohlm.	+	–	–	–	–	–	–	–	–	–
119	*Trematosphaeria* sp.	–	–	–	–	–	–	–	–	–	+
120	*Tirispora unicaudata* E.B.G. Jones & Vrijmoed, ≡*Aniptodera indica* Ananda & Sridhar,	–	–	–	–	–	–	–	–	–	–
121	*Tirispora mandoviana* V.V. Sarma & K.D. Hyde	–	–	–	–	–	–	–	–	–	–
122	*Tirispora* sp.	–	–	–	–	–	–	–		–	–
123	[c]*Tubeufia setosa* Sivan. & W.H. Hsieh	–	–	–	–	–	–	–		–	–
124	*Verrruculina enalia* (Kohlm.) Kohlm. & Volkm.–Kohlm.	+	+	+	+	+	+	+	+	–	+
125	*Zignoella* sp.	–	–	–	–	–	–	–	–	–	–
126	*Zopfiella* sp.	–	–	–	–	–	–	–	–	–	–

Basidiomycetes										
127 *Halocyphina villosa* Kohlm.	+	+	+	+	+	+	–	+	–	+
128 *Halocyphina* sp.	–	–	–	+	–	–	–	–	–	–
129 [e]*Calathella mangrovei* E.B.G. Jones & Agerer	–	–	–	–	–	–	–	–	–	–
130 *Calathella* sp.	–	–	–	+	–	–	–	–	–	–
131 *Nia vibrissa* R.T. Moore & Meyers	+	+	–	–	–	–	–	–	–	–
132 *Acremonium* sp.	–	–	–	–	–	–	–	–	–	–
133 *Alternaria* sp.	–	–	–	–	–	–	–	–	–	–
134 *Alveophoma sp.*	–	–	–	–	–	–	–	–	–	–
135 *Anguillospora marina* Nakagiri & Tubaki	–	–	–	–	–	–	–	–	–	–
136 *Arthrobotrys oligospora Fresen.*	–	–	–	–	–	–	–	–	–	–
137 *Bactrodesmium linderii* (Crane & Shearer) M. E. Palm & E.L. Stewart	–	–	–	–	–	–	–	–	–	–
138 *Bactrodesmium* sp./	–	–	–	+	–	–	–	–	–	–
139 *Brachysporiella gayana* Batista	–	–	–	–	–	–	–	–	–	–
140 *Cirrenalia basiminuta* Raghuk. & Zainal	–	–	–	–	–	–	–	–	–	+
141 *Cirrenalia macrocephala* (Kohlm.) Meyers & Mooore	+	–	–	–	–	–	–	–	–	+
142 *Cirrenalia pseudomacrocephala* Kohlm.	+	–	+	+	+	+	–	+	–	–
143 *Clavatospora bulbosa* (Anast.) Nakagiri & Tubaki	–	–	+	+	–	–	–	–	–	–
144 *Cumulospora marina* I. Schmidt	–	–	–	–	–	–	–	–	–	–
145 *Cytospora rhizophorae* Kohlm. & E. Kohlm.	+	–	–	–	–	–	–	–	+	–
146 *Cytospora* sp.	–	–	–	–	–	–	–	–	–	–
147 *Dactylella* sp.	–	–	–	–	–	–	–	–	–	–
148 *Delortia palmicola* Pat.	–	–	–	–	–	–	–	–	–	–
149 *Dendryphiella arenaria* Nicot	–	–	–	–	+	–	–	–	–	–
150 *Dendryphiella salina* (G.K. Sutherl.) G.J.F. Pugh & Nicot	+	–	–	–	–	–	–	–	–	–
151 *Dictyosporium elegans* Corda	–	–	–	–	–	–	+	–	–	–

(continued)

Table 13.1 (continued)

S. No.	Name of the species	Neotropics (N & S America)[1]	Belize, Fiji, Hawai[2]	Mandai, Siingapore[3]	Ranong, Thailand[4]	Kuala Selangor, Malaysia[5]	Morib, Malaysia[6]	Irimoto Island, Japan[7]	North Sumatra[8]	Queensland, Austrlaia[9]	Pichavaram, TN, India[10]
152	*Dictyosporium pelagicum* (Linder) G.C. Hughes	+	–	–	–	–	+	–	–	–	–
153	*Diplocladiela scalaroides* G. Arnaud ex M.B. Ellis	–	–	–	–	–	–	–	–	–	–
154	*Ellisembia vaga* (C.G. & T.F.L.) Subram.	–	–	–	–	–	–	–	–	–	–
155	*Endophragmia alternata* Tubaki & Saito	–	–	–	–	–	–	–	–	–	–
156	*Epicoccum purpurascens* Ehrenb.:Schlecht.	–	–	–	–	–	–	–	–	–	+
157	*Hansfordia* sp.	–	–	–	–	–	–	–	–	–	–
158	*Helicoma muelleri* Corda	–	–	–	–	–	–	–	–	–	–
159	*Helicomyces roseus* Link.	–	–	–	–	–	–	–	–	–	–
160	*Helicoon* sp.	–	–	–	+	–	–	–	–	–	–
161	*Helenospora varia* (Anastasiou) E.B.G. Jones ≡*Zalerion varium* Anastasiou	+	–	–	–	+	–	–	–	–	+
162	*Hydea pygmea* (Kohlm) K.L. Pang & E.B.G. Jones . ≡*Cirrenalia pygmea* Kohlm.	+	+	+	+	–	+	–	+	–	+
163	*Leptosphaeria salvinii* Cattaneo (anamorph, nakatea state)	–	–	–	–	–	–	–	–	–	–
164	*Matsusporium tropicale* (Kohlm.) E.B.G. Jones & K.L. Pang ≡*Cirrenalia tropicalis* Kohlm.	+	–	–	+	+	–	+	+	–	–
165	*Menispora cilliata* Corda	–	–	–	–	–	–	–	–	–	–
166	*Monodictys pelagica* Johnson & E.B.G. Jones	–	–	–	–	–	–	–	–	–	+
167	*M. putredinis* (Wallr.) S. Hughes	–	–	–	–	–	–	–	–	–	–
168	*Monodictys sp.*	–	–	–	–	–	–	–	–	–	–
169	*Mycoenterolobium platysporum* Goos	–	–	–	–	–	–	+	–	–	–
170	*Paramassariothea* sp.	–	–	–	–	–	–	–	–	–	+
171	*Phaeoisaria clematidis* (Fuckel) S. Hughes	–	–	–	–	–	–	–	–	–	–
172	*Periconia prolifica* Anastasiou	+	+	–	+	+	+	–	+	+	+
173	*Phragmospathula phoenicis* Subram. & Nair	–	–	–	–	–	–	+	–	–	–

174	*Phragmospathula* sp.	–	–		–	–	–	–	–	–	–
175	*Phialophorophoma* cf. *litoralis*	–	–	–	+	–	–	–	–	–	–
176	*Phoma suadae* Jaap	–	–	–	–	–	+	–	–	–	–
177	*Phoma sp.*	+	+	–	+	+	+	–	+	–	+
178	*Phomopsis mangrovei* K.D. Hyde	–	–	–	+	–	–	–	–	–	–
179	*Phomopsis sp.*	–	–	–	–	–	–	–	–	–	–
180	*Rhabdospora avicenniae* Kohlm. & E. Kohlm.	–	–	–	–	–	+	–	–	–	–
181	*Robillarda rhizophorae* Kohlm	+	–	–	–	–	–	–	–	–	–
182	*Sporidesmium* sp.1	–	–	–	–	–	–	–	–	–	–
183	*Sporidesmium* sp. 2	–	–	–	–	–	–	–	–	–	–
184	*Sporoschisma* sp.	–	–	–	–	–	–	–	–	–	–
185	*Stachybotrys mangiferae* Misra & Srivastava	–	–	–	–	–	–	+	–	–	–
186	*Stachylidium bicolor* Link.	–	–	–	–	–	–	–	–	–	–
187	*Stagonospora* sp.	–	–	–	–	–	–	–	–	–	–
188	*Taeniolella stricta* (Corda) S. Hughes	–	–	–	–	–	–	–	–	–	–
189	*Tetracrium* sp.	–	–	–	–	–	–	–	–	–	–
190	*Trichocladium achrasporum* (Meyers & R.T. Moore) Dixon	+	+	–	–	+	+	+	–	–	+
191	*T. alopallonellum* (Meyerss & R.T. Moore) Kohlm. & Volkm.–Kohlm.	+	+	–	+	+	+	–	–	+	+
192	*T. constrictum* I. Schmidt	–	–	–	–	–	–	–	–	–	–
193	*T. melhae* E.B.G. Jones, Abdel–Wahab & Vrijmoed	–	–	–	–	–	–	–	–	–	–
194	*Trichocladium opacum*-like	–	–	–	–	–	–	–	–	–	–
195	*Trichocladium* sp.	–	–	–	–	–	–	–	–	–	–
196	*Trimmatostroma sp.*	–	–	–	–	–	–	–	–	–	+
197	*Varicosporina ramulosa* Meyers & Kohlm.	+	–	–	–	–	–	–	–	–	–
198	*Verticillium* sp.	–	–	–	–	–	–	–	–	–	–
199[f]	[f]*Xylomyces rhizophorae* Kohlm. & Volkm.–Kohlm.	–	–	–	–	–	–	–	–	–	–
200	*Xylomyces* sp.	–	–	–	+	+	–	–	–	–	–
201	*Zalerion maritimum* (Linder) Anast.	–	–	–	–	–	–	–	–	–	–

Table 13.1 (continued)

S. No.	Name of the species	Krishna delta, AP, India[11]	Godavari delta, AP, India[12]	A & N Islands, India[13]	North Malabar, Kerala, India[14]	Nethravati, Karnataka, India[15]	Udyavara, Karnataka, India[16]	Mauritius[17]	Kampong Kapok, Brunei[18]	Total no. of mangrove sites in which a fungus was recorded[19]
1	*Acrocordiopsis patilii* Borse & K.D. Hyde	–	–	+	–	–	–	–	+	3
2	*Aigialus grandis* Kohlm. & S. Schatz	+	+	+	+	–	–	–	+	13
3	*A. mangrovei* Borse	–	–	+	–	+	+	+	–	8
4	*A. parvus* S. Schatz & Kohlm.	–	+	+	–	–	–	–	+	7
5	*A. striatispora* K.D. Hyde	–	–	–	–	–	–	–	–	1
6	*Aigialus* sp.	–	–	–	–	–	–	–	–	1
7	*Aniptodera chesapeakensis* Shearer & M.A. Mill.	+	+	+	+	–	+	+	+	11
8	*A. longispora* K.D. Hyde	–	–	–	–	–	–	–	+	5
9	*A. haispora* Vrijmoed, K.D. Hyde & E.B.G. Jones	–	+	–	+	–	–	–	–	2
10	*A. mangrovei* K.D. Hyde	+	+	+	–	+	+	–	+	8
11	*Aniptodera* sp.	–	–	–	–	–	+	+	+	4
12	*Antennospora quadricornuta* (Cribb & J.W. Cribb) T.W. Johnson	–	–	+	–	+	+	–	+	10
13	*Anthostomella* sp.	–	–	–	–	–	–	–	+	1
14	*Ascocratera manglicola* Kohlm.	+	–	+	–	–	–	–	–	6
15	*Ascocratera* cf. *manglicola* Kohlm.	–	–	–	–	–	–	–	–	1
16	*Astrosphaeriella striataspora* (K.D. Hyde) K.D. Hyde ≡*Trematosphaeria striatispora* K.D. Hyde	–	–	+	–	–	–	–	+	2
17	*Bathyascus grandisporus* K.D. Hyde & E.B. G. Jones	–	–	+	–	–	–	–	+	3
18	*Bathyascus mangrovei* Ravikumar & Vittal	–	–	–	–	–	–	–	–	1
19	*Bathyascus* sp.	–	–	–	–	–	–	–	–	1
20	*Belizeana tuberculata* Kohlm. & Volkm.–Kohlm.	–	–	+	–	–	–	–	+	5
21	*Biatriospora marina* K.D. Hyde & Borse	–	–	+	+	–	–	–	–	3

22	*Byssothecium obiones* (P. Crouan & H. Crouan) M.E. Barr	+	–	–	–	–	–	–	–	1
23	*Capillataspora corticola* K.D. Hyde	–	–	–	–	–	–	–	+	1
24	*Caryosporella rhizophorae* Kohlm.	–	–	+	–	+	+	+	+	9
25	*Coccostroma* sp.	–	–	–	–	–	–	–	–	1
26	*Corollospora pulchella* Kohlm., I. Schmidt & N.B. Nair	+	–	–	–	–	–	–	–	1
27	*Coronopapilla mangrovei* (K.D. Hyde) Kohlm. & Volkm.–Kohlm.	–	–	+	–	+	+	–	–	4
28	*Cryptosphaeria mangrovei* K.D. Hyde	+	+	–	–	–	–	–	–	3
29	*Cucullosporella mangrovei* K.D. Hyde & E.B.G. Jones	–	–	–	–	–	–	+	+	4
30	*Dactylospora haliotrepha* (Kohlm . & E. Kohlm.) Hafellner	+	+	+	+	–	+	+	+	16
31[a]	[a]*Diatrypasimilis australiensis* J. Zhou & Kohlm	–	–	–	–	–	–	–	–	1
32	[b]*Didymella avicenniae* Patil & Borse	–	–	–	–	–	–	–	–	1
33	*Durella* sp.	–	–	–	–	+	+	–	–	2
34	*Etheirophora bleapharospora* (Kohlm. & E. Kohlm.) Kohlm. & Volkm.–Kohlm.	–	–	+	–	–	–	–	+	5
35	[b]*Eutypa bathurstensis* K.D. Hyde & Rappaz	–	–	–	–	–	+	–	–	1
36	*Gnomonia* sp.-like	–	+	–	–	–	–	–	–	2
37	*Haiyanga salina* (Meyers) K.L. Pang & E. B. G. Jones, Yusoff & S. T.Moss ≡*Antennospora salina* (Meyers) Yusoff, E.B.G. Jones & S.T. Moss,	–	–	–	–	–	–	–	–	1
38	*Haligena* sp.	–	–	–	–	–	–	–	–	1
39	*Halomassarina thalassiae* (Kohlm. & Volkm.–Kohlm.) Suetrong, Sakayaroj, E.B.G. Jones, Kohlm., Volkm.–Kohlm. & C.L. Schoch, ≡*M. thalassiae* Kohlm. & Volkm.–Kohlm.	+	+	+	–	–	–	–	+	9
40	*Halorosellinia oceanica* Whalley, E.B.G. Jones, K.D. Hyde & Laessøe	+	+	+	+	+	+	+	+	15

(continued)

Table 13.1 (continued)

S. No.	Name of the species	Krishna delta, AP, India[11]	Godavari delta, AP, India[12]	A & N Islands, India[13]	North Malabar, Kerala, India[14]	Nethravati, Karnataka, India[15]	Udyavara, Karnataka, India[16]	Mauritius[17]	Kampong Kapok, Brunei[18]	Total no. of mangrove sites in which a fungus was recorded[19]
41	*Halosarpheia fibrosa* Kohlm. & E. Kohlm.	–	–	–	–	–	–	+	+	6
42	*H. marina* (Cribb & J.W. Cribb) Kohlm.	+	–	–	+	–	–	+	+	11
43	*H. minuta* Leong	–	–	+	–	–	–	–	+	4
44	*Halosarpheia* sp.	+	–	–	–	–	–	+	–	4
45	*Hapsidascus hadrus* Kohlm. & Volkm.–Kohlm.	–	–	–	–	–	–	–	–	1
46	*Heleococcum japonense* Tubaki	+	–	–	–	–	–	–	–	2
47	*Helicascus kanaloanus* Kohlm.	–	–	+	–	–	–	–	–	2
48	*Hypocrea* sp.	+	–	–	–	–	–	–	–	1
49	*Hypophloeda rhizospora* K.D. Hyde & E.B.G. Jones	–	–	–	–	–	–	–	+	4
50	*Hypxylon hypomiltum* Montagne	–	–	–	–	–	–	–	–	1
51	*Hypoxylon* sp.	+	+	–	–	+	+	–	–	5
52	[c]*Hysterium* sp.[b]	+	+	–	–	–	–	–	–	2
53	*Kallichroma tethys* (Kohlm. & E. Kohlm.) Kohlm. & Volkm.–Kohlm.	+	+	+	–	+	+	+	+	12
54	*Kirschsteiniothelia maritima* -like	–	+	–	–	–	–	–	–	1
55	[b]*Julella avicenniae* (Borse) K.D. Hyde	–	–	–	–	–	–	–	–	1
56	*Leptosphaeria australiensis* (Cribb & J.W. Cribb) G.C. Hughes	+	+	+	–	+	+	+	+	13
57	[b]*Leptosphaeria* cf. *avicenniae* Kohlm. & E. Kohlm.	–	–	–	–	–	–	–	+	1
58	[b]*Leptosphaeria*.cf. *peruviana* Speg	+	+	–	–	–	+	–	–	3
59	[b]*Leptosphaeria salvinii* Catt.	–	–	–	–	+	–	–	–	1
60	*Leptosphaeria* sp.	+	–	–	–	–	–	+	+	4
61	*Lignincola laevis* Höhnk	–	–	+	–	+	+	+	+	13
62	*L. tropica* Kohlm.	–	+	–	–	–	–	–	+	3
63	*Limacospora sundica* (Jørg. Koch & E.B.G. Jones) Jørg.	–	–	–	–	+	+	–	–	2

	≡*Ceriosporopsis sundica* Jørg Koch & E.B.G. Jones									
64	*Lineolata rhizophorae* (Kohlm.& E. Kohlm.) Kohlm.& Volkm.–Kohlm	+	–	+	–	–	–	+	–	9
65	*Lophiostoma armatisporum* (K.D. Hyde, Vrijmoed, Chinnaraj & E.B.G. Jones) Liew, Aptroot & K.D. Hyde, *Massarina armatispora* K.D. Hyde, Vrijmoed, Chinnaraj & E.B.G. Jones.	–	–	+	–	–	–	–	–	1
66[d]	[d]*Lophiostoma. rhizophorae* (Poonyth, K.D. Hyde, Aptroot & Peerally) Aptroot & K.D. Hyde	–	–	–	–	–	–	–	–	1
67	*Lulworthia grandispora* Meyers	+	+	+	+	+	+	–	+	11
68	*Lulworthia sp.*	+	+	+	–	+	+	+	+	16
69	*Manglicola guatemalensis* Kohlm. & E. Kohlm.	–	–	–	–	–	–	–	+	2
70	*Marinosphaera mangrovei* K.D. Hyde	–	–	+	–	–	–	+	+	8
71	*Massarina lacertensis* Kohlm. & Volkm.–Kohlm.	–	–	–	–	–	–	+	–	2
72	*Massarina* sp.	–	+	–	–	–	–	+	–	3
73	*Massariosphaeria typhicola* (P. Karst.) Barr	+	+	–	–	–	–	–	–	2
74	*Melaspilea mangrovei* Vrijmoed, K.D. Hyde & E.B.G. Jones	–	–	–	–	–	–	–	–	1
75	*Morosphaeria ramunculicola* (K.D. Hyde) Suetrong, Sakayaroj, E.B.G. Jones & C.L. Schoch ≡*Massarina ramunculicola* K.D. Hyde	–	–	–	–	–	–	–	–	3
76	*Morosphaeria velataspora* (K.D. Hyde & Borse) Suetrong, Sakayaroj, E.B.G. Jones & C.L. Schoch ≡*Massarina velataspora* K.D. Hyde & Borse	+	+	+	–	–	–	+	+	9

(continued)

Table 13.1 (continued)

S. No.	Name of the species	Krishna delta, AP, India[11]	Godavari delta, AP, India[12]	A & N Islands, India[13]	North Malabar, Kerala, India[14]	Nethravati, Karnataka, India[15]	Udyavara, Karnataka, India[16]	Mauritius[17]	Kampong Kapok, Brunei[18]	Total no. of mangrove sites in which a fungus was recorded[19]
77	[b]*Mycosphaerella pneumatophorae* Kohlm.	–	–	–	–	–	–	–	–	2
78	*Nais inornata* Kohlm.	–	–	–	–	–	–	–	–	2
79	*Nais* sp.	–	–	+	–	–	+	–	–	3
80	*Natantispora retorquens* (Shearer & J.L. Crane) J. Campb., J.L. Anderson & Shearer ≡*Halosarpheia retorquens* Shearer & J.L. Crane	–	–	–	–	–	–	–	–	1
81	*Neptunella longirostris* (Cribb & J.W. Cribb) K.L. Pang & E.B.G. Jones ≡*Lignincola longirostris* (Cribb & J.W. Cribb) Kohlm	+	–	–	–	–	–	–	–	4
82	*Okeanomyces cucullatus* (Kohlm.) K.L. Pang & E.B.G. Jones ≡*Halosphaeria cucullata* (Kohlm.) Kohlm	–	–	–	–	–	–	–	–	3
83	*Ophiodeira monosemeia* Kohlm. & Volkm.–Kohlm.-like	–	–	–	–	–	–	–	+	3
84	*Oceanitis cincinnatula* (Shearer & J.L. Crane) J. Dupont & E.B.G. Jones ≡*Halosarpheia cincinnatula* Shearer & J.L. Crane	–	–	–	–	+	+	–	–	4
85	*O. viscidula* (Kohlm. & E. Kohlm.) J. Dupont & E. B.G. Jones ≡*Haligena viscidula* Kohlm. & E. Kohlm., Nova Hedw. **9**: 92, 1965	–	–	–	–	–	–	–	–	1
86	*Panorbis viscosus* (I. Schmidt) J. Campb., J.L. Anderson & Shearer ≡*Halosarpheia viscosa* (I. Schmidt) Shearer & J.L. Crane	–	–	–	–	–	–	+	+	4
87	*Paraliomyces lentifer* Kohlm.	–	–	–	–	–	–	–	–	1

88	*Patellaria atra* (Hedw.) Fr., ≡*Lecanidion atratum* (Hedw.) Endl.	+	+	–	–	–	–	+	–	3
89	*Passeriniella savoryellopsis* K.D. Hyde & Mouzouras	–	–	–	–	–	–	–	+	4
90	*Passereniella mangrovei* G.L. Maria & K.R. Sridhar	–	–	–	–	+	+	–	–	2
91	*Payosphaeria minuta* W.F. Leong	–	–	–	–	–	–	–	–	1
92	*Payosphaeria* sp.	–	–	–	–	–	–	–	–	1
93	*Pedumispora rhizophorae* K.D. Hyde & E.B.G. Jones	+	+	–	–	–	–	–	–	2
94	*Phaeosphaeria neomaritima* (R.V. Gessner & Kohlm.) Shoemaker & C.E. Babc	–	–	–	–	–	–	–	–	1
95	*Phaeosphaeria oraemaris* (Linder) Khashn. & Shearer	–	–	–	–	–	–	–	–	1
96	*Polystigmina* sp.	–	–	–	–	–	–	–	+	1
97	*Pyrenographa xylographoides* Aptroot	–	–	–	–	–	–	–	–	1
98	*Quintaria lignatilis* (Kohlm.) Kohlm. & Volkm.–Kohlm.	+	–	+	–	–	–	–	+	11
99	*Rhizophila marina* K.D. Hyde & E.B.G. Jones	+	+	+	–	–	–	+	+	10
100	*Rimora mangrovei* (Kohlm. & Vittal) Kohlm.,Volkm–Kohlm., Suetrong, Sakayaroj & E.B.G. Jones ≡*Lophiostoma mangrovei* Kohlm. & Vital	+	+	+	–	+	+	–	–	9
101	*Saccardoella mangrovei* K.D. Hyde	–	–	–	–	–	–	–	–	1
102	*S. rhizophorae* K.D. Hyde	+	+	–	–	–	–	–	–	5
103	*Saagaromyces abonnis* (Kohlm.) K.L. Pang & E.B.G. Jones ≡*Halosarpheia abonnis* Kohlm.	+	+	+	–	–	–	+	–	8
104	*S. glitra* (J.L. Crane & Shearer) K.L. Pang & E.B.G. Jones, ≡*Nais glitra* J.L. Crane & Shearer	–	–	+	–	–	–	–	+	5

(continued)

Table 13.1 (continued)

S. No.	Name of the species	Krishna delta, AP, India[11]	Godavari delta, AP, India[12]	A & N Islands, India[13]	North Malabar, Kerala, India[14]	Nethravati, Karnataka, India[15]	Udyavara, Karnataka, India[16]	Mauritius[17]	Kampong Kapok, Brunei[18]	Total no. of mangrove sites in which a fungus was recorded[19]
105	*S. ratnagiriensis* (S.D. Patil & Borse) K.L. Pang & E.B.G. Jones ≡*Halosarpheia ratnagiriensis* S.D. Patil & Borse	+	+	+	–	+	+	–	+	8
106	*Savoryella lignicola* E.B.G. Jones & R.A. Eaton	+	–	+	+	+	+	+	+	9
107	*S. longispora* E.B.G. Jones & K.D. Hyde	–	–	–	–	+	+	–	–	3
108	*S. paucispora* (Cribb & J.W. Cribb) Jørg. Koch	–	–	+	+	+	+	–	–	5
109	*Splanchnonema brtizelmayriana*-like	–	+	–	–	–	–	–	–	1
110	*Splanchnonema* sp.	–	–	–	–	–	–	–	+	1
111	*Sporormiella grandispora* (Speg.) Ahmed & Cain	–	–	–	–	–	–	–	–	1
112	*S. minima* (Auersw.)S.I. Ahmed & Cain	–	–	–	–	–	–	–	–	1
113	*Swampomyces armeniacus* Kohlm. & Volkm.–Kohlm.	–	–	+	–	–	–	+	–	4
114	*S.wampomyces* cf. *armeniacus* Kohlm. & Volkm.–Kohlm.	–	–	–	–	–	–	–	+	3
115	*S. triseptatus* K.D. Hyde & Nakagiri	–	–	–	–	–	–	–	–	2
116	*Thalassogena sphaerica* Kolhm. & Volkm.–Kohlm.	–	–	–	–	–	–	+	+	4
117	*Torpedospora radiata* Meyers	–	–	–	–	–	–	–	+	4
118	*Trematosphaeria mangrovei* Kohlm.	–	–	–	–	–	–	–	–	1
119	*Trematosphaeria* sp.	–	–	–	–	–	–	–	+	2
120	*Tirispora unicaudata* E.B.G. Jones & Vrijmoed, ≡*Aniptodera indica* Ananda & Sridhar,	–	–	–	–	+	+	–	–	2
121	*Tirispora mandoviana* V.V. Sarma & K.D. Hyde	–	–	–	–	+	+	–	–	3

122	*Tirispora* sp.	–	–	–	–	+	+	–	–	2
123	[c]*Tubeufia setosa* Sivan. & W.H. Hsieh	–	+	–	–	–	–	–	–	1
124	*Verruculina enalia* (Kohlm.) Kohlm. & Volkm.–Kohlm.	+	+	+	+	+	+	–	+	16
125	*Zignoella* sp.	–	–	–	–	–	+	–	–	1
126	*Zopfiella* sp.	–	–	–	–	+	–	–	–	1
127	*Halocyphina villosa* Kohlm.	+	+	+	–	+	+	+	+	15
128	*Halocyphina* sp.	–	–	–	–	–	–	–	–	1
129	[e]*Calathella mangrovei* E.B.G. Jones & Agerer	–	–	–	–	–	–	–	–	1
130	*Calathella* sp.	–	–	–	–	–	–	–	–	1
131	*Nia vibrissa* R.T. Moore & Meyers	–	–	–	–	–	–	–	–	2
132	*Acremonium* sp.	–	–	–	–	+	+	–	–	2
133	*Alternaria* sp.	–	–	–	–	+	+	–	–	2
134	*Alveophoma sp.*	–	+	–	–	–	–	–	–	1
135	*Anguillospora marina* Nakagiri & Tubaki	–	–	–	–	+	+	–	–	2
136	*Arthrobotrys oligospora Fresen.*	–	–	–	–	+	+	–	–	2
137	*Bactrodesmium linderii* (Crane & Shearer) M.E. Palm & E.L. Stewart	+	+	–	–	+	+	–	+	5
138	*Bactrodesmium* sp./	–	–	–	–	–	–	–	–	1
139	*Brachysporiella gayana* Batista	–	–	–	–	+	+	–	–	2
140	*Cirrenalia basiminuta* Raghuk. & Zainal	–	+	–	–	–	–	+	–	3
141	*Cirrenalia macrocephala* (Kohlm.) Meyers & Mooore	+	–	–	–	+	+	–	–	4
142	*Cirrenalia pseudomacrocephala* Kohlm.	–	–	–	–	–	–	–	+	7
143	*Clavatospora bulbosa* (Anast.) Nakagiri & Tubaki	–	–	–	–	+	+	–	–	4
144	*Cumulospora marina* I. Schmidt	–	–	–	–	+	+	–	–	2
145	*Cytospora rhizophorae* Kohlm. & E. Kohlm.	+	–	+	–	–	–	+	–	5
146	*Cytospora* sp.	–	–	–	–	–	–	–	+	1

(continued)

Table 13.1 (continued)

S. No.	Name of the species	Krishna delta, AP, India[11]	Godavari delta, AP, India[12]	A & N Islands, India[13]	North Malabar, Kerala, India[14]	Nethravati, Karnataka, India[15]	Udyavara, Karnataka, India[16]	Mauritius[17]	Kampong Kapok, Brunei[18]	Total no. of mangrove sites in which a fungus was recorded[19]
147	*Dactylella* sp.	–	–	–	–	–	+	–	–	1
148	*Delortia palmicola* Pat.	–	–	–	–	+	+	–	–	2
149	*Dendryphiella arenaria* Nicot	–	–	–	–	–	–	–	–	1
150	*Dendryphiella salina* (G.K. Sutherl.) G.J.F. Pugh & Nicot	–	–	–	–	–	–	–	–	1
151	*Dictyosporium elegans* Corda	–	–	–	–	–	–	–	–	1
152	*Dictyosporium pelagicum* (Linder) G.C. Hughes	–	–	–	–	–	–	–	+	3
153	*Diplocladiela scalaroides* G. Arnaud ex M.B. Ellis	–	–	–	–	–	+	–	–	1
154	*Ellisembia vaga* (C.G. & T.F.L.) Subram.	+	+	–	–	–	–	–	–	2
155	*Endophragmia alternata* Tubaki & Saito	–	–	–	–	+	+	–	–	2
156	*Epicoccum purpurascens* Ehrenb.:Schlecht.	–	+	–	–	–	–	–	–	2
157	*Hansfordia* sp.	–	–	–	–	+	+	–	–	2
158	*Helicoma muelleri* Corda	–	–	–	–	–	+	–	–	1
159	*Helicomyces roseus* Link.	–	–	–	–	+	+	–	–	2
160	*Helicoon* sp.	–	–	–	–	–	–	–	–	1
161	*Helenospora varia* (Anastasiou) E.B.G. Jones ≡*Zalerion varium* Anastasiou	+	+	+	+	+	+	–	–	9
162	*Hydea pygmea* (Kohlm) K.L. Pang & E.B.G. Jones . ≡*Cirrenalia pygmea* Kohlm.	+	+	+	+	+	+	+	+	15
163	*Leptosphaeria salvinii* Cattaneo (anamorph, nakatea state)	–	–	–	–	–	+	–	–	1
164	*Matsusporium tropicale* (Kohlm.) E.B.G. Jones & K.L. Pang ≡*Cirrenalia tropicalis* Kohlm.	–	+	–	–	+	+	+	+	10
165	*Menispora cilliata* Corda	–	–	–	–	+	+	–	–	2

166	*Monodictys pelagica* Johnson & E.B.G. Jones	–	–	–	–	+	+	–	–	3
167	*M. putredinis* (Wallr.) S. Hughes	–	–	–	–	+	+	–	–	2
168	*Monodictys sp.*	+	–	–	–	–	–	–	–	1
169	*Mycoenterolobium platysporum* Goos	–	–	–	–	–	–	–	–	1
170	*Paramassariothea* sp.	–	–	–	–	–	–	–	–	1
171	*Phaeoisaria clematidis* (Fuckel) S. Hughes	–	–	–	–	+	+	–	–	2
172	*Periconia prolifica* Anastasiou	+	+	+	+	–	+	+	+	15
173	*Phragmospathula phoenicis* Subram. & Nair	–	–	–	–	–	–	–	–	1
174	*Phragmospathula* sp.	–	–	+	–	–	–	–	+	2
175	*Phialophorophoma* cf. *litoralis*	–	–	–	–	–	–	–	+	2
176	*Phoma suadae* Jaap	–	–	–	–	–	–	–	–	1
177	*Phoma sp.*	+	+	+	–	+	+	+	+	14
178	*Phomopsis mangrovei* K.D. Hyde	–	+	+	–	–	–	–	–	3
179	*Phomopsis sp.*	+	+	–	–	–	+	–	+	4
180	*Rhabdospora avicenniae* Kohlm. & E. Kohlm.	–	–	–	–	–	–	–	–	1
181	*Robillarda rhizophorae* Kohlm	–	–	–	–	–	–	–	–	1
182	*Sporidesmium* sp.1	–	–	–	–	–	+	–	–	1
183	*Sporidesmium* sp. 2	–	–	–	–	–	+	–	–	1
184	*Sporoschisma* sp.	–	–	–	–	+	–	–	–	1
185	*Stachybotrys mangiferae* Misra & Srivastava	–	–	–	–	–	–	–	–	1
186	*Stachylidium bicolor* Link.	–	–	–	–	+	+	–	–	2
187	*Stagonospora* sp.	–	–	–	–	–	–	–	–	1
188	*Taeniolella stricta* (Corda) S. Hughes	–	–	–	–	+	+	–	–	2
189	*Tetracrium* sp.	–	–	–	–	+	+	–	–	2
190	*Trichocladium achrasporum* (Meyers & R.T. Moore) Dixon	+	+	+	–	+	+	+	+	13
191	*T. alopallonellum* (Meyerss & R.T. Moore) Kohlm. & Volkm.–Kohlm.	+	+	+	–	–	+	–	+	12
192	*T. constrictum* I. Schmidt	–	–	–	+	–	–	–	–	1

(continued)

Table 13.1 (continued)

S. No.	Name of the species	Krishna delta, AP, India[11]	Godavari delta, AP, India[12]	A & N Islands, India[13]	North Malabar, Kerala, India[14]	Nethravati, Karnataka, India[15]	Udyavara, Karnataka, India[16]	Mauritius[17]	Kampong Kapok, Brunei[18]	Total no. of mangrove sites in which a fungus was recorded[19]
193	*T. melhae* E.B.G. Jones, Abdel–Wahab & Vrijmoed	–	–	–	–	+	+	–	–	2
194	*Trichocladium opacum*-like	–	–	–	–	–	–	–	+	2
195	*Trichocladium* sp.	–	–	–	–	–	+	–	+	2
196	*Trimmatostroma sp.*	+	+	–	–	–	–	–	–	3
197	*Varicosporina ramulosa* Meyers & Kohlm.	–	–	–	–	–	–	–	–	1
198	*Verticillium* sp.	–	–	–	–	+	+	–	–	2
199[f]	[f]*Xylomyces rhizophorae* Kohlm. & Volkm.–Kohlm.	–	–	–	–	–	–	–	–	1
200	*Xylomyces* sp.	–	–	–	–	–	–	–	+	3
201	*Zalerion maritimum* (Linder) Anast.	–	–	–	–	+	+	–	–	2

The fungal species occurring on aerial woody parts, rhizosphere, rhizoplane, phylloplane, phyllosphere and endophytic fungi *are not included* in the above list as the table was intended mainly for marine fungi colonizing dead and degrading samples of *Rhizophora* spp.

[a]Chalkley et al. (2010)

[b]Doubtful of the occurrence of the fungus on this host

[c]Doubtful whether it is truly marine fungus

[d]Hyde et al. (2002)

[e]From Chorao mangroves, Mandovi River, Goa, India included in CD–ROM on mangrove fungi, National Institute of Oceanography, Goa 403004, India

[f]Kohlmeyer and Volkmann-Kohlmeyer 1998

[Ω]No. of sites in which was fungus recorded: 1. Mostly from Neotropics (Kohlmeyer and Kohlmeyer 1979); 2. Belize, Fiji, Hawaii Islands, South America (B. Volkmann-Kohlmeyer and Kohlmeyer 1993; Kohlmeyer et al. 1995); 3. Mandai mangrove, Singapore (Leong et al. 1991); 4. Ranong mangrove, Thailand (Hyde et al. 1990); 5. Kuala Selangor & Kampong Sementa, Malaysia (Jones and Kuthubutheen 1989); 6. Morib, Malaysia (Alias and Jones 2000); 7. Irimoto island, Japan (Nakagiri 1993); 8. North Sumatra (Hyde 1989); 9. Queensland, Australia (Kohlmeyer and Volkmann-Kohlmeyer 1991); 10. Pichavaram, Tamil Nadu state, East coast of India (Ravikumar and Vittal 1996; Sridhar 2009); 11. Krishna delta, Andhra Pradesh state, east coast of India (Sarma and Vittal 1998–1999); 12. Godavari delta, Andhra Pradesh state, east coast of India (Sarma and Vittal 1998–1999); 13. Andaman and Nicobar Islands, East coast of India (Chinnaraj 1993); 14. Kerala state, West coast of India (Nambiar and Raveendran 2009); 15 Nethravati mangroves, Mangalore, Karnataka state, South West coast of India (Sridhar and Maria 2006); 16. Udyara, Karnataka state, South west coast of India (Maria and Sridhar 2003); 17. Mauritius (Poonyth et al. 1999); 18. Kampong Kapok, Brunei (Hyde 1990a, b; Hyde 1991); Kampong Danau (Hyde 1988); 19. No. of sites in which was fungus recorded

75 unitunicate ascomycetes have been reported so far from *Rhizophora* spp. The order Halosphaeriales alone consist 37 species, while 38 species belong to other unitunicate ascomycetes. More than 50 species possess active spore dispersal mechanism, either by having a pore in the ascus or by fissitunicate mechanism of dispersal.

On *Rhizophora* spp., most of the marine fungi belong to ascomycetes and sizeable portion to anamorphic fungi (35%) with only a few members belonging to basidiomycetes. Three marine basidiomycetes occur on *Rhizophora* spp. and all are tiny in size and belong to the family Lachnellaceae of the order Agaricales. All have reduced basidiomes, possibly as an adaptation to an aquatic environment where large fruit bodies would not survive (Jones 1988). Hibbett (2007) considers the minute forms of cyphelloid basidiomycetes to be related to selection of spore production from minimal substrates (Jones et al. 2009).

A close examination of the classification of the 85 marine fungal genera of Ascomycota occurring on *Rhizophora* spp. shows that though they belong to 6 of the 10 classes under the division Pezizomycotina of the Phylum Ascomycota, only two classes have large representation and they are Dothideomycetes (36 genera) and Sordariomycetes (42 genera). Whereas in the class Dothideomycetes, the order Pleosporales is represented by as many as 25 genera in the Sordariomycetes, the order Halosphaeriales is represented by 20 marine fungal genera. Three genera could not be assigned to any class and were kept incertae sedis under unitunicate ascomycetes. Including this, totally 85 marine fungal genera belonging to ascomycota has been recorded on *Rhizophora* spp.

A comparison between *Avicennia* spp. and *Rhizophora* spp., the two predominant mangrove tree species, shows that the following fungi are specific to *Avicennia*: *Adomia avicenniae*, *Camarosporium roumeguerii*, *Bathyascus avicenniae*, *Cryptovalsa* sp., *Didymella avicenniae*, *Eutypa bathurstensis*, *Eutypella naqsii*, *Julella avicenniae*, *Leptosphaeria avicenniae*, *Mycosphaerella pneumatophorae*, *Rhabdospora avicenniae*, *Zopfiella latipes*, *Zopfiella marina*. The following species have been found to be specific on *Rhizophora* spp: *Caryosporella rhizophorae*, *Cryptosporella mangrovei*, *Etheirophora blepharospora*, *Hypophloeda rhizospora*, *Lineolata rhizophorae*, *Lophiostoma rhizophorae*, *Pedumispora rhizophorae*, *Rhizophila marina*, *Sacardoella rhizophorae*, *Trematosphaeria mangrovei*, *Cytospora rhizophorae*, *Phomopsis mangrovei*, *Robillarda rhizophorae*, *Xylomyces rhizophorae*.

Many new genera and new species have been described after Hyde et al. (2000) in addition to transfers and/or creating new genera and new species. For an update on these new fungi/names, the following publications have been referred (Campbell et al. 2003; Pang et al. 2003; Jones et al. 2009; Suetrong et al. 2009; Abdel-Wahab et al. 2010) which are included in Table 13.1.

13.3 Geographical Distribution

Prior to 1980s, most of the studies on marine fungi in mangroves, particularly with reference to *Rhizophora* spp. was conducted in Neotropics covering mangrove formations in the coastal regions sandwiching North and South America including the coast of Florida, USA, Caribbean Islands, Belize and other surrounding sites (West Atlantic Ocean and Pacific Ocean regions). The host species studied were *Rhizophora mangle*, *R. racemosa* and *R. stylosa* (Kohlmeyer and Kohlmeyer 1979). A few scattered studies were also conducted from West African region particularly Sierra Leone (Aleem 1980). Around 42 mangrove fungi were reported by Kohlmeyer and Kohlmeyer (1979) in their treatise on higher marine fungi, of which 33 were on *Rhizophora* spp.

Subsequent studies of marine fungi in mangroves in the following two decades were from South East Asia, South Asia, Far East, Seychelles, Mauritius and Australia. Among these, South East Asia including Brunei, Malaysia, Thailand, Philippines and Singapore, has been thoroughly sampled. A large number of marine fungi have been recorded on mangroves from this region (Table 13.1). Many mangrove formations are yet to be explored including Mahanadi mangroves and Sunderbans on the east coast of India, Indonesian mangroves and the African and South American continents.

Fungi occurring on *Rhizophora* spp. from eighteen sites are listed in Table 13.1. These sites may be divided into three regions, viz., (a) Neotropics covering the mangrove formations in between North and South America and Belize, Fiji and Hawaii, (b) South East Asia covering Singapore, Thailand, Malaysia and Brunei, (c) Indian Ocean coasts particularly South and East coast of India and (d) Miscellaneous containing Japan, Mauritius, Australia and North Sumatra. From the available data, it could be seen (Table 13.1) that 60 marine fungi have been recorded on *Rhizophora* spp. from Neotropics and Belize and Hawaii islands belonging to 44 ascomycetes, 2 basidiomycetes and 14 anamorphic fungi. The following fungi were not recorded in the other 3 regions of the 4 mentioned above and they are *Aigialus* sp., *Haligena* sp., *Halosarpheia* sp., *Hapsidascus hadrus*, *Helicascus kanaloanus*, *Oceanitis viscidula*, *Paraliomyces lentifer*, *Trematosphaeria mangrovei*, *Dendryphiella salina*, *Robillarda rhizophorae* and *Varicosporina ramulosa*. Though these fungi have not been recorded on *Rhizophora* spp., there is a possibility of their presence on other hosts.

Trichocladium is represented by 6 species; *Aigialus*, *Aniptodera* and *Leptosphaeria* are represented by 5 species each, while *Halosarpheia*, *Saagaromyces*, *Savoryella*, *Tirispora* and *Cirrenalia* are represented by 3 species each. Most of the marine fungal genera recorded on the host *Rhizophora* spp. are monospecific or have two species. From India a total of 142 marine fungal species (comprising 89 Ascomycota, 1 Basidiomycota and 21 anamorphic fungi) have been recorded. Out of these, 94 marine fungi comprising 40 Ascomycota, 8 Basidiomycota 1 and 30 Anamorphic fungi have been recorded from west coast of India. While 48 fungi (40 Ascomycetes and 8 Anamorphic fungi) were not recorded from the west coast, around 47 marine fungi (16 ascomycetes, 1 basidiomycete and 30 anamorphic fungi) recorded from west coast of India have not been recorded from east coast of India. Between the coasts, 47 marine fungi (33 Ascomycetes, 1 basidiomycete and 13 anamorphic fungi) are common. Jaccard's Index expressed in percentage showed 33% similarity between the two coasts of India. This shows that there is considerably less commonality between the two coasts and more fungi are unique in the each coast. The reason for variation in fungal diversity could be attributed to the temperature and salinity differences between the two coasts. On the east coast the salinity is normally less and temperatures are high and vice versa are true in the case of west coast (Vinayachandran and Kurian 2008). Of the fungi recorded from east coast of India on *Rhizophora* spp., 77 % are ascomycetes, whereas only 52% of ascomycetes are found in west coast of India. Similarly, the anamorphic fungi recorded on *Rhizophora* spp. from east coast of India is only 22%, whereas it is 46% from west coast of India. However, the total number of marine fungi recorded on *Rhizophora* from both the coasts did not vary much (95 and 94%, respectively).

From the South East Asian region, 103 marine fungi have been reported to occur on *Rhizophora* spp. (76 ascomycetes, 3 basidiomycetes and 24 anamorphic fungi). Though the number of marine fungi recorded on *Rhizophora* spp. from this region is low when compared to the Indian coastline, South East Asia could well be the richest in terms of marine fungal diversity in mangroves. For example, Alias et al. (2010) have reported more than 139 marine fungi covering all mangrove plants from Malaysia with many waiting for formal identification. Similarly, 154 fully identified marine fungi have been reported from Thailand by Jones et al. (2006) that include hosts *Nypa fruticans* and *Rhizophora* spp.

A comparison between the marine fungi reported from the South East Asia and Indian coasts showed that 40 marine fungi reported from the South East Asia are not found in the Indian coasts, while 73 marine fungi found in the Indian coasts are not reported from the S.E. Asia. In between these two regions, 63 fungi have been found to be commonly occurring on *Rhizophora* spp. While 55 fungi are common to the South East Asian region and the East coast of India, 44 marine fungi have been found to be common on *Rhizophora* spp. occurring in the South East Asia and the west coast of India. The reason that relatively more number of marine fungi are common between the east coast of India and the South East Asia could be due to the geographical proximity where the Bay of Bengal waters extend and mix into waters of the Indian Ocean in the South East Asia.

Most of the marine fungi reported from the Mauritius have been reported from the Indian mangroves. Same is the case with North Sumatra and Australia. In Japan, with exception of three or four species, all other marine fungi reported are the commonly occurring on *Rhizophora* spp. reported from all other regions.

13.3.1 Cosmopolitan Fungi

Out of 201 marine fungi recorded on *Rhizophora* spp. from 18 different mangrove formation/sites throughout the world, only 30 species have been recorded in more than 9 out of the 18 sites (Table 13.1). *Dactylospora haliotrepha*, *Lulworthia* sp. and *Verruculina enalia* from 16 sites, *Halorosellinia oceanica*, *Halocyphina villosa*, *Hydea pygmea* and *Periconia prolifica* were reported from 15 sites; *Phoma* sp. 14 sites; *A. grandis*, *Leptosphaeria australiensis*, *Lignincola laevis* and *Trichocladium achrasporum* in 13 sites; *Kallichroma tethys* and *Trichocladium alopalonellum* in 12 sites; *Aniptodera chesapeakensis*, *Halosarpheia marina*, *Lulworthia grandispora*, *Quintaria lignatilis* in 11 sites; *Antennosopora quadricornuta*, *Rhizophila marina* and *Matsusporium tropicale* 10 sites; *Caryosporella rhizophorae*, *Halomassarina thalassiae*, *Lineolata rhizophorae*, *Morosphaeria velatospora*, *Rimora mangrovei*, *Savoryella lignicola*, *Helenospora varia* in 9 sites; *Aigialus mangrove*, *Aniptodera mangrovei*, *Halomassarina thalassiae*, *Marinosphaera mangrovei*, *Saagaromyces abonnis* and *Saagaromyces ratnagiriensis* in 8 sites. These fungi could be characterized as having cosmopolitan occurrence and hence could be described as having "high ecological amplitude."

13.3.2 Marine Fungi with Restricted Distribution on Rhizophora spp.

Of the 201 marine fungi reported on *Rhizophora* spp. from 18 mangroves sites listed (Table 13.1), seventy-one marine fungal species occurred only in any one particular mangrove sites; forty-four in any two mangrove sites; twenty-four in any three mangroves sites and 17 in any four mangroves sites. Though fungi with restricted distribution add to the diversity of the species to a particular site, it also shows the vulnerability of the species to extinction due to their endemic occurrence and selection pressure. Though the exact reason for their restricted distribution is not known, it could be hypothesized that they are of recent origin in geological timescale and speciation, and hence have not reached other mangrove sites.

13.4 Adaptations

On driftwood not many groups of ascomycetes are found but are mostly dominated by Halosphaeriales. It is different in the case of mangroves where marine fungi belonging to several groups of fungi could be recorded. The reasons could be (a) mangroves have regular freshwater in-surge from rivers which not only reduces salinity but also brings propagules of various groups of fungi, (b) intermittent exposure and inundation by sea water during high tides, (c) special adaptations of the mangrove plants, (d) fixed plant parts of some of the mangroves that provide decaying substrata for colonization by marine fungi, (e) less or no turbulence in mangroves when compared to high seas or beaches. The above factors might have provided a natural selection to develop a unique marine mycota in mangroves.

While Halosphaeriales prefer lower parts of the mangroves where inundation takes place, the loculoascomycetes (Pleosporales, Dothideomycetes) prefer the intertidal parts. This clearly shows that mangroves in general, and *Rhizophora* spp. in particular, are influenced by tidal amplitude where the sea water inundation promote fungi belonging to Halosphaeriales and other unitunicate ascomycetes, wheras the intertidal region and the exposure of the plant parts promote the loculoascomycetes with fissitunicate mechanism (Hyde 1988, 1990a; Hyde et al. 1993; Alias and Jones 2000; Sarma and Vittal 2002). Many representatives of Halosphaeriales and other unitunicates having an apical ring at the apex of the asci and bitunicates having fissitunicate mechanism are suited for an active dispersal of the spores in the mangrove environment. Though the fungi have adaptations to withstand exposure to sunlight, desiccation and dryness, they also have features of typical aquatic mode of life, where the spores of many of the marine ascomycete genera, also have appendages or mucilaginous sheaths for attachment to the surfaces. While the appendages is found more in the genera of Halosphaeriales, the mucilaginous sheaths are found more in the order Pleosporales.

Fungi that occur above the water mark have different strategies. The apothecoid fungi viz., *Lecanidion atratum* and *Dactylospora haliotrepha* that occur superficially produce superficial ascomata and have a gelatinous substance in their fruit bodies, particularly the latter fungus. In these two fungi, as they are exposed to sunlight and heat, they form melanized fruit bodies to escape from the solar radiation. Another fungus in this category is *Hysterium* sp. which produces hysterothecia and produces superficial fruit bodies. Further, many of the fungi growing above tidal mark mainly belong to mitosporic fungi, e.g., *Epicoccum purpurascens*, *Trimmatostroma* sp. (Vittal and Sarma 2006).

The status of a few fungi occurring on this host is doubtful and a few others are actually terrestrial fungi but occur occasionally in the intertidal region. These are highlighted in Table 13.1 and they require further studies to establish whether they are marine fungi or "salt-tolerant fungi." Typical mangrove plants in general accumulate salt through their roots and excrete it out through salt glands on the leaves. That would mean the aerial parts have considerable amount of salt concentrations. Hence the fungi growing on the aerial parts of the mangrove plants have to cope up with this excreted salt. From this angle, it could be said that the fungi occurring on aerial parts of the mangroves are "salt-tolerant fungi."

Plant parts in mangroves are exposed intermittently to sea water for different periods of time or upper parts mostly exposed to sun light and radiation without inundation by water. This kind of a condition has allowed more number of loculoascomycetes with active spore dispersal mechanism to thrive in mangroves. Most of these loculoascomycetes are not found on driftwood samples. But these submerged parts are also exposed for short periods or longer periods of time if the tidal range is small. Hence the marine fungi (Halosphaeriales and other Unitunicate ascomycetes) are exposed to sun light, radiation and desiccation, even if it is for shorter periods of time. Due to these reasons the Halosphaeriales commonly recorded from mangroves are not the ones found on driftwood samples. Similarly, typical Halosphaerialean fungi with elaborate appendages, bristles and sheaths are not found in mangroves or have low percentage occurrence in mangroves. The main factor that influences mycota seems to be the continuous connection of the drift wood samples with the sea water and their non-exposure even for shorter periods of time, which distinguishes "driftwood-marine fungi" from "mangrove wood-marine fungi."

Within mangroves, once again, each plant has its own special adaptations. For example, *Rhizophora* spp. have prop roots that branch out from the main trunk and are fixed into the marshy mangrove habitat and characteristic viviparous seedlings, whereas *Avicennia* spp. possess negatively geotropic pneumatophores and a large trunk with aerial branches which are not subjected to inundation. Hence in the case of *Avicennia* spp. the dead and fallen woody litter on the mangrove floor is the available substratum for marine fungi to colonize. Rarely the roots of *Avicennia* spp. on the banks of the canals and tributaries are exposed and are available for fungal colonization. These factors contribute to the difference in marine mycota occurring on *Avicennia* spp. or other mangrove species. However, there could be an overlap of marine fungi between *Rhizophora* spp. and *Avicennia* spp., indicating their wide host range and high ecological amplitude.

13.5 Conclusions

Rhizophora spp. are important hosts for marine fungi. Accommodating more than 201 fungi is a unique feature of this host genus. No other genus supports so many fungi. *Nypa fruticans* is another host which accommodates more than 100 marine fungi. This is followed by *Avicennia* spp. However, the advantage of *Rhizophora* spp. is that in addition to being a host for rich fungal diversity, it is also one of the ideal plants for ecological experimentation including vertical zonation, frequency of occurrence, substrate preferences and various other ecological observations.

We need to study the relationship between the chemistry/biochemistry of the host and the number of fungi growing on *Rhizophora* spp. and compare fungal diversity and the chemistry of prop roots, wood and leaves to substantiate the hypothesis that "the chemistry of the host plays a major role in supporting the diversity of fungi on a particular host." Further, there are more than 10 fungi that are specific to *Rhizophora* spp. and around 40 fungi are considered as "core group fungi" (most frequent) that thrive on this host. In future, it would be worthwhile to direct studies on these fungi, at molecular level, and also trace the biochemical pathways, to observe which particular carbohydrate or other compounds they prefer to thrive on this host. It is recommended that marine mycologists provide data on marine fungi occurring on each mangrove host separately in future so that maintaining a database will be easy.

Acknowledgements The author would like to thank Prof. BPR Vittal, CAS in Botany, University of Madras, Chennai, India; Dr. S Raghukumar, MykoTech Pvt. Ltd. 313, Vainguinnim Valley, Dona Paula, Goa - 403 004, India, for encouragement. Prof. EBG Jones of Biotech, Thailand, for generously providing me copies of some of his invaluable publications, for his suggestions and encouragement.

References

Abdel-Wahab MA, Pang K-L, Nagahama T, Abdel-Aziz FA, Jones EBG (2010) Phylogenetic evaluation of anamorphic species of *Cirrenalia* and *Cumulospora* with the description of eight new genera and four new species. Mycol Prog 9:537–558

Aleem AA (1980) Distribution and ecology of marine fungi in Sierra Leone (Tropical West Africa). Bot Mar 23:679–688

Alias SA, Jones EBG (2000) Vertical distribution of marine fungi in *Rhizophora apiculata* at Morib mangrove, Selangor, Malaysia. Mycoscience 41:431–436

Alias SA, Zainuddin N, Jones EBG (2010) Biodiversity of marine fungi in Malaysian mangroves. Bot Mar 53:545–554

Campbell J, Anderson JL, Shearer CA (2003) Systematics of *Halosarpheia* based on morphological and molecular data. Mycologia 95:530–552

Chalkley DB, Suh S-O, Volkmann-Kohlmeyer B, Kohlmeyer J, Zhou JJ (2010) *Diatrypasimilis australiensis*, a novel xylarialean fungus from mangrove. Mycologia 102:430–437

Chinnaraj S (1993) Higher marine fungi from mangroves of Andaman and Nicobar islands. Sydowia 45:109–115

Fell JW, Master IM (1980) The association and potential role of fungi in mangrove detrital systems. Bot Mar 23:257–263

Fell JW, Cefalu RC, Master IM, Tallman AS (1975) Microbial activities in the mangrove (*Rhizophora mangle*) leaf detrital system. In: Walsh GE, Snedaker SC, Teas HJ (eds) Biology and management of mangrove, Proceedings of the international symposium on biology and management of mangrove. University of Florida, Gainesville, FL, pp 661–679

Heald EJ, Odum WE (1970) The contribution of mangrove swamps to Florida fisheries. Proc Gulf and Caribb Fish Inst 22:130–135

Hibbett DS (2007) After the gold rush, or before the flood? Evolutionary morphology of mushroom-forming fungi (Agaricomycetes) in the early 21st century. Mycol Res 111:1001–1018

Hyde KD (1988) Observations on the vertical distribution of marine fungi in *Rhizophora* spp. at Kampong Danau mangrove, Brunei. Asian Mar Biol 5:77–81

Hyde KD (1989) Intertidal mangrove fungi from north Sumatra. Can J Bot 67:3078–3082

Hyde KD (1990a) A study of the vertical zonation of intertidal fungi on *Rhizophora apiculata* at Kampong Kapok mangrove, Brunei. Aquat Bot 36:255–262

Hyde KD (1990b) A comparison of intertidal mycota of five mangrove tree species. Asian Mar Biol 7:93–107

Hyde KD (1991) Fungal colonization of *Rhizophora apiculata* and *Xylocarpus granatum* poles in Kampong Kapok mangrove, Brunei. Sydowia 43:31–38

Hyde KD (1996) Marine fungi. In: Grgurinovic C, Mallett K (eds) Fungi of Australia, vol 1B. ABRS/CSIRO, Canberra, pp 39–64

Hyde KD, Chalennpongse A, Boonthavikoon T (1990) Ecology of intertidal fungi at Ranong mangrove, Thailand. Trans Mycol Soc Jpn 31:17–27

Hyde KD, Chalermpongse A, Boonthavikoon T (1993) The distribution of intertidal fungi on *Rhizophora apiculata*. In: Morton B (ed) The marine biology of the South China Sea, Proceedings of the first international conference on the marine biology of Hong Kong and South China Sea. Hong Kong University Press, Hong Kong, pp 643–652

Hyde KD, Jones EBG, Leano E, Pointing SB, Poonyth AD, Vrijmoed LLP (1998) Role of marine fungi in marine ecosystems. Biodivers Conserv 7:1147–1161

Hyde KD, Sarma VV, Jones EBG (2000) Morphology and taxonomy of higher marine fungi. In: Hyde KD, Pointing SB (eds) Marine mycology – a practical approach, Fungal diversity research series 1. Fungal Diversity Press, Hong Kong, pp 172–204

Hyde KD, Wong WS, Aptroot A (2002) Marine and estuarine species of *Lophiostoma* and *Massarina*. In: Hyde KD (ed) Fungi in marine environments, vol 7, Fungal diversity research series. Fungal Diversity Press, Hong Kong, p 93

Jones EBG (1988) Do fungi occur in the sea? Mycologist 2:150–157

Jones EBG, Alias SA (1997) Biodiversity of mangrove fungi. In: Hyde KD (ed) Biodiversity of tropical microfungi. Hong Kong University Press, Hong Kong, pp 71–92
Jones EBG, Kuthubutheen AJ (1989) Malaysian mangrove fungi. Sydowia 41:160–169
Jones EBG, Pilantanapak A, Chatmala I, Sakayaroj J, Phongpaichit S, Choeyklin R (2006) Thai marine fungal diversity. Songklanakarin J Sci Technol 28:687–708
Jones EBG, Sakayaroj J, Suetrong S, Somrithipol A, Pang KL (2009) Classification of marine Ascomycota, anamorphic taxa and Basidiomycota. Fungal Divers 35:1–187
Kohlmeyer J (1969) Ecological notes on fungi in mangrove forests. Trans Br Mycol Soc 53:237–250
Kohlmeyer J, Kohlmeyer E (1979) Marine mycology: the higher fungi. Academic, New York, 690 pp
Kohlmeyer J, Volkmann-Kohlmeyer B (1991) Marine fungi of Queensland, Australia: *Massarina lacertensis* Kohlm. and Volkm.-Kohlm. Aust J Mar Freshw Res 42:91–99
Kohlmeyer J, Volkmann-Kohlmeyer B (1998) A new marine *Xylomyces* on *Rhizophora* from the Caribbean and Hawaii. Fungal Divers 1:159–164
Kohlmeyer J, Bebout B, Volkmann-Kohlmeyer B (1995) Decomposition of mangrove wood by marine fungi and teredinids in Belize. Mar Ecol (PSZNI) 16:27–39
Kuthubutheen AJ (1981) Fungi associated with the aerial parts of Malaysian mangrove plants. Mycopathol 76:33–43
Leong WF, Tan TK, Jones EBG (1991) Fungal colonization of submerged *Bruguiera cylindrical* and *Rhizophora apuculata* wood. Bot Mar 34:69–76
Lugo AE, Snedaker SC (1974) The ecology of mangroves. Annu Rev Ecol Syst 5:39–64
Maria GL, Sridhar KR (2003) Diversity of filamentous fungi on woody litter of five mangrove plant species from the southwest coast of India. Fungal Divers 14:109–126
Nakagiri A (1993) Intertidal mangrove fungi from Irimote Island. IFO Res Commun 16:24–62
Nambiar GR, Raveendran K (2009) Manglicolous marine fungi on *Avicennia* and *Rhizophora* along Kerala coast (India). Middle-East J Sci Res 4:48–51
Odum EP (1971) Fundamentals of mycology. W.B. Saunders Company, Philadelphia, USA
Pang K-L, Vrijmoed LLP, Kong RYC, Jones EBG (2003) *Lignincola* and *Nais*, polyphyletic genera of the Halosphaeriales (Ascomycota). Mycol Prog 2:29–36
Poonyth AD, Hyde KD, Peerally A (1999) Intertidal fungi in Mauritian mangroves. Bot Mar 42:243–252
Raghukumar S, Sharma S, Raghukumar C, Sathe Pathak V, Chandramohan D (1994) Thraustochytrid and fungal component of marine detritus. 4. Laboratory studies on decomposition of leaves of the mangrove *Rhizophora apiculata* Blume. J Exp Mar Biol Ecol 183:113–131
Ravikumar DR, Vittal BPR (1996) Fungal diversity on decomposing biomass of mangrove plant Rhizophora in Pichavaram estuary, east coast of India. Ind J Mar Sci 25:142–144
Sarma VV, Vittal BPR (1998–1999) Ecological studies on mangrove fungi from east coast of India. Observations on seasonal occurrence. Kavaka 26–27:105–120
Sarma VV, Vittal BPR (2002) Observations on vertical distribution of manglicolous fungi on prop roots of *Rhizophora apiculata* Blume at Krishna delta, east coast of India. Kavaka 30:21–29
Schmidt JP, Shearer CL (2003) A checklist of mangrove-associated fungi, their geographical distribution and known host plants. Mycotaxon 85:423–477
Sridhar KR (2009) Fungal diversity of Pichavaram mangroves, Southeast coast of India. Nat Sci 7:67–75
Sridhar KR, Maria GL (2006) Fungal diversity on mangrove woody litter *Rhizophora mucronata* (Rhizophoraceae). Ind J Mar Sci 35:318–325
Suetrong S, Schoch CL, Spatafora JW, Kohlmeyer J, Volkmann-Kohlmeyer B, Sakayaroj J, Phongpaichit S, Tanaka K, Hirayama K, Jones EBG (2009) Molecular systematics of the marine *Dothideomycetes*. Stud Mycol 64:155–173

Vinayachandran PN, Kurian J (2008) Modeling Indian ocean circulation: Bay of Bengal fresh plume and Arabian sea mini warm pool. Proceedings of the 12th Asian Congress of Fluid Mechanics, 18–21 August 2008, Daejeon, Korea

Vittal BPR, Sarma VV (2006) Diversity and ecology of fungi on mangrove in Bay of Bengal region – an overview. Ind J Mar Sci 35:308–317

Volkmann-Kohlmeyer B, Kohlmeyer J (1993) Biogeographic observations on Pacific marine fungi. Mycologia 85:337–346

Chapter 14
Biotechnology of Marine Fungi

Samir Damare, Purnima Singh, and Seshagiri Raghukumar

Contents

Abstract Filamentous fungi are the most widely used eukaryotes in industrial and pharmaceutical applications. Their biotechnological uses include the production of enzymes, vitamins, polysaccharides, pigments, lipids and others. Marine fungi are a still relatively unexplored group in biotechnology. Taxonomic and habitat diversity form the basis for exploration of marine fungal biotechnology. This review covers what is known of the potential applications of obligate and marine-derived fungi obtained from coastal to the oceanic and shallow water to the deep-sea habitats. Recent studies indicate that marine fungi are potential candidates for novel enzymes, bioremediation, biosurfactants, polysaccharides, polyunsaturated fatty acids and secondary metabolites. Future studies that focus on culturing rare and novel marine fungi, combined with knowledge of their physiology and biochemistry will provide a firm basis for marine mycotechnology.

S. Damare (✉)
Marine Biotechnology Laboratory, CSIR-National Institute of Oceanography, Dona Paula, Goa 403004, India
e-mail: samir@nio.org

C. Raghukumar (ed.), *Biology of Marine Fungi*,
Progress in Molecular and Subcellular Biology 53,
DOI 10.1007/978-3-642-23342-5_14, © Springer-Verlag Berlin Heidelberg 2012

14.1 Introduction

Freshwater and terrestrial bacteria and fungi have played a major role in biotechnology by virtue of their biochemical diversity and adaptations to a variety of environmental conditions (Bennett 1998). In comparison, marine microorganisms, particularly marine fungi are yet to find a prominent place in biotechnology. The oceans are home to a vast biodiversity (Ray 1988; Norse 1993). Most of marine microbial diversity has not been discovered and characterized, both taxonomically, as well as biochemically. It is logical, therefore, to conclude that marine microorganisms hold vast promise for improved and novel biotechnologies. Marine bacteria have been major targets for biotechnology research. On the contrary, marine fungi have attracted attention from biotechnologists only in recent years. This review focuses on what is known of marine fungal biotechnology so far, their potentials and future directions.

14.2 Diversity and Habitats of Fungi in the Marine Environment

Taxonomic and phylogenetic diversity of organisms, as well as their occurrence and adaptation to various habitats often provide clues to their potential applications in biotechnology (Raghukumar 2008a). Marine fungi may fall under 'true fungi' of the Kingdom Fungi or the 'straminipilan fungi' of the Kingdom Straminipila. Taxonomic diversity of marine fungi can be assessed by direct observations of sporulating structures, culturing and metagenomics. Culturable and taxonomically characterized marine fungi belong either to obligate or facultative (marine-derived) fungi (Kohlmeyer and Kohlmeyer 1979). The former are not found in freshwater and terrestrial environments, and grow and reproduce exclusively in the sea. Most of these are morphologically distinct. Some, such as the Halosphaeriales and Lulworthiales also constitute taxonomically distinct orders among the Ascomycetes (Jones et al. 2009). Current evidence points out that these are terrestrial fungi that have colonized marine habitats and have become adapted to those conditions (Spatafora et al. 1998). About 530 species of obligate marine fungi are known so far, most of these occurring on decomposing lignocellulosic material in the coastal environment, although a few have been described from decaying wood in the deep sea (Jones et al. 2009). The very fact that these are obligately marine, points out to their novelty compared to terrestrial fungi and makes their study attractive for biotechnological applications. In contrast to the obligate forms, 'facultative' or marine-derived fungi belong to terrestrial species. These have been isolated from a wide variety of marine habitats and it is strongly believed that these terrestrial species have adapted to marine conditions and probably have special biochemical properties (Damare et al. 2006a, b). The straminipilan, marine fungi, the Labyrinthulomycetes were once considered 'true' fungi, but are now known to belong to the Kingdom Stramenopila (Raghukumar 2002). Metagenomic studies of marine habitats have revealed a large diversity of hitherto uncultured fungi (Edgcomb et al. 2002; Burgaud et al. 2009; Singh et al. 2011). Most of these do not appear to correspond to known fungal taxa and are of much interest to marine biotechnologists.

Salinity is the most defining feature of the oceanic environment. Mechanisms of salinity tolerance in fungi have been discussed by Jennings (1986) and Blomberg and Adler (1992). Physiological and biochemical adaptation of fungi to salinity can be of relevance to biotechnology. The ability to withstand salinity has been addressed from the point of view of biotechnology in the yeast *Saccharomyces cerevisiae*. Overexpression of the *HAL1* gene in yeast has a positive effect on salt tolerance by maintaining a high internal K^+ concentration and decreasing intracellular Na^+ during salt stress. Gisbert et al. (2000) have successfully introduced *HAL1* gene from this yeast into tomato. Higher level of salt tolerance was recorded in the progeny of two different transgenic plants bearing four copies or one copy of the *HAL1* gene. This has the scope for designing salt-tolerant crops.

In terms of nutrients in the marine environment, lignocellulosic substrates from coastal environments have been known for a long time to harbour a distinct diversity of obligate marine fungi, the lignicolous fungi (Kohlmeyer and Kohlmeyer 1979; Sridhar 2005). Among these, the manglicolous fungi that grow on decaying woody material in mangroves comprise an important group. Mangroves are detritus-based ecosystems and substantial fungal populations are involved in detritus processing (Raghukumar 2004). Mangrove fungi are the second largest group among the marine fungi.

Marine endophytic fungi that live inside the living tissues of seagrasses and macroalgae constitute an interesting ecological group of fungi (Raghukumar et al. 1992; Sathe-Pathak et al. 1993; Thirunavukkarasu et al. 2011).

Coral reefs provide an excellent habitat for fungi. Fungi grow both on the surface and interior of scleractinian corals. They are found both within the polyps and calcium carbonate skeleton of such corals. In the latter case, they exist as endoliths by penetrating the calcium carbonate structures. Coral reef fungi are also found as endobionts, being associated with living organisms such as coral polyps, sponges and holothurians, or as saprotrophs in coral mucus, plant detritus, sediments and water column. They are not only responsible for occasional coral diseases but may also have a mutualistic role in healthy corals (Le Campion-Alsumard et al. 1995). The report of aspergillosis disease in the sea fan *Gorgonia* (Smith et al. 1996) has kindled a worldwide interest in mycoflora of soft corals. Fungi have been isolated from the gorgonian soft coral, *Gorgonia ventalina* collected from Puerto Rico (Toledo-Hernández et al. 2007). Koh et al. (2000) isolated 16 fungal genera and 51 species, including 2 yeasts from 10 species of gorgonian corals in Singapore by culture-dependent technique. The microbial metagenome of *Porites astreoides* collected from Bocal del Toro, Panama, showed fungi to be the dominant community, contributing 38% to the genome of the coral (Wegley et al. 2007).

The deep sea is an extreme environment, characterized by temperatures around 2°C and high hydrostatic pressures. Detailed studies on deep-sea sediments of

Central Indian Basin suggest that fungi are an important component therein (Damare et al. 2006a; Damare and Raghukumar 2008; Singh et al. 2010). Most of the fungi isolated from these sediments belong to terrestrial species. Burgaud et al. (2009) have successfully cultured fungi from various benthic organisms in deep-sea hydrothermal vents. Physiological studies showed that these were adapted to deep-sea conditions. Several of these appear to be undescribed forms. Such fungi are important tools for biotechnology research. Raghukumar et al. (2008) isolated fungi and thraustochytrids for the first time from a shallow water hydrothermal vent site at the D. João de Castro Seamount (DJCS) located in the North Atlantic, between the Azorean islands São Miguel and Terceira lying on the Terceira Rift. Thraustochytrids isolated from this habitat had a high tolerance to heavy metals and secreted protease even under metal stress. Bass et al. (2007) have reported the dominance of yeasts in the deep sea. Singh et al. (2011) applied a metagenomics approach to detect fungi from the deep-sea sediments of the Central Indian Basin, using a multiple primer approach that included 18 S rDNA and ITS primers. The sea is also a window for 'virtual time travel', making it possible to culture fungi from a past of different environmental conditions. For example, Raghukumar et al. (2004) cultured fungi that had been presumably buried for nearly 420,000 years at about 365 cm below the sea floor at 5,000 m depth in the Central Indian Ocean. Such fungal 'dinosaurs' and 'wooly mammoths' might have their own uses.

Salt pans are an attractive source for extremophilic microorganisms. The high salinity in hypersaline environments exerts an osmotic effect and also adversely affects protein structure and enzyme function (Cooke and Whipps 1993). No obligate marine fungi have been detected so far in salt pans. However, other fungi, such as black yeasts, *Hortaea werneckii*, *Phaeotheca triangularis*, *Aureobasidium pullulans* and *Trimatostroma salinum* have been isolated from salt pans of 15–30% salinity (Zalar et al. 1999a, b; Gunde-Cimerman et al. 2000). Out of these, the black yeast and *Phaeotheca triangularis* appear to be obligate halophiles, while the rest are facultative halotolerant.

Yet another extreme environment for marine fungi are the anoxic habitats. Fungi are primarily aerobic heterotrophs and were thought to play a negligent role in the ecosystem processes of anoxic environments (Dighton 2003). However, many fungi were recently shown to possess metabolic adaptations to utilize nitrate and (or) nitrite as an alternative for oxygen (Shoun et al. 1992). Jebaraj and Raghukumar (2009) have shown that fungi isolated from anoxic marine waters of the Arabian Sea are capable of growth under oxygen-deficient conditions while performing anaerobic denitrification, a characteristic that is bound to interest biotechnologists.

14.3 Enzymes from Marine Fungi

Fungi produce a variety of extracellular enzymes, such as proteases, laccase, amylases, xylanases and cellulases that have found applications in all fields of life, including food, beverages, detergents and medicines (Table 14.1). Enzymes

Table 14.1 Enzymes from marine fungi with potential applications

Enzyme	Source of isolation	Classification of compound	Substrate	References
Laccase	Basidiomycetous fungi	Copper containing protein	Phenolic compounds	Raghukumar et al. (1999)
Manganese peroxidase	Basidiomycetous fungi	Heam protein	Phenolic compounds	Raghukumar et al. (1999)
Lignin peroxidase	Basidiomycetous fungi	Heam protein	Phenolic compounds	Raghukumar et al. (1999)
Alkaline Protease	*Aspergillus ustus*	Serine protease	Azocasein	Damare et al. (2006a)
Cellulase	Lignicolous mangrove fungi	–	Cellulose	Eriksson et al. (1990)
L-Glutaminase	*Beauveria bassiana*	–	L-Glutamine	Keerthi et al. (1999)
Xylanase	*Aspergillus niger*	Glycoside hydrolase	Paper pulp	Raghukumar et al. (2004)
Gelatinase	*Halosphaeria mediosetigera*	Metalloproteinase	Corn meal	Pisano et al. (1964)
Amylase	*Aspergillus flaviceps*	Glycoside hydrolase	Starch	Frolova et al. (2001)

obtained from marine fungi are likely to differ from those produced by terrestrial fungi because of their differences in taxonomic diversity and environmental adaptations (Abe et al. 2001, 2006; Damare 2007).

Most research on enzymes from marine fungi has focused on lignocellulases obtained from fungi in the coastal environment. Fungal degradation of lignin has attracted the attention of researchers because lignin-degrading enzymes have applications in using lignocellulose as a renewable resource for the production of paper products, feeds, chemicals and fuels (Kuhad et al. 1997). Fungi degrade lignin by production of the extracellular enzymes, lignin peroxidase (LiP), manganese peroxidase (MnP) and laccase. These enzymes degrade or modify not only lignin, but also several aromatic, recalcitrant environmental pollutants such as those that occur in crude oil wastes, textile effluents and organochloride agrochemicals which are a cause of serious environmental pollution (Mtui and Nakamura 2004; Kiiskinen et al. 2004). Marine mangrove fungi have proven to be an important source of lignocellulose degrading enzymes (Raghukumar et al. 1994; Grant et al. 1996; Pointing et al. 1998; Pointing and Hyde 2001). Pointing and Hyde (2001) suggested that most marine fungal strains capable of decay activity are likely to be soft-rot fungi, with relatively few capable of white-rot decay. Raghukumar et al. (1994) demonstrated the presence of laccase, xylanase and cellulase activities in several obligate marine as well as marine-derived fungi isolated from mangroves. Luo et al. (2005) have reported cellulolytic, xylanolytic and ligninolytic enzymes from fungi obtained from tropical and sub-tropical mangroves. Most of the fungi were ascomycetes, except for one basidiomycete, *Calathella mangrovei* and a mitosporic fungus, *Cirrenalia tropicalis*. A marine-derived fungus, designated NIOCC #312 produced all the three lignin degradative enzymes, namely LiP, MnP and laccase to varying extents when grown in sea water medium containing sugarcane bagasse fibres, pine and poplar wood shavings as carbon and nitrogen source (Raghukumar et al. 1999). Yet another marine-derived, white-rot fungus isolated from decaying mangrove wood, designated NIOCC #2a produced enhanced levels of laccase in the presence of several phenolics and lignin-derivatives (D'Souza-Ticlo et al. 2006). The enzyme showed optimum activity at 70°C, with half-life for 90 min at 70°C. It was active in the presence of 1 mmol NaCl and was not inhibited by Pb, Fe, Ni, Li, Co and Cd at 1 mmol. Verma et al. (2010) reported laccase production in marine-derived asco- and basidiomycetes, and demonstrated decolorization and detoxification of raw textile mill effluents by these fungi.

Deep-sea fungi are likely to produce interesting enzymes. Thus, a deep-sea yeast isolated from the Japan Trench at a depth of 4,500–6,500 m, showed the presence of an endopolygalacturonase, active at 0–10°C with no loss in activity up to 100 MPa at 24°C (Abe et al. 2006). High level of superoxide dismutase activity was detected in a *Cryptococcus* strain isolated from deep-sea sediments (Abe et al. 2001). A deep-sea isolate of *Aspergillus ustus*, cultured from sediments of the Central Indian Basin from the depths of 5,100 m, produced a low-temperature active protease (Damare et al. 2006b). The enzyme had a broad pH range of 6–10, with an optimum at pH 9. The optimum temperature for protease activity was 45°C and approximately 10% of the activity was retained at 2°C. Its activity remained unaffected in the presence of 0.5 M NaCl, equivalent to seawater salinity of 29 psu (practical salinity units). Daniel et al. (2006) proposed that it might be possible to perform a reaction requiring high temperature at lower temperature under elevated hydrostatic pressure. If so, enzymes of marine microorganisms adapted to such conditions would be ideal candidates.

Enzymes degrading polysaccharides are of great interest (Schaumann and Weide 1990; Pointing et al. 1998). To increase the yield of desired products, glucanases are used for destroying the cellular walls of plants and microorganisms. Using these enzymes, the structure and biological role of glucans can be established. Highly purified b-1, 3-glucanase has been successfully employed for the enzyme transformation of laminaran into a biologically active glucan, translam (Zvyagintseva et al. 1995).

The Labyrinthulomycetes, by virtue of their ubiquitous and abundant occurrence in the oceans are considered to play an important role in mineralization processes and are, therefore, potential sources of useful enzymes (Raghukumar 2008b). Although the presence of several degradative enzymes have been described in these organisms, such studies have been sporadic and of preliminary nature (Sharma et al. 1994; Bongiorni et al. 2005; Raghukumar et al. 2010; Nagano et al. 2011). Salt-dependent alkaline lipase production in two *Thraustochytrium* isolates was reported recently (Kanchana et al. 2011). Enzymes from these fungi deserve more attention.

14.4 Bioremediation Using Marine Fungi

Terrestrial and freshwater microorganisms, particularly bacteria, are now extensively used to treat and remedy a number of polluted environments (Brar et al. 2006). Marine microorganisms in general, and fungi in particular, have not been studied adequately for such applications. A good example of the potentials of marine-derived fungi for application in bioremediation of industrial pollutants is that of a ligninolytic marine-derived, mangrove fungus, designated NIOCC #2a (Verma et al. 2010, 2011). This fungus has been shown to decolorize and detoxify a wide variety of textile mill and molasses-based raw industrial effluents accompanied by reduction in COD and phenolics.

14.4.1 Hydrocarbon Degradation

The marine environment in many parts of the world suffers from chronic oil pollution (GESAMP 2007). Catastrophic oil pollutions such as the BP disaster in the Gulf of Mexico have often resulted in tremendous environmental damage. Marine fungi have not been adequately studied as candidates for bioremediation of marine oil spills. Ahearn and Meyers (1972) were amongst the first to discuss hydrocarbon degradation by marine fungi. Some of the most important and ecologically damaging components of oil pollution are the polycyclic aromatic hydrocarbons (PAHs). Ligninolytic fungi are interesting from this point of view, because they can oxidize PAHs by producing a non-specific enzymatic extracellular complex that is normally used for lignin depolymerization. These lignin-degrading enzymes are comprised of lignin-peroxidase (LiP), MnP and laccase (Hamman 2004; Peng et al. 2008). White-rot fungi possess powerful extracellular lignin-degrading enzymatic systems that can degrade a broad variety of different pollutants and have been extensively studied for this purpose (Junghanns et al. 2005; Sette et al. 2008). A novel PAH metabolic pathway described in fungi involves hydroxylation by cytochrome P-450 monooxygenase through a sequence of reactions that is similar to the reactions that are involved in mammalian metabolism (Capotorti et al. 2004). This pathway is shared by many non-ligninolytic fungi that could effectively degrade PAHs (Krivobok et al. 1998; Ravelet et al. 2000). The use of marine fungi in the bioremediation of polluted marine environments may be achieved because of their tolerance to saline conditions. Passarini et al. (2011) have recently reported that a marine *Aspergillus sclerotiorum* CBMAI 849 (isolated from cnidarians) was able to reduce 99.7% pyrene and 76.6% benzo[a] pyrene after 8 and 16 days, respectively. They also reported substantial amounts of benzo[a]pyrene (>50.0%) depletion by a *Mucor racemosus* CBMAI 847. They attributed these reductions showed by *A. sclerotiorum* CBMAI 849 and *M. racemosus* CBMAI 847 to their ability to metabolize pyrene to the corresponding pyrenylsulfate and benzo[a]pyrene to benzo[a]pyrenylsulfate, suggesting that the mechanism of hydroxylation is mediated by a cytochrome P-450 monooxygenase, followed by conjugation with sulfate ions.

Cooney et al. (1993) have reported obligately marine fungi growing in artificial sea water with single hydrocarbon as their sole source of carbon and energy. They observed that the unsaturated compound 1, 1,4-tetradecadiene and the methyl-branched compound pristane were used by several fungi, while none of the fungi used aromatic hydrocarbons as their sole source of carbon. The isolates reported by them were *Corollospora lacera*, *C. maritima* and *Lulworthia* sp. They found that four of the five fungi examined form microbodies when the fungi were grown on hydrocarbons as sole C-source but not when they were grown on glucose.

Asphaltenes constitute the most recalcitrant components of oil. Certain parts of the world's beaches, such as those in Goa, India, are polluted by tarballs formed from asphaltenes on a regular, annual basis. Organisms that are capable of degrading these are of tremendous biotechnological interest. Raikar et al. (2001) reported extensive degradation of tarballs by thraustochytrids isolated from coastal waters and sediments of Goa coast. They found that there was a 71% reduction of tarball contents by the thraustochytrids inoculated to tarball-enriched sediment after a month's incubation. Up to 30% of tarballs added to peptone broth were degraded in 7 days. This showed for the first time that thraustochytrids play a definite role in tarball degradation in sediments.

14.4.2 Heavy Metals

Various toxic heavy metals are released into the environment through natural as well as anthropogenic sources (Shukla and Singhal 1984). These cause serious environmental and health problems, because of their bioaccumulation in the food web. Heavy metals have been reported from the sea sediments (Karageorgis et al. 2005), resulting from river run off and anthropogenic activities. Conventional physicochemical treatment technologies (Dermont et al. 2008) become less effective and more expensive when metal concentrations are in the higher range (1–100 ppm). Many biological sources including marine algae, fungi, yeast and bacteria remove toxic metals from the environment through adsorption as well as their metabolic activity (Davis et al. 2003; Malik 2004; Kadukova and Vircikova 2005).

The potential of living and dead fungal biomass has been recognized for the removal of heavy metals through absorption (Bishnoi and Garima 2005). The latter seems to be a preferred alternative due to the absence of toxicity limitations, absence of requirements of growth media and nutrients in the treatment and also due to the fact that the biosorbed metals can be recovered and the biomass reused. There are not many reports available on the use of marine fungi for biosorption of metals in bioremediation application. Babich and Stotzky (1983) reported that the growth of some marine fungi exposed to nickel was less depressed in the presence of magnesium compared to their growth in the presence of nickel alone. Hicks and Newell (1984) found that exposing *Phaeospharia typharum*, a salt marsh fungus, to mercury at metal concentration of 0.74 mg L^{-1} resulted in no significant change in growth as compared to the cultures grown in the absence of mercury. Taboski et al. (2005) discovered the effect of lead and cadmium on the growth of two species of marine fungi, *Corollospora lacera* and *Monodictys pelagic*. A marine fungus *Aspergillus candidus* has been shown to grow successfully in the presence of arsenic (25 and 50 mg L^{-1}). The fungus decreased the amounts of the metal by means of bioaccumulation. Raghukumar et al. (2008) have shown the capability of thraustochytrids from shallow water hydrothermal vents to withstand high levels of heavy metals, as described in Sect. 14.2. A metal-tolerant yeast *Lodderomyces elongisporus* was isolated from metal-contaminated site and found to show considerable tolerance towards various heavy metals (Rehman et al. 2008). Production of the antioxidant enzyme, superoxide dismutase was found to increase as a defensive mechanism against high concentration of $CuSO_4$ in *Cryptococcus* sp., isolated from deep-sea sediments of the Japan Trench (Abe et al. 2001). Marine fungi deserve to be studied in greater detail for applications concerning heavy metal removal.

14.5 Biosurfactants and Polysaccharides from Marine Fungi

Biosurfactants were used as hydrocarbon dissolution agents during the late 1960s, but their applications have been greatly extended recently as an improved alternative to chemical surfactants (carboxylates, sulphonates and sulphate acid esters), especially in food, pharmaceutical and oil industry, because of their diversity, environmentally friendly nature, possibility of large-scale production, selectivity, performance under extreme conditions and potential applications in environmental protection (Karanth et al. 1999; Banat et al. 2000; Rahman et al. 2002).

A lipid–carbohydrate–protein complex from tropical marine yeast *Yarrowia lipolytica* was extracted from cell wall (Zinjarde and Pant 2002). Recently, a biosurfactant (novel glycolipid) producing yeast was isolated from *Calyptogena soyoae* (deep-sea cold-seep clam) in the deep sea (Konishi et al. 2010). This yeast showed abundant production of glycolipid biosurfactants at around 30 g L^{-1} for 96 h. The isolation of novel forms of such biosurfactants may prove promising for degradation of complex hydrocarbons contaminating marine environment.

An endosymbiotic isolate of *Aspergillus ustus* (MSF3) from the marine sponge *Fasciospongia cavernosa*, collected from the peninsular coast of India, was reported to produce high amounts of biosurfactants (Kiran et al. 2009). The biosurfactant produced by MSF3 was partially characterized as glycoprotein. The partially purified biosurfactant showed a broad spectrum of antimicrobial activity. It was also suggested that the biosurfactant could be used for the microbially enhanced oil recovery process. Sun et al. (2009) have reported 3 different polysaccharides PS1-1, PS1-2 and PS2-1 from a marine fungus *Penicillium* sp. F23-2. All three polysaccharides primarily consisted of mannose with variable amounts of glucose, whereas their glucoronic acid contents, molecular weights and glycosidic linkage patterns were different. All the three polysaccharides showed good antioxidant properties, especially scavenging abilities on superoxide radicals and hydroxyl radicals.

14.5.1 Production of Omega-3 Polyunsaturated Fatty Acids

The omega-3 polyunsaturated fatty acid (ω-3 PUFA), docosahexaenoic acid (DHA; 22:6ω-3) is important in human health and aquaculture (Horrocks and Yeo 1999).

While fish oil has been the commercial source of DHA for many years, microbial DHA from thraustochytrids has now captured a large market in the world, both as an adult and infant nutritional supplement and as ingredients in aquaculture feeds (Fan and Chen 2007; Raghukumar 2008b). Martek Biosciences, USA, are the major commercial manufacturers of DHA from *Schizochytrium.* Commercial processes have been improved and optimized to yield high biomass in excess of 100 g dry weight of the organism per litre culture, of which at least 50% constitute lipids, and a further 25% of this as DHA. There is a tremendous international interest in obtaining better strains, improved processes and also the molecular biology of the fatty acid synthesis pathways of thraustochytrids to enable the development of recombinant methods to produce DHA (Lippmeier et al. 2009). Thraustochytrids have now become very important organisms for marine biotechnology, in addition to their usefulness for DHA production. Thus, carotenoid and enzyme production of these organisms, besides synthesis of various essential fatty acids other than DHA are attracting attention.

14.6 Secondary Metabolites from Marine Fungi

Microorganisms have historically been a rich source of leads for pharmaceutical development, particularly for antibiotics. Over 20,000 microbial metabolites have been described, most of which were isolated from the terrestrial environment. The trend of tapering off in drug discovery from terrestrial sources has led to exploration of marine organisms for the purpose. Marine fungi have become an important component of this search.

As early as 1968, a marine isolate of the fungus *Cephalosporium acremonium* obtained from the sea near a sewage outfall of the coast of Sardina was reported to produce a number of antibiotic substances (Godzeski et al. 1968). A penicillinase-sensitive antibiotic substance named ‘Antibiotic N’, active against gram-negative bacteria, was isolated from this source. Szaniszlo et al. (1968) reported production of capsular polysaccharide by marine filamentous fungi. In recent years, marine fungi have proven to be a rich and promising source of novel anticancer, antibacterial, antiplasmodial, anti-inflammatory and antiviral agents and many new metabolites have been isolated from them (Kobayashi and Ishibashi 1993; Davidson 1995; Liberra and Lindequist 1995; Daferner et al. 2002; Bhadury et al. 2006; Ebel 2006). Bhadury et al. (2006) have reviewed novel antibacterial, antiviral, antiprotozoal compounds isolated from marine-derived fungi and their future commercial exploitation as drugs, using metabolic engineering and post-genomics approaches.

Some of the secondary metabolites produced by marine fungi and their characteristics are listed in Table 14.2. Amongst the fungi from different marine environments, marine mangrove fungi have proved to be most significant source of new bioactive compounds (Lin et al. 2001).

More than 700 unique molecular structures had been discovered from marine fungi by the end of 2008. These are divisible into most of the major classes of natural products, mirroring the situation for cultured terrestrial-derived fungi.

Ebel (2006) has provided strong evidence that 'marine-derived' fungi, although taxonomically belonging to terrestrial fungi, are an interesting ecological group that produce unique compounds. In his review, he reported 240 new compounds described from marine-derived fungi described between 2002 and 2004. A variety of secondary metabolites not found in terrestrial fungi possibly act as a chemical defence, enabling marine-derived fungi to survive competition with native microorganisms (Fenical and Jensen 1993; Gallo et al. 2004). It is believed that the use of seawater for isolation and growth of marine-derived terrestrial species enhances recovery of fungi that yield secondary metabolites, as compared to the use of media prepared in distilled water. Compounds such as the neomangicols, which are structurally unprecedented halogenated sesterterpenes that possess cytotoxic and antibacterial properties and various metabolites that exhibit cytotoxicity, anti-inflammatory and antifungal activities, as well as inhibitors of viral topoisomerase and protein tyrosine phosphatase enzymes have been cited as examples (Alvi et al. 1998; Toske et al. 1998; Schlingmann et al. 1998; Hwang et al. 1999; Belofsky et al. 2000; Rowley et al. 2003). One possible example is *Emericella variecolor* (anamorph: *Aspergillus variecolor*). Strains of this fungus have been the source of a variety of natural products, mainly sesterterpenes with unusual polycyclic skeletons, for example, astellatol (Sadler and Simpson 1989) or variecolin (Hensens et al. 1991), and prenylated xanthones (Chexal et al. 1974; Kawahara et al. 1988). Brauers et al. (2000) found a remarkable diversity of secondary metabolites in the preliminary HPLC screening produced by a strain of *E. variecolor* isolated from the sponge *Haliclona valliculata*. Detailed chemical analysis revealed evariquinone (1, 2, 3- trihydroxy-6-methyl-8-methoxyanthraquinone), which is a new anthraquinone, besides the known 1,2,3,8-tetrahydroxy-6-methylanthraquinone (7-hydroxyemodin). Other new compounds were prenylxanthone and isoemericellin, accompanied by the biosynthetically related known metabolite shamixanthone. Furthermore, the C-glycosidic depside stromemycin was identified and patented for its metalloproteinase-inhibiting activity (Hopmann et al. 2001).

Coral reefs are a rich source of fungi that produce secondary metabolites. These may be produced as a result of association with other biotic components present in the reefs. Namikoshi et al. (2000) have reported two new compounds, paecilospirone and phomopsidin, and seven known compounds, chaetoglobosin A, griseofulvin, fusarielin A, fusapyrone, deoxyfusapyrone, verrucarins J and L acetate from marine-derived fungi collected in tropical and sub-tropical coral reef environments. An anticancer compound, Sorbicillactone A has been isolated from a strain of *Penicillium chrysogenum* cultured from the Mediterranean sponge *Ircinia fasciculate*. This compound exhibited a highly selective activity against the murine leukemic lymphoblast cell line L5178y, and has antiviral and neuro-protective properties. Owing to its excellent antileukemic properties, this compound has also qualified for human trials (Bringmann et al. 2007; Bhatnagar and Kim 2010).

Table 14.2 Secondary metabolites produced by marine fungi

Secondary metabolite	Source of isolation and habitat	Type of compound	Activity	References
Fusidic acid	*Stilbella aciculosa*	Steroid	Antibacterial	Kuznetsova et al. (2001)
Aspergillitine	*Aspergillus versicolor* (isolated from the sponge *Xestospongia exigua*)	Chromone-derivative	Antibacterial	Lin et al. (2003)
Guisinol	*Emericella unguis* (obtained from a mollusc)	Depside	Antibacterial	Nielsen et al. (1999)
Penicillic acid	*Aspergillus* sp. (Coral reef of South china Sea)	–	Mycotoxin, potential carcinogen and mutagen	Hou-jin et al. (2010)
Varixanthone	*Emericella variecolor* (sponge derived)	–	Antibacterial	Malstrom et al. (2002)
7-deacetoxyyanuthone A	*Penicillium* sp.	Polyoxygenatedfarnesyl cyclohexenones	In vitro activity against methicillin and multidrug-resistant *S. aureus*	Li et al. (2003)
Microsphaeropsin	*Microsphaeropsis* sp. (derived from the sponge *Myxilla incrustans*)	Eremophilane derivative	Antifungal	Holler (1999)
Penicitides A	*Penicillium chrysogenum* (endophytic fungus from a marine red algal species of the genus *Laurencia)*	Polyketide derivatives	Inhibits pathogenic fungus *A. brassicae*, human hepatocellular liver carcinoma cell line	Gao et al. (2011)
Penicimonoterpene	*Penicillium chrysogenum* (endophytic fungus from a marine red algal species of the genus *Laurencia)*	Monoterpene derivative	Inhibits pathogenic fungus *A. brassicae*	Gao et al. (2011)
Zopfiellamides A and B	*Zopfiella latipes*	Pyrrolidinone derivative	Inhibits *Nematospora coryli* and *Saccharomyces cerevisiae*	Daferner et al. (2002)
Seragikinone A	Unidentified marine fungus (derived from the Rhodophyte *Ceratodictyon spongiosum*)	Anthracycline-related pentacyclic compound	Weak antifungal activity against *C. albicans*	Shigemori et al. (1999)
	Keissleriella sp. YS4108	–		Liu et al. (2002)

3,6,8-trihydroxy-3-[3,5-dimethyl-2-oxo-3(E)-heptenyl]-2,3-dihydronaphthalen-1 (4H)-one			Inhibits *C. albicans*, *T. rubrum* and *A. niger*	
Enniatin B	*Fusarium* sp.	Cyclodepsipeptide	Exhibits antibiotic activity against *S. aureus* and vancomycin-resistant enterococci	Jiang et al. (2002)

Metabolites exploited in pharmaceutical and agricultural industries are widespread among endophytic fungi (Petrini et al. 1992). These fungi produce an extraordinary diversity of metabolites some of which have therapeutic value as novel antibiotics or anticancer chemicals (Gunatilaka 2006; Suryanarayanan et al. 2009; Weber 2009). Thirunavukkarasu et al. (2011) have recently reported production of L-asparaginase from a *Fusarium* sp. isolated from the thallus of *Sargassum wightii* and a sterile mycelial form isolated from the thallus of *Chaetomorpha* sp. showed maximum activity of the enzyme.

With more access to the deeper parts of oceans, the information about the fungi from deep sea is increasing rapidly and the forms obtained would prove very good source of novel biomolecules.

Most of the herbicides in use today are chemical moieties. Decreasing chemical heterogeneity of herbicides targeting fewer mechanisms of action is increasing the prevalence of herbicide resistance (Lein et al. 2004). Motti et al. (2007) screened extracts of 449 marine-derived fungi for inhibition of pyruvate phosphate dikinase (PPDK), which hinders growth in C4 plants. This enzyme occurs primarily in plants, but not reported so far from vertebrate or invertebrate animals, with the exception of protozoan, *Giardia*, potentially minimizing the risk of PPDK inhibitors exhibiting adverse toxicological effects. They isolated unguinol, a known compound and found it to inhibit PPDK via a novel mechanism of action which also translates to an herbicidal effect on whole plants.

14.7 Future Directions

Marine mycotechnology has already revealed much promise. Interesting enzymes, novel metabolic properties and secondary metabolites have been discovered from obligate and marine-derived fungi from a variety of marine habitats, ranging from coastal to oceanic and shallow water to the deep sea. There is scope to examine them for many other useful compounds, such as extracellular polysaccharides. The premise that genetic diversity based on taxonomy and adaptations to environmental conditions is the source of novel applications needs to be examined more thoroughly in future for marine fungi. Further studies on the biology of marine fungi may reveal interesting physiological and biochemical characteristics that will be useful in novel biotechnology applications. Culturing of hitherto uncultured marine fungi and more detailed studies on the physiology and biochemistry of rare and interesting ones will truly lay a foundation for the marine fungal technology.

Acknowledgements The first author is thankful to Director, NIO, for the support for the research work. The second author wishes to acknowledge UGC for the Research Fellowship provided to carry out the work. This is NIO's contribution number 5012.

References

Abe F, Miura T, Nagahama T, Inoue A, Usami R, Horikoshi K (2001) Isolation of a highly copper-tolerant yeast, *Cryptococcus* sp., from the Japan Trench and the induction of superoxide dismutase activity by Cu^{2+}. Biotechnol Lett 23:2027–2034

Abe F, Minegishi H, Miura T, Nagahara T, Usami R, Horikoshi K (2006) Characterization of cold- and high-pressure-active polygalacturonases from a deep-sea yeast, *Cryptococcus liquefaciens* strain N6. Biosci Biotechnol Biochem 70:296–299

Ahearn DG, Meyers SP (1972) The role of fungi in the decomposition of hydrocarbons in the marine environment. In: Walters AH, Vander Hueck, Plas EH (eds) Biodeterioration of materials. Applied Science, London, pp 12–18

Alvi KA, Casey A, Nair BG (1998) Pulchellalactam: a CD45 protein tyrosine phosphatase inhibitor from the marine fungus *Corollospora pulchella*. J Antibiot 51:515–517

Babich H, Stotzky G (1983) Nickel toxicity to estuarine/marine fungi and its amelioration by magnesium in sea water. Water Air Soil Poll 19:193–202

Banat IM, Makkar RS, Cameotra SS (2000) Potential commercial applications of microbial surfactants. Appl Microbiol Biot 53:495–508

Bass D, Howe A, Brown N, Barton H, Demidova M, Michelle H, Li L, Sanders H, Watkinson SC, Willcock S, Richards TA (2007) Yeast forms dominate fungal diversity in the deep oceans. Proc R Soc B 22(274):3069–3077

Belofsky GN, Anguera M, Jensen PR, Fenical W, Kock M (2000) Oxepinamides A–C and fumiquinazolines H–I: bioactive metabolites from a marine isolate of a fungus of the genus *Acremonium*. Eur J Chem 6:1355–1360

Bennett JW (1998) Mycotechnology: the role of fungi in biotechnology. J Biotechnol 11:101–107

Bhadury P, Mohammad BT, Wright C (2006) The current status of natural products from marine fungi and their potential as anti-infective agents. J Ind Microbiol Biotechnol 33:325–337

Bhatnagar I, Kim Se-Kwon (2010) Immense essence of excellence: marine microbial bioactive compounds. Mar Drugs 8:2673–2701

Bishnoi NR, Garima A (2005) Fungus – an alternative for bioremediation of heavy metal containing wastewater: a review. J Sci Ind Res 64:93–100

Blomberg A, Adler L (1992) Physiology of osmotolerance in fungi. Adv Microb Physiol 33:145–212

Bongiorni L, Pusceddu A, Danovaro R (2005) Enzymatic activities of epiphytic and benthic thraustochytrids involved in organic matter degradation. Aquat Microb Ecol 41:299–305

Brar SK, Verma M, Surampalli RY, Misra K, Tyagi RD, Meunier N, Blais JF (2006) Bioremediation of Hazardous Wastes – A Review. J Hazard Toxic Radioactive Wastes 10:59–72

Brauers G, Edrada RA, Ebel R, Proksch P, Wray V, Berg A, Gräfe U, Schächtele C, Totzke F, Finkenzeller G, Marme D, Kraus J, Münchbach M, Michel M, Bringmann G, Schaumann K (2000) Two new betaenone derivatives and three new anthraquinones from the sponge-associated fungus *Microsphaeropsis* sp. J Nat Prod 63:739–745

Bringmann G, Gulder TA, Lang G, Schmitt S, Stöhr R, Wiese J, Nagel K, Imhoff JF (2007) Large-scale biotechnological production of the antileukemic marine natural product sorbicillactone A. Mar Drugs 5:23–30

Burgaud G, Calvez T, Arzur D, Vandenkoornhuyse P, Barbier G (2009) Diversity of culturable marine filamentous fungi from deep-sea hydrothermal vents. Environ Microbiol 11:1588–1600

Capotorti G, Digianvincenzo P, Cesti P, Bernardi A, Guglielmetti G (2004) Pyrene and benzo(a) pyrene metabolism by an *Aspergillus terreus* strain isolated from a polycylic aromatic hydrocarbons polluted soil. Biodegradation 15:79–85

Chexal KK, Fouweather C, Holker JSE, Simpson TJ, Young K (1974) Structure of shamixanthone and tajixanthone, metabolites of *Aspergillus variecolor*. J Chem Soc Perkin Trans 1:1584–1593

Cooke RC, Whipps JM (1993) Ecophysiology of fungi. Blackwell Scientific Publication, London, pp 324–345

Cooney JJ, Doolittle MM, Grahl-Nielsen O, Haaland IM, Kirk PW (1993) Comparison of fatty acids of marine fungi using multivariate statistical analysis. J Ind Microbiol 12:373–378

D'Souza-Ticlo D, Verma AK, Mathew M, Raghukumar C (2006) Effect of nutrient nitrogen on laccase production, its isozyme pattern and effluent decolorization by the fungus NIOCC No. 2a, isolated from mangrove wood. Ind J Mar Sci 35:364–372

Daferner M, Anke T, Sterner O (2002) Zopfiellamides A and B, antimicrobial pyrrolidinone derivatives from the marine fungus *Zopfiella latipes*. Tetrahedron 58:7781–7784

Damare S (2007) Deep-sea fungi: occurrence and adaptations. PhD thesis, Goa University, India

Damare S, Raghukumar C (2008) Fungi and macroaggregation in deep-sea sediments. Microb Ecol 27:168–177

Damare S, Raghukumar C, Raghukumar S (2006a) Fungi in deep-sea sediments of the Central Indian Basin. Deep-Sea Res I 53:14–27

Damare S, Raghukumar C, Muraleedharan UD, Raghukumar S (2006b) Deep-sea fungi as a source of alkaline and cold-tolerant proteases. Enzyme Microbiol Technol 39:172–181

Daniel I, Oger P, Winter R (2006) Origins of life and biochemistry under high-pressure conditions. Chem Soc Rev 35:858–875

Davidson BS (1995) New dimensions in natural products research: cultured marine microorganisms. Curr Opin Biotechnol 6:284–291

Davis TA, Volesky B, Mucci A (2003) A review of biochemistry of heavy metal biosorption by brown algae. Water Res 37:4311–4330

Dermont G, Bergeron M, Mercier G, Richer-Lafleche M (2008) Soil washing for metal removal: a review of physical/chemical technologies and field applications. J Hazard Mater 152:1–31

Dighton J (2003) Fungi in ecosystem processes. Marcel Dekker Inc., New York

Ebel R (2006) Secondary metabolites from marine-derived fungi. In: Proksch P, Müller WEG (eds) Frontiers in marine biotechnology. Horizon Bioscience, England, pp 73–143

Edgcomb VP, Kysela DT, Teske A, de Vera GA (2002) Benthic eukaryotic diversity in the Guaymas Basin hydrothermal vent environment. Proc Natl Acad Sci U S A 99:7658–7662

Eriksson K-E, Blanchette RA, Ander P (1990) Microbial and enzymatic degradation of wood and wood components. Springer, Berlin, p 407

Fan KW, Chen F (2007) Production of high-value products by marine microalgae thraustocytrids. In: Yang S-T (ed) Bioprocessing for value-added products from renewable resources. New Technologies and Applications, Amsterdam, Elsevier, pp 293–324

Fenical W, Jensen PR (1993) Marine microorganisms: a new biomedical resource. In: Attaway DH, Zaborsky OR (eds) Marine biotechnology, vol 1. Plenum Press, New York, pp 419–457

Frolova GM, Sil'chenko AS, Pivkin MV, Mikhailov VV (2001) Amylases of the fungus *Aspergillus flavipes* associated with *Fucus evanescens*. Appl Biochem Microbiol 38:134–138

Gallo ML, Seldes AM, Cabrera GM (2004) Antibiotic long-chain and α, β-unsaturated aldehydes from the culture of the marine fungus Cladosporium sp. Biochem Syst Ecol 32:545–551

Gao S, Li X, Du F, Li C, Proksch P, Wang B (2011) Secondary metabolites from a marine-derived endophytic fungus *Penicillium chrysogenum* QEN-24 S. Mar Drugs 9:59–70

GESAMP (IMO/FAO/UNESCO-IOC/UNIDO/WMO/IAEA/UN/UNEP Joint Group of Experts on the Scientific Aspects of Marine Environmental Protection) (2007) Estimates of oil entering the marine environment from sea-based activities. Rep Stud GESAMP No. 75, 96 pp

Gisbert C, Rus AM, Bolarín MC, López-Coronado JM, Arrillaga I, Montesinos C, Caro M, Serrano R, Moreno V (2000) The yeast *HAL1* gene improves salt tolerance of transgenic tomato. Plant Physiol 123:393–402

Godzeski CWJ, Kobayashi J, Ishibashi M (1968) Bioactive metabolites of symbiotic marine microorganisms. Chem Rev 93:1753–1769

Grant WD, Atkinson M, Burke B, Molly C (1996) Chitinolysis by the marine ascomycete *Corollospora maritima* Werdermann: purification and properties of chitobiosidase. Bot Mar 39:177–186

Gunatilaka AAL (2006) Natural products from plant-associated microorganisms: distribution, structural diversity, bioactivity and implications of their occurrence. J Nat Prod 69:509–526

Gunde-Cimerman N, Zalar P, de Hoog S, Plemenitas A (2000) Hypersaline waters in saltern-natural ecological niches for black halophilic yeast. FEMS Microbiol Ecol 32:235–240
Hamman S (2004) Bioremediation capabilities of white rot fungi. BI570 – review article Spring
Hensens OD, Zink D, Williamson JM, Lotti VJ, Chang RSL, Goetz MA (1991) Variecolin, a sesterterpenoid of novel skeleton from *Aspergillus variecolor* MF138. J Org Chem 56:3399–3403
Hicks RE, Newell SY (1984) The growth of bacteria and the fungus *Phaeosphaeriatypharum* (Desm.) Holm (Eumycota: Ascomycotina) in salt-marsh microcosms in the presence and absence of mercury. J Exp Mar Biol Ecol 78:143–155
Holler U (1999) Isolation, biological activity and secondary metabolite investigations of marine derived fungi and selected host sponges. PhD thesis, Universitat Carolo-Wilhelmina
Hopmann C, Knauf MA, Weithmann K, Wink J (2001) Aventis Pharma Deutschland GmbH, Germany, 2001. Preparation of Stromemycins as stromelysin inhibitors. PCT International Patent Application No. WO 01/44264 A2
Horrocks LA, Yeo YK (1999) Health benefits of docosahexaenoic acid (DHA). Pharmacol Res 40:211–225
Hou-jin L, Yong-tong C, Yun-yun C, Chi-keung C, Wen-jian L (2010) Metabolites of marine fungus *Aspergillus* sp. collected from soft coral *Sarcophyton tortuosum*. Chem Res Chinese U 26:415–419
Hwang Y, Rowley D, Rhodes D, Gertsch J, Fenical W, Bushman F (1999) Mechanism of inhibition of a poxvirus topoisomerase by the marine natural product sansalvamide A. Mol Pharmacol 55:1049–1053
Jebaraj CS, Raghukumar C (2009) Anaerobic denitrification in fungi from the coastal marine sediments off Goa, India. Mycol Res 113:100–109
Jennings DH (1986) Fungal growth in the sea. In: Moss ST (ed) The biology of marine fungi. Cambridge University Press, London, pp 1–10
Jiang Z, Barret MO, Boyd KG, Adams DR, Boyd ASF, Burgess JG (2002) JM47, a cyclic tetrapeptide HC-toxin analogue from a marine *Fusarium* species. Phytochemistry 60:33–38
Jones EBG, Sakayaroj J, Suetrong S, Somrithipol S, Pang KL (2009) Classification of marine Ascomycota, anamorphic taxa and Basidiomycota. Fungal Divers 35:1–189
Junghanns C, Moeder M, Krauss G, Martin C, Schlosser D (2005) Degradation of the xenoestrogen nonylphenol by aquatic fungi and their laccases. Microbiology 151:45–57
Kadukova J, Vircikova E (2005) Comparison of differences between copper bioaccumulation and biosorption. Environ Int 31:227–232
Kanchana R, Muraleedharan U, Raghukumar S (2011) Alkaline lipase activity from the marine protists, thraustochytrids. World J Microbiol Biotechnol. doi:doi:10.1007/s11274-011-0676-8
Karageorgis AP, Anagnostou CL, Kaberi H (2005) Geochemistry and mineralogy of the NW Aegean Sea surface sediments: implications for river runoff and anthropogenic impact. Appl Geochem 20:69–88
Karanth NGK, Deo PG, Veenanadig NK (1999) Microbial production of biosurfactants and their importance. Curr Sci 77:116–123
Kawahara N, Nozawa K, Nakajima S, Kawai K (1988) Isolation and structure determination of arugosin E from *Aspergillus silvaticus* and cycloisoemericellin from *Emericellastriata*. J Chem Soc, Perkin Trans 1:907–911
Keerthi TR, Suresh PV, Sabu A, Rajeevkumar S, Chandrasekaran M (1999) Extracellular production of L-glutaminase by alkalophilic *Beauveria bassiana* BTMF S10 isolated from marine sediment. World J Microbiol Biotechnol 15:751–752
Kiiskinen LL, Rättö M, Kruus K (2004) Screening for novel laccase producing microbes. J Appl Microbiol 97:640–646
Kiran GS, Hema TA, Gandhimathi R, Selvin J, Thomas TA, Rajeetha Ravji T, Natarajaseenivasan K (2009) Optimization and production of a biosurfactant from the sponge-associated marine fungus *Aspergillus ustus* MSF3. Colloids Surf B Biointerfaces 73:250–256
Kobayashi J, Ishibashi M (1993) Bioactive metabolites of symbiotic marine microorganisms. Chem Rev 93:1753–1769

Koh LL, Goh NKC, Chou LM, Tan YW (2000) Chemical and physical defenses of Singapore gorgonians (Octocorallia: Gorgonacea). J Exp Mar Biol Ecol 251:103–115

Kohlmeyer J, Kohlmeyer E (1979) Marine mycology: the higher fungi. Academic Press, New York, 690

Konishi M, Fukuoka T, Nagahama T, Morita T, Imura T, Kitamoto D, Hatada Y (2010) Biosurfactant-producing yeast isolated from *Calyptogena soyoae* (deep-sea cold-seep clam) in the deep sea. J Biosci Bioeng 110:169–175

Krivobok S, Miriouchkine E, Seigle-Murandi F, Benoit-Guyod JL (1998) Biodegradation of anthracene by soil fungi. Chemosphere 37:523–530

Kuhad RC, Singh A, Eriksson KEL (1997) Microorganisms and enzymes involved in the degradation of plant fiber cell walls. Adv Biochem Eng Biotechnol 57:47–125

Kuznetsova TA, Smetanina OF, Afiyatullov SS, Pivkin MV, Denisenko VA, Elyakov GB (2001) The identification of fusidic acid, a steroidal antibiotic marine isolate of the fungus *Stilbella aciculosa*. Biochem Syst Ecol 29:873–874

Le Campion-Alsumard T, Golubic S, Priess K (1995) Fungi in corals: symbiosis or disease? Interaction between polyps and fungi causes pearl-like skeleton biomineralization. Mar Ecol Prog Ser 117:137–147

Lein W, Bornke F, Reindl A, Ehrhardt T, Stitt M, Sonnewald U (2004) Target-based discovery of novel herbicides. Curr Opin Plant Biol 7:219–225

Li X, Choi HD, Kang JS, Lee CO, Son BW (2003) New polyoxygenated farnesylcyclohexenones, deacetoxyyanuthone A and its hydro derivative from the marine-derived fungus *Penicillium* sp. J Nat Prod 66:1499–1500

Liberra K, Lindequist U (1995) Marine fungi-a prolific resource of biologically active natural products. Pharmazie 50:583–588

Lin Y, Wu X, Feng S, Jiang G, Luo J, Zhou S, Vrijmoed LLP, Jones EBG, Krohn K, Steongröver K, Zsila F (2001) Five unique compounds: xyloketales from mangrove fungus *Xylaria* sp. from the South China Sea coast. J Org Chem 66:6252–6256

Lin W, Brauers G, Ebel R, Wray V, Berg A, Sudarsono PP (2003) Novel chromone derivatives from the fungus *Aspergillus versicolor* isolated from the marine sponge *Xestospongia exigua*. J Nat Prod 66:57–61

Lippmeier JC, Crawford KS, Owen CB, Rivas AA, Metz JG, Apt KE (2009) Characterization of both polyunsaturated fatty acid biosynthetic pathways in *Schizochytrium* sp. Lipids 44:221–230

Liu CH, Meng JC, Zou WX, Huang LL, Tang HQ, Tan RX (2002) Antifungal metabolite with a new carbon skeleton from *Keissleriella* sp YS4108, a marine filamentous fungus. Planta Med 68:363–365

Luo W, Vrijmoed LLP, Jones EBG (2005) Screening of marine fungi for lignocellulose-degrading enzyme activities. Bot Mar 48:379–386

Malik A (2004) Metal bioremediation through growing cells. Environ Int 30:261–278

Malstrom J, Christophersen C, Barrero AF, Oltra JE, Justicia J, Rosales A (2002) Bioactive metabolites from a marine derived strain of the fungus *Emericella variecolor*. J Nat Prod 65:364–367

Motti CA, Bourne DG, Burnell JN, Doyle JR, Haines DS, Liptrot CH, Llewellyn LE, Ludke S, Muirhead A, Tapiolas DM (2007) Screening marine fungi for inhibitors of the C_4 plant enzyme pyruvate phosphate dikinase: unguinol as a potential novel herbicide candidate. Appl Environ Microbiol 73:1921–1927

Mtui G, Nakamura Y (2004) Lignin-degrading enzymes from mycelial cultures of basidiomycetes fungi isolated in Tanzania. J Chem Eng Jpn 37:113–118

Nagano N, Matsui S, Kuramura T, Taoka Y, Honda D, Hayashi M (2011) The distribution of extracellular cellulase activity in marine eukaryotes, thraustochytrids. Mar Biotechnol 13:133–136

Namikoshi M, Kobayashi H, Yoshimoto T, Meguro S, Akano K (2000) Isolation and characterization of bioactive metabolites from marine-derived filamentous fungi collected from tropical and sub-tropical coral reefs. Chem Pharm Bull 48:1452–1457

Nielsen J, Nielsen PH, Frisvad JC (1999) Fungal depside, guisinol, from a marine derived strain of *Emericella unguis*. Phytochemistry 50:263–265

Norse EA (1993) Global marine biological diversity: a strategy for building conservation into decision making. Island Press, Washington, DC, p 383

Passarini MZR, Rodrigues MVN, da Silva M, Sette LD (2011) Marine-derived filamentous fungi and their potential application for polycyclic aromatic hydrocarbon bioremediation. Mar Poll Bull 62:364–370

Peng RH, Xiong AS, Xue Y, Fu XY, Gao F, Zhao W, Tian YS, Yao QH (2008) Microbial biodegradation of polyaromatic hydrocarbons. FEMS Microbiol Rev 32:927–955

Petrini O, Sieber TN, Toti L, Vivet O (1992) Ecology, metabolite production and substrate utilisation in endophytic fungi. Nat Toxins 1:185–196

Pisano MA, Mihalik JA, Catalano GR (1964) Gelatinase activity by marine fungi. Appl Microbiol 12:470–474

Pointing SB, Hyde KD (eds) (2001) BioExploitation of filamentous fungi. Fungal Divers Res Ser 6:1–467

Pointing SB, Vrijmoed LLP, Jones EBG (1998) A qualitative assessment of lignocellulose degrading activity in marine fungi. Bot Mar 41:290–298

Raghukumar S (2002) Ecology of the marine protists, the Labyrinthulomycetes (Thraustochytrids and Labyrinthulids). Eur J Protistol 38:127–145

Raghukumar S (2004) The role of fungi in marine detrital processes. In: Ramaiah N (ed) Marine microbiology: facets and opportunities. NIO, Dona Paula, Goa, India, pp 125–140

Raghukumar C (2008a) Marine fungal biotechnology: an ecological perspective. Fungal Divers 31:19–35

Raghukumar S (2008b) Thraustochytrid marine protists: production of PUFAs and other emerging technologies. Mar Biotechnol 10:631–640

Raghukumar C, Nagarkar S, Raghukumar S (1992) Association of thraustochytrids and fungi with living marine algae. Mycol Res 96:542–546

Raghukumar C, Raghukumar S, Chinnaraj S, Chandramohan D, DeSouza TM, Reddy CA (1994) Laccase and other lignocellulose modifying enzymes of marine fungi isolated from the coast of India. Bot Mar 37:515–523

Raghukumar C, D'Souza TM, Thorn RG, Reddy CA (1999) Lignin-modifying enzymes of *Flavodon flavus*, a Basidiomycete isolated from a coastal marine environment. Appl Environ Microbiol 65:2103–2111

Raghukumar C, Raghukumar S, Sheelu G, Gupta SM, Nagender Nath B, Rao BR (2004) Buried in time: culturable fungi in a deep-sea sediment core from the Chagos Trench, Indian Ocean. Deep-sea Res I 51:1759–1768

Raghukumar C, Mohandass C, Cardigos F, DeCosta PM, Santos RS, Colaco A (2008) Assemblage of benthic diatoms and culturable heterotrophs in shallow-water hydrothermal vent of the D. Joao de Castro Seamount; Azores in the Atlantic Ocean. Curr Sci 95:1715–1723

Raghukumar C, Damare SR, Singh P (2010) A review on deep-sea fungi: occurrence, diversity and adaptations. Bot Mar 53:479–492

Rahman KSM, Thahira-Rahman J, McClean S, Marchant R, Banat IM (2002) Rhamnolipid biosurfactants production by strains of *Pseudomonas aeruginosa* using low cost raw materials. Biotechnol Prog 18:1277–1281

Raikar MT, Raghukumar S, Vani V, David JJ, Chandramohan D (2001) Thraustochytrid protists degrade hydrocarbons. Ind J Mar Sci 30:139–145

Ravelet C, Krivobok S, Sage L, Steiman R (2000) Biodegradation of pyrene by sediment fungi. Chemosphere 40:557–563

Ray GC (1988) Ecological diversity in coastal zones and oceans. In: Willson EO (ed) Biodiversity. National Academy Press, Washington, DC, pp 36–50

Rehman A, Farooq H, Hasnain H (2008) Biosorption of copper by yeast, *Lodderomyces elongisporus*, isolated from industrial effluents: its potential use in wastewater treatment. J Basic Microbiol 48:195–201

Rowley DC, Kelly S, Kauffman CA, Jensen PR, Fenical W (2003) Halovirs A-E, new antiviral agents from a marine-derived fungus of the genus *Scytalidium*. Bioorg Med Chem 11:4263–4274

Sadler IH, Simpson TJ (1989) The determination by NMR methods of the structure and stereochemistry of astellatol, a new and unusual sesterterpene. J Chem Soc Chem Commun 21:1602–1604

Sathe-Pathak V, Raghukumar S, Raghukumar C, Sharma S (1993) Thraustochytrid and fungal component of marine detritus. 1. Field studies on decomposition of the brown alga *Sargassum cinereum* J. Ag. Indian J Mar Sci 22:159–167

Schaumann K, Weide G (1990) Enzymatic degradation of alginate by marine fungi. Hydrobiologia 205:589–596

Schlingmann G, Milne L, Williams DR, Carter GT (1998) Cell wall active antifungal compounds produced by the marine fungus *Hypoxylon oceanicum* LL-15 G256. II. Isolation and structure determination. J Antibiot 51:303–316

Sette LD, Oliveira VM, Rodrigues MFA (2008) Microbial lignocellulolytic enzymes: industrial applications and future perspectives. Microbiol Aus 29:18–20

Sharma S, Raghukumar C, Raghukumar S, Sathe-Pathak V, Chandramohan D (1994) Thraustochytrid and fungal component of marine detritus II. Laboratory studies on decomposition of the brown alga *Sargassum cinereum* J. Ag. J Exp Mar Biol Ecol 175:227–242

Shigemori H, Komatsu K, Mikami Y, Kobayashi J (1999) Seragakinone A, a new pentacyclic metabolite from a marine derived fungus. Tetrahedron 55:14925–14930

Shoun H, Kim DH, Uchiyama H, Sugiyama J (1992) Denitrification by fungi. FEMS Microbiol Lett 94:277–282

Shukla GS, Singhal RL (1984) The present status of biological effects of toxic metals in the environment: lead, cadmium, and manganese. Can J Physiol Pharmacol 62(8):1015–1031

Singh P, Raghukumar C, Verma P, Shouche Y (2010) Phylogenetic diversity of culturable fungi from the deep-sea sediments of the Central Indian Basin and their growth characteristics. Fungal Divers 40:89–102

Singh P, Raghukumar C, Verma P, Shouche Y (2011) Fungal community analysis in the deep-sea sediments of the Central Indian Basin by culture-independent approach. Microb Ecol 61:507–517

Smith GW, Nagelkerken IA, Ritchie KB (1996) Caribbean sea-fan mortalities. Nature 383:487

Spatafora JW, Volkmann-Kohlmeyer B, Kohlmeyer J (1998) Independent terrestrial origins of the Halosphaeriales (marine Ascomycota). Am J Bot 85:1569–1580

Sridhar KR (2005) Diversity of fungi in mangrove ecosystems. In: Satyanarayana T, Johri BN (eds) Microbial diversity: current perspectives and potential applications. I.K. International Pvt. Ltd., New Delhi, pp 129–147

Sun HH, Mao WJ, Chen Y, Guo SD, Li HY, Qi XH, Chen YL, Xu J (2009) Isolation, chemical characteristics and antioxidant properties of the polysaccharides from marine fungus *Penicillium* sp. F23-2. Carbohydr Polym 78:117

Suryanarayanan TS, Thirunavukkarasu N, Govindarajalu MB, Sasse F, Jansen R, Murali TS (2009) Fungal endophytes - Mycosphere and bioprospecting. Fung Biol Rev 23:9–19

Szaniszlo PJ, Carl Wirsen JR, Mitchell R (1968) Production of a capsular polysaccharide by a marine filamentous fungus. J Bacteriol 96:1474–1483

Taboski MAS, Rand TG, Piorko A (2005) Lead and cadmium uptake in the marine fungi *Corollospora lacera* and *Monodictys pelagica*. FEMS Microbiol Ecol 53:445–453

Thirunavukkarasu N, Suryanarayanan TS, Murali TS, Ravishankar JP, Gummadi SN (2011) L-Asparaginase from marine derived fungal endophytes of seaweeds. Mycosphere (Online). J Fung Biol 2(2):147–155

Toledo-Hernández C, Sabat AM, Zuluaga Montero A (2007) Density, size structure and aspergillosis prevalence in *Gorgonia ventalina* at six localities in Puerto Rico. Mar Biol 152:527–535

Toske SG, Jensen PR, Kauffman CA, Fenical W (1998) Aspergillamidales A and B: Modified cytotoxic tripeptides produced by a marine fungus of the genus *Aspergillus*. Tetrahedron 54:13459–13466

Verma AK, Raghukumar C, Verma P, Shouche YS, Naik CG (2010) Four marine-derived fungi for bioremediation of raw textile mill effluents. Biodegradation 21:217–233

Verma AK, Raghukumar C, Naik CG (2011) A novel hybrid technology for remediation of molasses-based raw effluents. Bioresour Technol 102:2411–2418

Weber D (2009) Endophytic fungi, occurence and metabolites. In: Anke T, Weber D (eds) The Mycota XV Physiology and Genetics. Springer-Verlag, Berlin, pp 153–195

Wegley L, Edwards R, Rodriguez-Brito B, Liu H, Rohwer F (2007) Metagenomic analysis of the microbial community associated with the coral *Porites astreoides*. Environ Microbiol 9:2707–2719

Zalar P, de Hoog GS, Gunde-Cimerman N (1999a) Ecology of halotolerant dothideaceous black yeasts. Stud Mycol 43:38–48

Zalar P, de Hoog GS, Gunde-Cimerman N (1999b) *Trimmatostroma salinum*, a new species from hypersaline water. Stud Mycol 43:57–62

Zinjarde SS, Pant A (2002) Emulsifier from a tropical marine yeast, *Yarrowia lipolytica* NCIM 3589. J Basic Microbiol 42:67–73

Zvyagintseva TN, Elyakova LA, Isakov VV (1995) The enzymatic transformations of laminarans in 1fi3; 1fi6-b-D-glucans with immunostimulating activity. Bioorg Khim 21:218–225

Chapter 15
Degradation of Phthalate Esters by *Fusarium* sp. DMT-5-3 and *Trichosporon* sp. DMI-5-1 Isolated from Mangrove Sediments

Zhu-Hua Luo, Ka-Lai Pang, Yi-Rui Wu, Ji-Dong Gu, Raymond K.K. Chow, and L.L.P. Vrijmoed

Contents

L.L.P. Vrijmoed (✉)
Department of Biology and Chemistry, City University of Hong Kong, 83 Tat Chee Avenue, Kowloon Tong, Hong Kong SAR, PR China
e-mail: bhlilian@cityu.edu.hk

C. Raghukumar (ed.), *Biology of Marine Fungi*,
Progress in Molecular and Subcellular Biology 53,
DOI 10.1007/978-3-642-23342-5_15,

Abstract Phthalate esters (PAEs) are important industrial compounds mainly used as plasticizers to increase flexibility and softness of plastic products. PAEs are of major concern because of their widespread use, ubiquity in the environment, and endocrine-disrupting toxicity.

In this study, two fungal strains, *Fusarium* sp. DMT-5-3 and *Trichosporon* sp. DMI-5-1 which had the capability to degrade dimethyl phthalate esters (DMPEs), were isolated from mangrove sediments in the Futian Nature Reserve of Shenzhen, China, by enrichment culture technique. These fungi were identified on the basis of spore morphology and molecular typing using 18S rDNA sequence. Comparative investigations on the biodegradation of three isomers of DMPEs, namely dimethyl phthalate (DMP), dimethyl isophthalate (DMI), and dimethyl terephthalate (DMT), were carried out with these two fungi. It was found that both fungi could not completely mineralize DMPEs but transform them to the respective monomethyl phthalate or phthalate acid. Biochemical degradation pathways for different DMPE isomers by both fungi were different. Both fungi could transform DMT to monomethyl terephthalate (MMT) and further to terephthalic acid (TA) by stepwise hydrolysis of two ester bonds. However, they could only carry out one-step ester hydrolysis to transform DMI to monomethyl isophthalate (MMI). Further metabolism of MMI did not proceed. Only *Trichosporon* sp. was able to transform DMP to monomethyl phthalate (MMP) but not *Fusarium* sp. The optimal pH for DMI and DMT degradation by *Fusarium* sp. was 6.0 and 4.5, respectively, whereas for *Trichosporon* sp., the optimal pH for the degradation of all the three DMPE isomers was at 6.0. These results suggest that the fungal esterases responsible for hydrolysis of the two ester bonds of PAEs are highly substrate specific.

15.1 What Do We Know About Phthalate Esters and Their Microbial Degradation?

15.1.1 World-Wide Consumption

Phthalate esters (PAEs), or phthalates, are industrially synthetic compounds principally functioning as plasticizers to provide flexibility, workability, and distensibility for consumer products such as building materials, home furnishings, clothing, and packing of food and medicine (Staples et al. 1997a; Cartwright et al. 2000; Wang et al. 2004; Xu et al. 2005a, b). These are also used as additives in the manufacture of paints, adhesives, cardboard, lubricants, and fragrances (Gu et al. 2005). *Ortho*-PAEs alone account for 80% of plasticizers in the world (Baikova et al. 1999; Wang et al. 2003). As plasticizers, PAEs comprise as much as 30% of the polymer matrices in plastic products (Wang et al. 2003). The consumption of PAEs reaches millions of tonnes world-wide every year (Harris and Sumpter 2001). Dimethyl phthalate esters (DMPEs), with the simplest structures amongst PAEs, have three isomers, namely dimethyl phthalate (*ortho*-DMP; *ortho*-arrangement of the two carboxyl groups), dimethyl isophthalate (DMI; *meta*-), and dimethyl terephthalate (DMT; *para*-).

DMP is used as a plasticizer in cellulose ester-based plastics including cellulose acetate and cellulose butyrate (Staples et al. 1997a). DMI is mainly used to synthesize sodium dimethyl isophthalate-5-sulfonate to improve the dyeing properties of polyesters (Gu et al. 2005). DMT is generally applied in the production of polyesters, such as polyethylene terephthalate (PET) and polytrimethylene terephthalate (PTT).

The phthalate plasticizers do not bind covalently to the plastic resin and thus are able to migrate into the environment during manufacture, usage, and disposal (Bauer and Herrmann 1997; Staples et al. 1997a). Due to the large world-wide consumption of PAEs, they have been widely detected in both aquatic and terrestrial environments, including agricultural soils (Xu et al. 2008), river water and sediments (Tan 1995; Long et al. 1998), marine water and sediments (Peterson and Freeman 1982; Xie et al. 2005), sewage sludge and effluents (Fromme et al. 2002), and landfill leachates (Paxéus 2000; Jonsson et al. 2003).

15.1.1.1 Toxicity

Toxicity of PAEs has attracted increasing attention due to their widespread use and ubiquitous occurrence in the environment. Known as endocrine-disrupting chemicals, PAEs may affect the normal function of the reproductive system and development of humans and animals at very low concentrations (Jobling et al. 1995; Colón et al. 2000; Gu et al. 2005; Xu et al. 2005b). Some PAEs, such as di-2-ethylhexylphthalate (DEHP) and di-isononyl phthalate (DINP), are suspected to be mutagens and carcinogens (David and Gans 2003; Chang et al. 2004; Xu et al. 2006). In addition, several low molecular weight PAEs (with alkyl side chains of less than six carbon atoms) have been demonstrated to have acute or chronic toxicity on aquatic organisms (microorganisms, algae, aquatic invertebrates, and fish) at concentrations below their water solubility (Staples et al. 1997b). As a result, six PAEs, including DMP, have been designated as priority pollutants by the United States Environmental Protection Agency due to their endocrine-disrupting activity (Gu et al. 2005).

15.1.2 Bacterial Degradation

Microbial degradation is suggested to be the principal mechanism for removal of PAEs in the environment (Staples et al. 1997a). The biochemical pathways and molecular mechanism of degradation of PAEs by bacteria have been well documented. A large number of bacterial species capable of degrading PAEs were isolated from activated sludge, mangroves, soils, rivers, and the marine environment (Chatterjee and Dutta 2003; Chang et al. 2004; Li et al. 2005a, b; Nalli et al. 2005; Xu et al. 2005b; Wang and Gu 2006a, b; Fang et al. 2007). It was demonstrated in some studies that construction of a bacterial consortia was more efficient to completely mineralize PAEs than single species (Vega and Bastide 2003; Wang et al. 2003; Li and Gu 2007). Degradation of PAEs by bacteria is initiated by esterases which catalyze stepwise hydrolysis of both carboxylic ester linkages to form phthalate monoester and then phthalic acid (PA) (Gu et al. 2005; Li et al. 2005b; Li and Gu 2006). Two distinct pathways have been elucidated for the transformation of PA to protocatechuate by Gram-positive and Gram-negative bacteria, respectively (Chang and Zylstra 1998; Stingley et al. 2004). Protocatechuate is further metabolized by *ortho*-cleavage pathway or *meta*-cleavage pathway (Keyser et al. 1976; Eaton and Ribbons 1982; Williams et al. 1992; Harwood and Parales 1996; Eaton 2001). Some functional operons involved in the metabolism of PA have been identified and characterized in *Arthrobacter keyseri* 12B, *Mycobacterium vanbaalenii* PYR-1, *Burkholderia cepacia* DBO1, *Pseudomonas putida* NMH102-2, and other bacterial strains (Chang and Zylstra 1998; Eaton 2001; Stingley et al. 2004).

15.1.3 Fungal Degradation

On the other hand, degradation of PAEs by fungi has received surprisingly little attention. The potential role of fungi in degradation of a wide range of recalcitrant compounds, including polycyclic aromatic hydrocarbons (PAHs), benzene-toluene-ethylbenzene-xylenes (BTEX), chlorophenols, polychlorinated biphenyl, munitions waste, and pesticides has been demonstrated (Tortella et al. 2005). The limited reports for fungal degradation of PAEs are mainly grouped into two categories: degradation of PAEs by pure fungal cultures (Table 15.1) (Sivamurthy et al. 1991; Ganji et al. 1995; Pradeepkumar and Karegoudar 2000; Lee et al. 2004; Hwang et al. 2008) and by purified fungal enzymes (Table 15.2) (Kim et al. 2002, 2003, 2005; Kim and Lee 2005; Ahn et al. 2006). The role of fungi on the environmental fate of PAEs remains largely unknown.

Most reports for degradation of PAEs by pure fungal cultures focused on two functional groups of filamentous fungi, namely the ligninolytic fungi (also commonly referred as the white-rot fungi) and non-ligninolytic fungi, respectively.

Many ligninolytic fungi have shown the ability to metabolize PAEs. The fungus *Sclerotium rolfsii* was able to transform DMT to terephthalic acid (TA) through the formation of monomethyl phthalate (MMT) as the intermediate. However, TA was the dead-end product and further metabolism was not observed. Hwang et al. (2008) investigated the degradation of three PAEs (DMP, diethyl phthalate (DEP), and butyl benzyl phthalate (BBP)) by several white-rot fungi, including *Irpex lacteus*, *Merullius tremellosus*, *Polyporus brumalis*, *Trametes versicolor*, *Schizophyllum commune*, *Formitella fraxinea*, *Pleurotus ostreatus*, and *T. versicolor* MrP1, MrP13, and MnP26 (transformant of the Mn-repressed peroxidase gene (*MrP*) and Mn-dependent peroxidase gene (*MnP*) of *T. versicolor*). All these fungi were able to degrade the three PAEs and *P. ostreatus* showed the highest degradation rates among the tested fungi. Degradation of BBP by *P. ostreatus* showed that fungal mycelium was much more efficient to degrade the substrate than the supernatant of the liquid culture, indicating that the initial metabolism of BBP by

Table 15.1 Degradation of PAEs by fungal cultures

Fungal species	Substrates[a]	Degradation	References
Ligninolytic fungi			
Sclerotium rolfsii	DMT	Incomplete	Sivamurthy et al. (1991)
Irpex lacteus	DMP, DEP, BBP	NM	Hwang et al. (2008)
Merullius tremellosus	DMP, DEP, BBP	NM	Hwang et al. (2008)
Polyporus brumalis	DMP, DEP, BBP	NM	Hwang et al. (2008)
Trametes versicolor	DMP, DEP, BBP	NM	Hwang et al. (2008)
Schizophyllum commune	DMP, DEP, BBP	NM	Hwang et al. (2008)
Formitella fraxinea	DMP, DEP, BBP	NM	Hwang et al. (2008)
Pleurotus ostreatus	DMP, DEP, BBP	NM	Hwang et al. (2008)
Phanerochaete chrysosporium	DBP	NM	Lee et al. (2004)
Trametes versicolor	DBP	NM	Lee et al. (2004)
Daldinia concentrica	DBP	Incomplete	Lee et al. (2004)
Polyporus brumalis	DBP	Incomplete	Lee et al. (2007)
Non-ligninolytic fungi			
Aspergillus oryzae	DEHP	NM	Chai et al. (2008)
Aspergillus sydowii	DEHP	NM	Chai et al. (2008)
Aspergillus ustus	DEHP	NM	Chai et al. (2008)
Fusarium graminearum	DEHP	NM	Chai et al. (2008)
Fusarium morniforme	DEHP	NM	Chai et al. (2008)
Fusarium sporotrichioides	DEHP	Incomplete	Chai et al. (2008)
Penicillium citrinum	DEHP	NM	Chai et al. (2008)
Penicillium expansum	DEHP	NM	Chai et al. (2008)
Penicillium frequentans	DEHP	NM	Chai et al. (2008)
Curvularia lunata	DEHP	NM	Chai et al. (2008)
Trichoderma viride	DEHP	NM	Chai et al. (2008)
Rhizopus stolonifer	DEHP	NM	Chai et al. (2008)
Aspergillus niger	DMP	Complete	Pradeepkumar and Karegoudar (2000)
Aspergillus niger	DMT	Complete	Ganji et al. (1995)
Yeasts			
Rhodotorula rubra	DOP, DOTP	Complete	Gartshore et al. (2003)
Saccharomyces cerevisiae	DEP, DBP, BBP	NM	Begum et al. (2003)

NM not mentioned

[a]Abbreviations are explained in the text

P. ostreatus may be catalyzed by mycelium-associated enzymes or intracellular enzymes rather than extracellular enzymes. Lee et al. (2004) reported that *Phanerochaete chrysosporium*, *Trametes versicolor*, and *Daldinia concentrica* were able to completely degrade dibutyl phthalates (DBP) with the formation of a large number of products, including *o*-phenyl acetic acid, isobenzofuranone, *O*-anisic acid, phenylethyl alcohol, di-butyl-4-mehtoxy phenol and 1,2-benzenedicarboxylic acid, diisooctyl ester. Another study on the degradation of DBP by *Polyporus brumalis* recorded that DBP was metabolized to phthalic acid anhydride (PAA) through trans-esterification and de-esterification pathways with DEP and mono-butyl phthalate

Table 15.2 Enzymatic degradation of several PAEs by two purified fungal enzymes, *Fusarium oxysporum* f. sp. *pisi* cutinase and *Candida cylindracea* esterase

PAEs	Enzymes	Degradation rate	Degradation products	Toxic effects of degradation products	References
Di-2-ethylhexyl-phthalate (DEHP)	Cutinase	Fast	1,3-Isobenzofurandione (IBF)	Non-toxic	Kim et al. (2003)
	Esterase	Slow	1,3-Isobenzofurandione (IBF), unidentified compound (X)	Toxic hazard, causing oxidative stress and damage to protein synthesis	
Dipropyl phthalate (DPrP)	Cutinase	Fast	1,3-Isobenzofurandione (IBF)	Non-toxic	Kim et al. (2005)
	Esterase	Slow	1,3-Isobenzofurandione (IBF), propyl methyl phthalate (PrMP	Toxic hazard, causing oxidative stress and damage to protein synthesis	
Dipentyl phthalate (DPeP)	Cutinase	Fast	1,3-Isobenzofurandione (IBF)	Non-toxic	Ahn et al. (2006)
	Esterase	Slow	1,3-Isobenzofurandione (IBF) , pentyl methyl phthalate (PeMP)	Toxic hazard, causing oxidative stress and damage to protein synthesis	
Butyl Benzyl phthalate (BBP)	Cutinase	Fast	1,3-Isobenzofurandione (IBF)	Non-toxic	Kim et al. (2002)
	Esterase	Slow	1,3-Isobenzofurandione (IBF), butyl methyl phthalate (BMP)	Toxic hazard, causing oxidative stress and damage to protein synthesis	
Dibutyl phthalate (DBP)	Cutinase	Fast	1,3-Isobenzofurandione (IBF)	Non-toxic	Kim and Lee (2005)
	Esterase	Slow	1,3-Isobenzofurandione (IBF), butyl methyl phthalate (BMP)	Toxic hazard, causing oxidative stress and damage to protein synthesis	

(MBuP) as intermediates, respectively (Lee et al. 2007). At the same time, some derivates of PAA were also detected as by-products in trace amounts, including α-hydroxyphenylacetic acid, benzyl alcohol, and *o*-hydroxyphenylacetic acid.

Degradation of PAEs by non-ligninolytic fungi has also been documented. Chai et al. (2008) reported the degradation of DEHP by 14 non-ligninolytic fungi, including *Aspergillus oryzae*, *A. sydowii*, *A. ustus*, *Fusarium graminearum*, *F. morniforme*, *F. sporotrichioides*, *Penicillium citrinum*, *P. expansum*, *P. frequentans*, *Curvularia lunata*, *Trichoderma viride*, and *Rhizopus stolonifer* species. All these fungal strains were able to metabolize DEHP. *Fusarium* species demonstrated the highest degradation efficiencies among the tested fungi. Degradation products of DEHP by *F. sporotrichioides* were identified as *o*-phthalic acid, 2-ethylhexanol, and 2-ethylhexanoic acid. In addition, degradation of DMPE isomers was observed in *Aspergillus niger*. Pradeepkumar and Karegoudar (2000) reported that *A. niger* was capable of completely metabolizing DMP through MMP, PA, and protocatechuate. Protocatechuate was further oxidized via *ortho*-cleavage pathway. Another study investigated the degradation of DMT by *A. niger* (Ganji et al. 1995). DMT was completely metabolized by the fungus with the formation of MMT, TA, and protocatechuate as the intermediates. The aromatic metabolite, protocatechuate, was further metabolized by *A. niger* via *meta*-cleavage pathway in this report, totally different from the study of DMP degradation reported by Pradeepkumar and Karegoudar (2000), in which protocatechuate was degraded by the *ortho*-cleavage pathway.

Compared to filamentous fungi, degradation of PAEs by yeasts has been largely overlooked; only limited information is available. The yeast *Rhodotorula rubra* could completely mineralize bis-2-ethylhexyl adipate (B(EH)A), di-octyl phthalate (DOP), and dioctyl terephthalate (DOTP) while partially transformed dipropylene glycol dibenzoate (D(PG)DB) and diethylene glycol dibenzoate (D(EG)DB) (Gartshore et al. 2003). Incomplete degradation of two dibenzoate plasticizers, D(PG)DB and D(EG)DB, resulted in the production of two toxic metabolites, dipropylene glycol monobenzoate and diethylene glycol monobenzoate, respectively. By contrast, no metabolites were observed and the toxicity of media did not increase during the degradation of B(EH)A, DOP, and DOTP by *R. rubra*. It should be noted that glucose was required as a co-substrate to degrade PAEs by this yeast because PAEs could not support substantial growth of the yeast when serving as the sole source of carbon and energy. In addition, *Saccharomyces cerevisiae* also displayed the capability to degrade DEP, DBP, and BBP (Begum et al. 2003), following an efficiency order of DBP > BBP > DEP.

Other studies focused on the degradation of PAEs using purified fungal enzymes. A Korean research group undertook comparative investigations on enzymatic degradation of PAEs using two purified fungal enzymes, *Fusarium oxysporum* f. sp. *pisi* cutinase and *Candida cylindracea* esterase (Kim et al. 2002, 2003, 2005; Kim and Lee 2005; Ahn et al. 2006). They compared the degradation efficiency of these two enzymes on several PAEs, including DBP, BBP, DEHP, di-*n*-propyl phthalate (DPrP), and di-*n*-pentyl phthalate (DPeP), and obtained similar results. Degradation of PAEs by fungal cutinase had some advantages over yeast esterase: (1) fungal cutinase showed high stability in ester-hydrolytic activity and was able to degrade PAEs much faster than the yeast esterase, (2) enzymatic degradation products of PAEs by fungal cutinase, mainly 1,3-isobenzofurandione (IBF) were non-toxic, whereas the hydrolysis of PAEs (such as BBP and DBP, DPrP, DPeP, and DEHP) by yeast esterase produced toxic compounds, including IBF, butyl methyl phthalate (BMP), propyl methyl phthalate (PrMP), pentyl methyl phthalate (PeMP), and an unidentified compound (X) which could lead to oxidative stress and damage to protein synthesis in bacteria.

15.1.4 Bacterial Degradation of PAEs in Mangrove Sediments

Mangrove ecosystems, commonly found in the inter-tidal estuarine zone of tropical and subtropical regions, are preferential sinks of man-made pollutants from fresh waters as well as tidal waters (Tam 2006). Mangrove microorganisms may play an important role in the degradation of recalcitrant pollutants; examples of bacteria isolated from mangrove sediments capable of completely mineralizing PAEs are *Pseudomonas fluorescens* B-1, *Pasturella multocida* Sa, and *Sphingomonas paucimobilis* Sy (Li et al. 2005b; Xu et al. 2005b, 2006; Li and Gu 2006). Others such as *Rhodococcus ruber* Sa exhibited only partial degradation (Li et al. 2005a). Li and Gu (2007) also reported complete degradation of DMI by the cooperation between two mangrove sediment bacteria *Klebsiella oxytoca* Sc and *Methylobacterium mesophilicum* Sr.

15.1.5 Investigation on Fungal Degradation of PAEs in Mangrove Sediments

Very few reports are available on the degradation of PAEs by mangrove fungi to date. An investigation was undertaken recently to explore if these fungi could play a similar role as bacteria (Luo 2010). The site of investigation was a mangrove wetland, the Futian National Nature Reserve (22°32′N, 114°05′E) with an area of 304 ha located at the border of Shenzhen, one of the cities with top economic growth in China. The Reserve is suffering from increasing pollution burden derived from the rapid urbanization and industrialization of the city which is mainly related to the discharge of domestic sewage, livestock, and industrial wastewater (Tam et al. 1995; Tam 2006). High concentrations of PAHs and heavy metals were found in the sediments (Zhang et al. 2004) and elevated levels of bisphenol A (BPA) were also recorded in the tidal waters (Li et al. 2009). These findings indicate that the sediments of this Reserve were sinks of pollutants from all possible sources, including PAH-related chemicals, such as PAEs. Therefore, this site was selected for the investigation with the aims to (a) isolate PAE – degrading fungi from sediments of this Reserve, (b) to assess the degradation ability of isolated fungi, and (c) to compare the degradation pathways of different isomers of the PAEs by the isolated fungi (Luo et al. 2009; Luo 2010; Luo et al., 2011).

15.2 Isolation and Screening of Dimethyl-Phthalate Ester (DMPE): Degrading Fungi from Mangrove Sediments

15.2.1 Enrichment Cultures and Isolation of Fungi

DMPE – degrading fungi were isolated from mangrove sediments by an enrichment culture technique. Enrichment cultures were initiated by inoculating ca. 5.0 g wet mangrove sediments into 100 ml of mineral salt medium (MSM) with 100 mg l^{-1} of either one of DMPE isomers (DMP, DMI, and DMT) as the sole source of carbon and energy in 250 ml Erlenmeyer flasks. The MSM was modified from Hartmans et al. (1989) supplemented with vitamin solution (Kao and Michayluk Vitamin Solution, Sigma-Aldrich, Germany) and antibiotics (penicillin G (sodium salt) 0.5 g l^{-1}, and streptomycin sulfate 0.5 g l^{-1}) (Luo et al. 2009). The initial pH of the culture medium was adjusted to 5.5 $\pm$ 0.1. The flasks inoculated with mangrove sediments were incubated at 25°C on a rotary shaker at 150 rpm in the dark. After 1 week of incubation, the fungal mycelia/yeast biomass were observed and 5 ml of aliquot cultures were extracted and transferred to a new 250 ml Erlenmeyer flask containing 100 ml fresh culture medium identical to the initial one. DMPE – degrading cultures were obtained through six serial enrichment transfers.

The final enrichment culture was diluted 100 times with MSM and then plated on MSM agar plates. Single colonies were picked up and transferred to fresh MSM plates as pure cultures. A total of 30 fungi were isolated from the enrichment cultures, including 12 DMP-degrading fungi, 6 DMI-degrading fungi, and 10 DMT-degrading fungi. The fungal isolates were maintained on MSM slants and stored at 25°C. MSM agar plates and slants were similar to enrichment cultures but supplemented with 1.2% of agar.

15.2.2 Characterization of the Selected Fungal Isolates

The fungal isolates were identified on the basis of microscopic spore morphology. Most of the DMP and DMI-degrading isolates, with three exceptions could not be identified because they were non-sporulating cultures. The three isolates, DMP-3-1, DMI-5-1, and DMI-5-2 were identified as *Aureobasidium pollulans* c.f., *Trichosporon* sp., and *Penicillium* sp., respectively. Ten DMT-degrading isolates belong to two genera, eight of which were *Fusarium* spp. and the remaining two were *Aureobasidium* spp.

For selected strains used for detailed investigation of DMPE degradation (DMT-5-3 and DMI-5-1), 18S rDNA gene sequencing was further employed to confirm their taxonomic position determined by examination with an optical microscope (Luo 2010, Luo et al. 2011). The results are illustrated in Figs. 15.1–15.4.

15.2.3 Preliminary Investigation of DMPE Degradation Capability of Selected DMP-/DMI-/DMT-Degrading Fungi

The DMPE degradation capability of the 30 fungal isolates was indirectly tested through another growth assay in MSM amended with each DMPE as sole source of carbon and energy as described previously. Fungal isolates which showed a fast and good growth in the liquid medium after 1 week incubation were then subjected to further DMPE-degrading screening. Three such isolates were selected which were later identified to be *Fusarium* sp. DMT-5-3, *Trichosporon* sp. DMI-5-1, and *Penicillium* sp. DMI-5-2. As shown in Fig. 15.5, *Trichosporon* sp. and *Fusarium* sp. showed much higher DMPE degradation ability compared with *Penicillium* sp. 80% of 84 mg l^{-1} of DMI was metabolized by *Trichosporon* sp. over the 8-day of

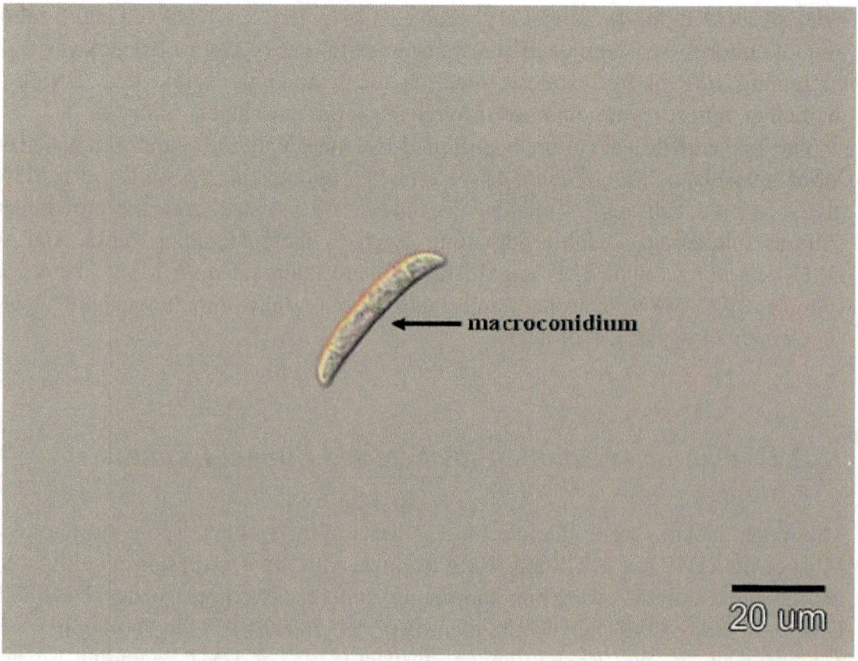

Fig. 15.1 Optical micrograph showing macroconidium of *Fusarium* sp. DMT-5-3 grown on corn meal agar

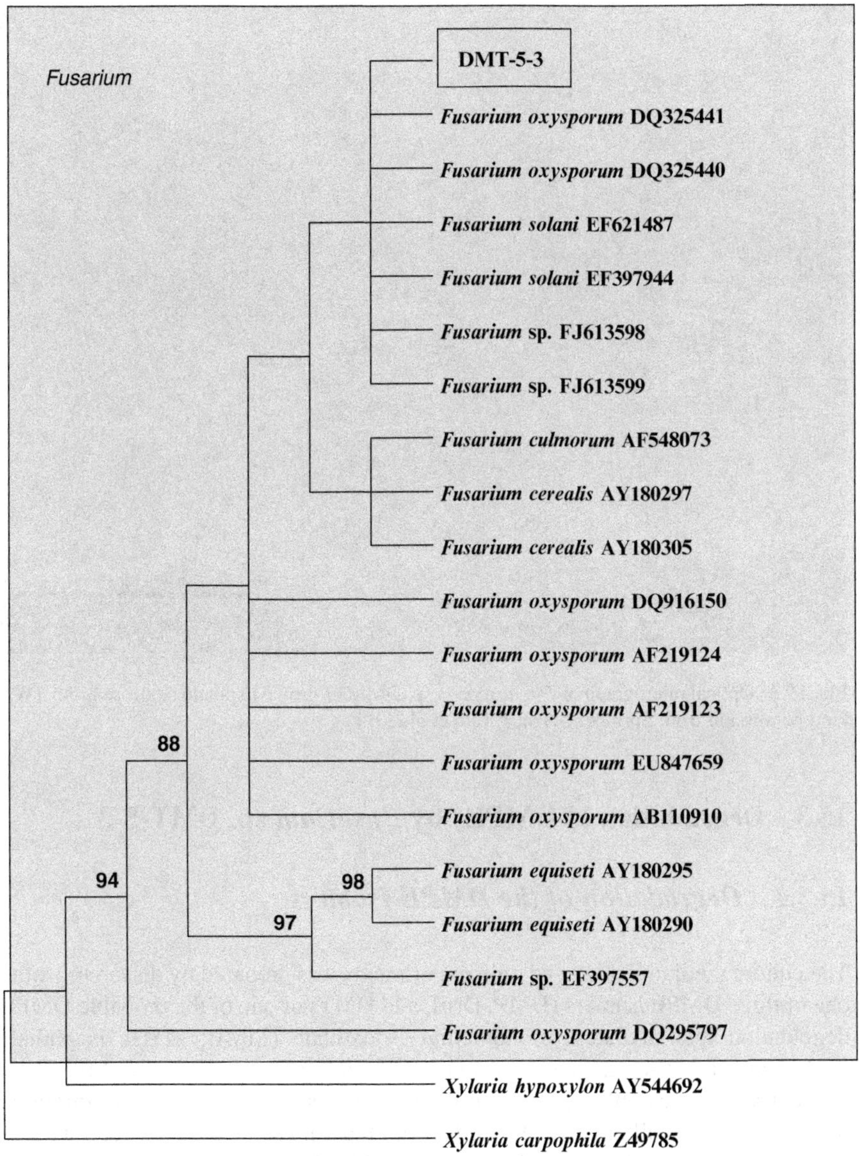

Fig. 15.2 The phylogenetic 18S rDNA-based tree showing the relationship between the fungal strain DMT-5-3 and selected members of genus *Fusarium*

incubation (Fig. 15.5a). By contrast, only 15% of 80 mg l^{-1} of DMI was degraded by *Penicillium* sp. after 8 days (Fig. 15.5b). *Fusarium* sp. was able to completely degrade 22 mg l^{-1} of DMT after 4 days of incubation (Fig. 15.5c). Therefore, *Trichosporon* sp. and *Fusarium* sp. were chosen for further investigation.

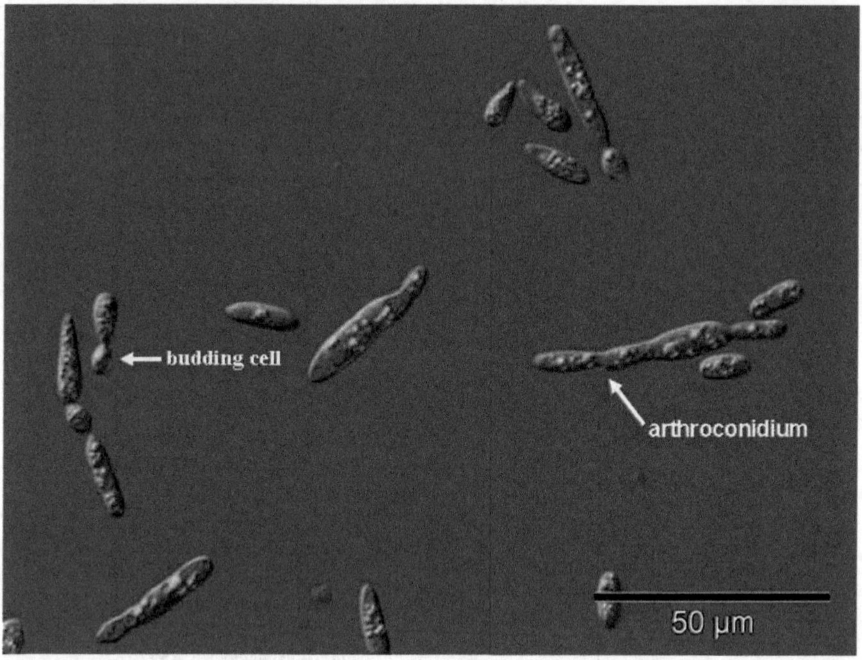

Fig. 15.3 Optical micrograph of *Trichosporon* sp. DMI-5-1 grown in potato dextrose broth. (With kind permission from Springer Science: Luo et al. 2011)

15.3 Degradation of DMPEs by *Fusarium* sp. DMT-5-3

15.3.1 Degradation of the DMPE Isomers

The culture medium for degradation experiments was prepared by dissolving either one of three DMPE isomers (DMP, DMI, and DMT) or one of the probable DMPE-degradation intermediates (monomethyl isophthlate (MMI), MMT, isophthalic acid (IA), and TA) as the sole source of carbon and energy in MSM at a concentration of ca. 30 mg l^{-1}. The medium was sterilized by passing through a 0.2-μm membrane filter (Toyo Roshi Kaisha, Japan). Active inocula were prepared by subculturing the fungus DMT-5-3 onto a MSM agar plate supplemented with 50 mg l^{-1} of the three DMPE isomers (DMP, DMI, and DMT). Six agar plugs (6 mm in diameter) with active fungal mycelia were aseptically inoculated into 125 ml Erlenmeyer flasks with 50 ml of culture medium. The control was composed of culture medium without inoculum (for detecting abiotic degradation). Triplicate flasks were set up for each treatment. The flasks were incubated on an orbital shaker at 150 rpm and 25°C. At scheduled times, three flasks for each treatment were extracted as follows: the fungal mycelia were harvested for cellular

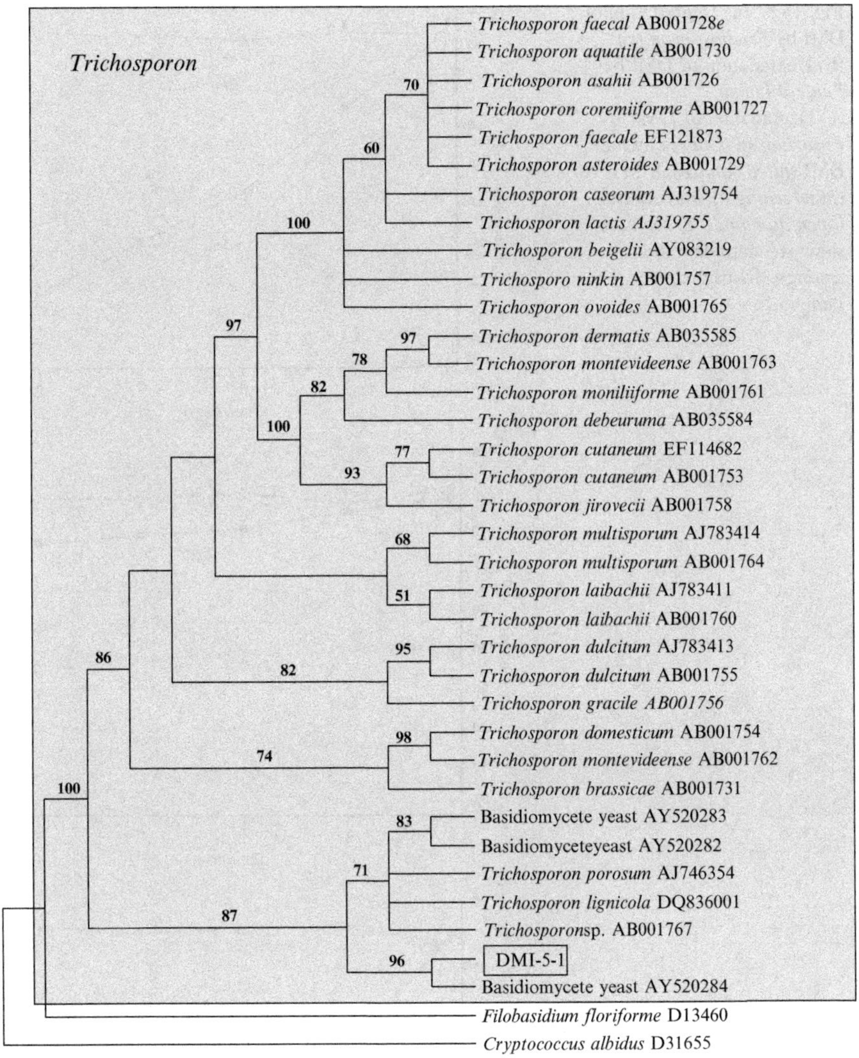

Fig. 15.4 The phylogenetic 18S rDNA-based tree showing the relationship between the yeast strain DMI-5-1 and selected members of genus *Trichosporon* (with kind permission from Springer Science: Luo et al. 2011)

protein measurement by a method adapted from Philips and Gordon (1989); and an aliquot of 2 ml of the cell-free supernatant was filtered and stored frozen (−20°C) for HPLC analysis adapted from Wang and Gu (2006a). The results are illustrated in Fig. 15.6. *Fusarium* sp. was able to degrade DMI and DMT, but not DMP (Luo et al. 2009).

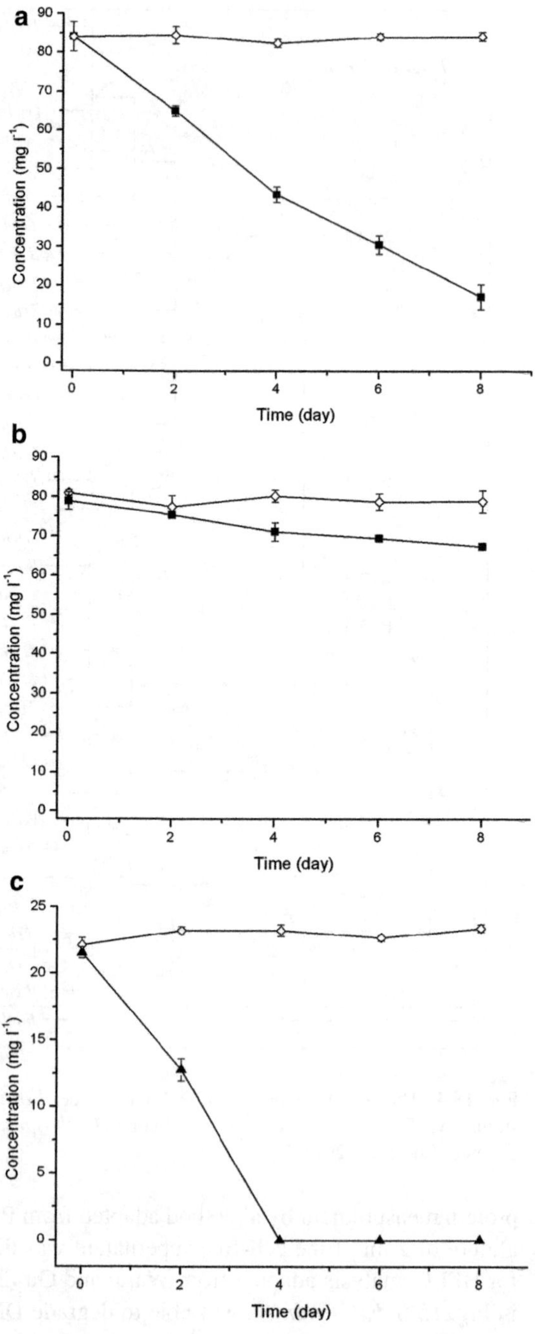

Fig. 15.5 (**a**) Degradation of DMI by *Trichosporon* sp. (**b**) Degradation of DMI by *Penicillium* sp. (**c**) Degradation of DMT by *Fusarium* sp. Curves include DMI (*filled square*), MMT (*filled triangle*), and control (*open diamond*). *Error bars* show standard deviations amongst the triplicate samples

15.3.2 Degradation of DMP

The concentration of DMP remained unchanged over the 24-day incubation period, and none of the expected intermediates (MMP and PA) appeared in the culture medium (data not shown), indicating that *Fusarium* sp. was not able to metabolize DMP (Fig. 15.7a).

15.3.3 Degradation of DMI

As shown in Fig. 15.6a, 24 mg l^{-1} of DMI completely disappeared in 24 days. MMI appeared and accumulated to a peak level of 28 mg l^{-1} after 24 days. IA was not observed in the culture medium over the 24-day incubation period, indicating that further metabolism of MMI did not proceed. The degradation of DMI was accompanied by an increase in cellular protein concentrations, indicating the growth of the fungus during the degradation process. In order to confirm the proposed degradation pathway, degradation of the two suspected intermediates (MMI and IA) was also assessed using this fungus. The results show that neither MMI nor IA was metabolized (data not shown). All these results suggest that *Fusarium* sp. could only transform DMI to MMI through one step of ester hydrolysis, but lacked the ability to hydrolyze the second carboxylic ester-linked methyl group from MMI (Fig. 15.7b).

15.3.4 Degradation of DMT and MMT

Similarly, the fungus was capable of transforming DMT. As shown in Fig. 15.6b, DMT was quickly metabolized by this fungus from 20 mg l^{-1} to 0.7 mg l^{-1} in 12 days. Meanwhile, the intermediate MMT gradually accumulated to the maximum level of 20.3 mg l^{-1} in 8 days, after which it remained constant. Another intermediate, TA, was also detected in the culture medium from day 4 and its concentration remained at 5.3 mg l^{-1} after 24 days. Cellular protein concentrations increased with the depletion of DMT, suggesting that DMT was utilized by this fungus as the sole source of carbon and energy. At the same time, MMT and TA were also used as the initial substrates for the fungus to confirm the degradation pathway.

MMT, with an initial concentration of 29.5 mg l^{-1} was degraded to 24.1 mg l^{-1} over 24 days of incubation. Simultaneously, TA was observed as a degradation intermediate in the culture and accumulated to 5.3 mg l^{-1} after 24 days (Fig. 15.6c). However, further degradation of TA was not observed (data not shown). Based on these results, it can be concluded that *Fusarium* sp. was capable of transforming

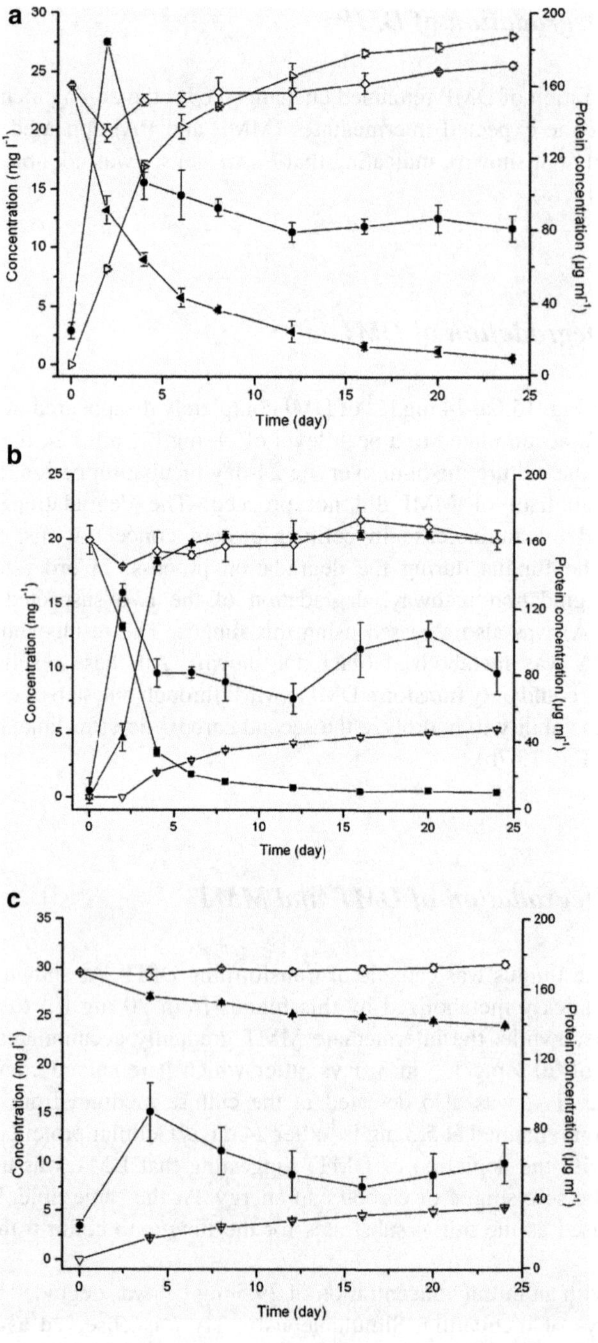

Fig. 15.6 Degradation of DMPEs and suspected intermediates by *Fusarium* sp. over 24 days: (**a**) DMI, (**b**) DMT, and (**c**) MMT. Curves include DMI (*left pointed filled triangle*), MMI (*right pointed open triangle*), DMT (*filled square*), MMT (*filled triangle*), TA (*open inverted triangle*), protein (*filled circle*), and control (*open diamond*). *Error bars* show standard deviations among the triplicate samples (reprinted from Marine Pollution Bulletin with permission from Elsevier: Luo et al. 2009)

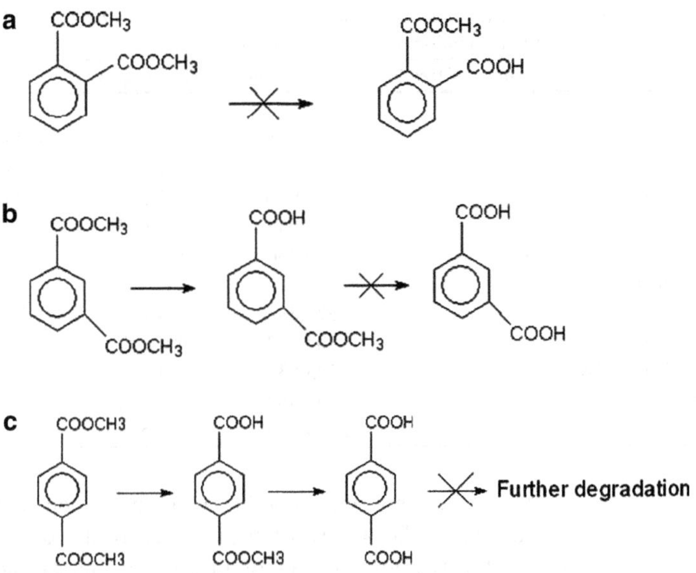

Fig. 15.7 Proposed biochemical degradation pathways for dimethyl phthalate esters (DMPEs) by *Fusarium* sp. (**a**) dimethyl phthalate (DMP), (**b**) dimethyl isophthalate (DMI), and (**c**) dimethyl terephthalate (DMT) (reprinted from Marine Pollution Bulletin with permission from Elsevier Luo et al. 2009)

DMT to MMT and then to TA by stepwise hydrolysis of two carboxylic ester bonds (Fig. 15.7c).

15.3.5 Effect of pH on the Degradation of DMPEs

Preparation of the degradation experiment was similar to the procedures described in Sect. 15.3.1. Initial pH of culture medium was adjusted to four levels (4.5, 5.0, 5.5, and 6.0) using appropriate buffers to investigate the effect on the degradation of DMPEs by the test fungus. At days 4 and 10, an aliquot of 2 ml of cell-free supernatant was extracted from the culture medium, filtered, and stored frozen (−20°C) for HPLC analysis. The removal percentages of DMPEs after 4 and 10 days of incubation were calculated. The differences of removal percentages of each isomer of DMPEs among four pH levels and between two incubation days were compared by two-way ANOVA (SigmaStat 3.0). The differences of removal percentages of each DMPE isomers under different pH levels on the same incubation time were compared by Tukey Test (SigmaStat 3.0).

As shown in Table 15.3, both pH levels and incubation time were able to significantly influence the degradation of DMPEs ($p < 0.05$). The interaction between pH and incubation time was not significant for DMI degradation ($p = 0.435$) but was statistically significant for DMT degradation ($p < 0.05$).

Table 15.3 A comparison of removal percentages of DMPEs by *Fusarium* sp. DMT-5-3 under different pH levels after two incubation days using two-way ANOVA

Source of variation	Degrees of freedom	Sum of square	Mean square	*F*	*P*
DMI degradation					
pH	3	1294.725	431.575	56.762	<0.001
Incubation days	1	2012.458	2012.458	164.683	<0.001
pH × incubation days	3	21.920	7.307	0.961	0.435
Residual	16	121.653	7.603		
Total	23	3450.756	150.033		
DMT degradation					
pH	3	973.883	324.628	119.719	<0.001
Incubation days	1	907.847	907.847	334.804	<0.001
pH × incubation days	3	545.533	181.844	67.062	<0.001
Residual	14	37.962	2.712		
Total	21	2404.823	114.515		

The significant interactive effect between pH and incubation time on DMT degradation was probably related to the ability of the fungus to almost completely degrade DMT at all pH levels after 10 days of incubation, indicating that the incubation time was too long and the influence of pH on DMT degradation cannot be demonstrated clearly (Fig. 15.8b).

However, comparison of removal percentages of each DMPE isomer under different pH levels on the same incubation day by Tukey Test demonstrated that the optimal pH for DMI degradation by *Fusarium* sp. was 6.0, while that for DMT degradation was 4.5 (Fig. 15.8a, b).

15.4 Degradation of DMPEs by *Trichosporon* sp. DMI-5-1

15.4.1 Degradation of DMPE Isomers

The procedures for this experiment were as described for *Fusarium* sp. in Sect. 15.3.1. The results are illustrated in Fig. 15.9. *Trichosporon* sp. was not able to completely mineralize DMPEs but transform them to respective phthalate monoester or PA (Luo et al. 2011).

15.4.2 Degradation of DMP

As shown in Fig. 15.9a, in 24 days, 21.5 mg l^{-1} of DMP was reduced to 6.7 mg l^{-1}, MMP was detected as an intermediate in the culture medium and accumulated to 17.4 mg l^{-1}. Another suspected intermediate, PA, was not observed in the culture medium, indicating that further metabolism of MMI did not proceed. Degradation

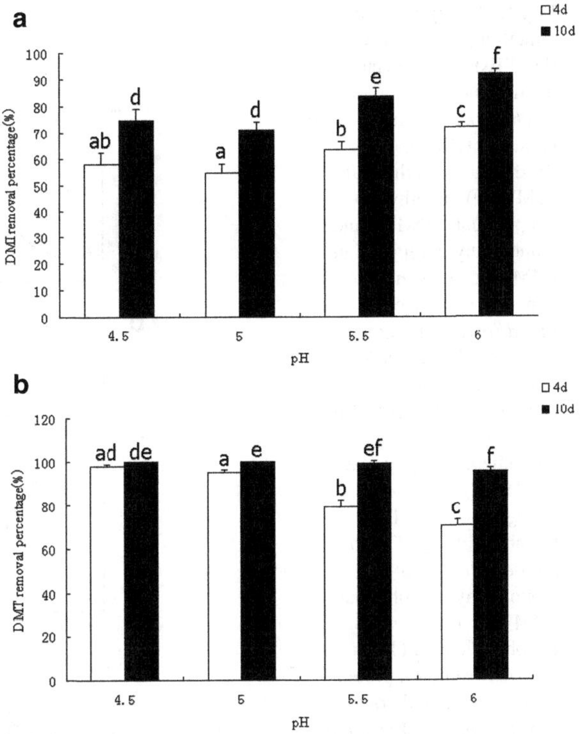

Fig. 15.8 Effect of pH levels on the degradation of DMPEs by *Fusarium* sp. after 4 days and 10 days of incubation: (**a**) DMI degradation and (**b**) DMT degradation. *Bars* with different letters are significantly different at $p < 0.05$ (Tukey Test). *Error bars* show standard deviations among the triplicate samples (reprinted from Marine Pollution Bulletin with permission from Elsevier: Luo et al. 2009)

of DMP was accompanied by an increase in cellular protein concentrations, indicating the growth of the yeast during the degradation process. In order to confirm the degradation pathway, degradation of two suspected intermediates (MMP and PA) was carried out and neither MMP nor PA was degraded (data not shown). These results collectively suggest that *Trichosporon* sp. could only carried out one-step ester hydrolysis to transform DMP to MMP, but lacked the ability to remove the second carboxylic ester-linked methyl group of DMP (Fig. 15.10a).

15.4.3 Degradation of DMI

DMI at 20.2 mg l^{-1} completely disappeared after 16 days of incubation (Fig. 15.9b). The intermediate MMI gradually accumulated to peak level of 27.4 mg l^{-1} at the same time and remained constant afterwards. No expected intermediate IA appeared in the medium over the 24-day incubation period, suggesting that MMI was the last end product of DMI degradation. Cellular protein concentration increased with the depletion of DMI, indicating that DMI was utilized by this yeast as the sole carbon source. No MMI or IA was utilized when they were used as the initial substrates (data not shown). Based on these results,

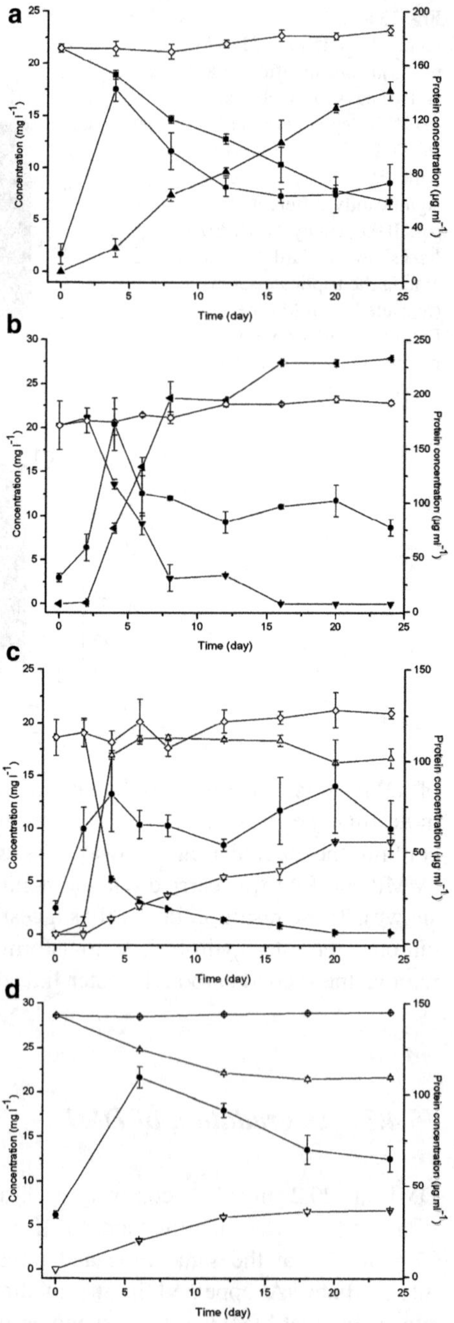

Fig. 15.9 Degradation of dimethyl phthalate esters (DMPEs) and suspected intermediates by *Trichosporon* sp. (**a**) dimethyl phthalate (DMP), (**b**) dimethyl isophthalate (DMI), (**c**) dimethyl terephthalate (DMT), and (**d**) monomethyl terephthalate (MMT). *Curves* include dimethyl phthalate (DMP) (*filled square*), monomethyl phthalate (MMP) (*filled triangle*), dimethyl isophthalate (DMI) (*filled inverted triangle*), monomethyl isophthalate (MMI) (*left pointed filled triangle*), dimethyl terephthalate (DMT) (*right pointed filled triangle*), monomethyl terephthalate (MMT) (*open triangle*), terephthalic acid (TA) (*open inverted triangle*), protein (*filled circle*), and control (*open diamond*). *Error bars* show standard deviations among the triplicate samples (with kind permission from Springer Science: Luo et al. 2011)

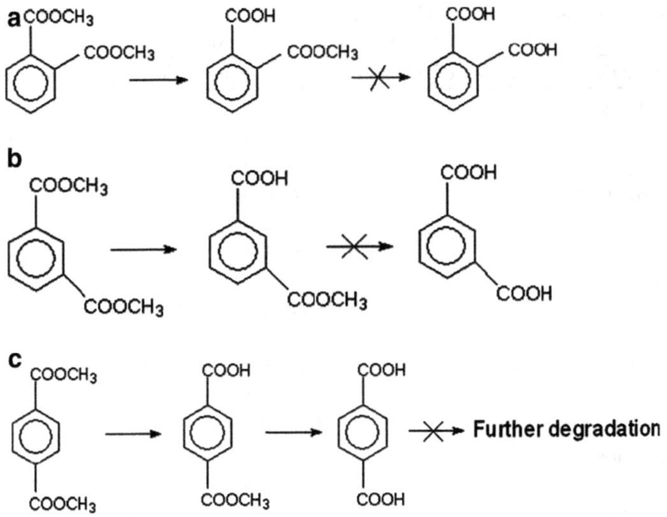

Fig. 15.10 Proposed biochemical degradation pathways for dimethyl phthalate esters (DMPEs) by *Trichosporon* sp. (**a**) dimethyl phthalate (DMP), (**b**) dimethyl isophthalate (DMI), and (**c**) dimethyl terephthalate (DMT) (with kind permission from Springer Science: Luo et al. 2011)

Trichosporon sp. also carried out one-step ester hydrolysis of DMI to MMI and could not further degrade MMI, which was similar with DMP degradation (Fig. 15.10b).

15.4.4 Degradation of DMT

DMT was gradually metabolized from 18.6 mg l^{-1} to 0.2 mg l^{-1} within 20 days of incubation (Fig. 15.9c). At the same time, MMT was observed as an intermediate and quickly accumulated to a maximum level of 18.5 mg l^{-1} in 6 days, after which it remained constant. TA was also detected as another intermediate in the culture medium from day 4 and gradually accumulated to 8.8 mg l^{-1} over the 24-day experimental period. Cellular protein concentration of the fungus increased with the degradation of DMT, suggesting the increase of yeast biomass during the DMT degradation. Confirmation experiment using MMT and TA as the initial substrates showed that the yeast was able to degrade MMT, but not TA (data not shown).

15.4.5 Degradation of MMT

MMT was degraded from 28.6 mg l^{-1} to 21.6 mg l^{-1} over the 24-day incubation period, and expected intermediate TA accumulated simultaneously in the culture

Table 15.4 A comparison of removal percentages of DMPEs by *Trichosporon* sp. DMI-5-1 under different pH levels after 2 incubation days using two-way ANOVA

Source of variation	Degrees of freedom	Sum of square	Mean square	*F*	*P*
DMP degradation					
pH	3	413.131	137.710	16.886	<0.001
Incubation days	1	3513.925	3513.925	430.888	<0.001
pH × incubation days	3	25.848	8.616	1.057	0.395
Residual	16	130.481	8.155		
Total	23	4083.386	177.539		
DMI degradation					
pH	3	15596.098	5198.699	265.519	<0.001
Incubation days	1	4138.700	4138.700	211.380	<0.001
pH × incubation days	3	483.352	161.117	8.229	0.002
Residual	15	293.691	19.579		
Total	22	21986.723	999.396		
DMT degradation					
pH	3	16746.105	5582.035	291.384	<0.001
Incubation days	1	625.324	625.324	32.642	<0.001
pH × incubation days	3	147.577	49.192	2.568	0.093
Residual	15	287.355	19.157		
Total	22	18450.535	838.661		

medium and finally reached to 6.7 mg l^{-1} after 24 days (Fig. 15.9d). Further metabolism of TA did not proceed by this yeast (data not shown). These results indicate that *Trichosporon* sp. carried out stepwise ester hydrolysis to transform DMT to TA through MMT (Fig. 15.10c).

15.4.6 Effect of pH on the Degradation of DMPEs

The procedures for this experiment were as described for *Fusarium* sp. and the results are illustrated in Table 15.4 and Fig. 15.11. As shown in Table 15.4, both pH levels and incubation time were able to influence the degradation of DPMEs significantly ($p < 0.05$). The interaction between pH and incubation time was not significant for degradation of DMP ($p = 0.395$) and DMT ($p = 0.093$), but was statistically significant for degradation of DMI ($p < 0.05$). The latter was probably related to the ability of the fungus to almost completely remove DMI at three pH levels (5, 5.5, and 6.0) after 10 days of incubation, indicating that the incubation time was too long and the influence of pH on DMI degradation could not be demonstrated clearly under the experimental conditions (Fig. 15.11b). Moreover, removal percentages of each isomer of DMPE under four pH levels at the same incubation day were compared by Tukey Test and the results indicated that the optimal pH for degradation of all the three isomers was at 6.0 (Fig. 15.11).

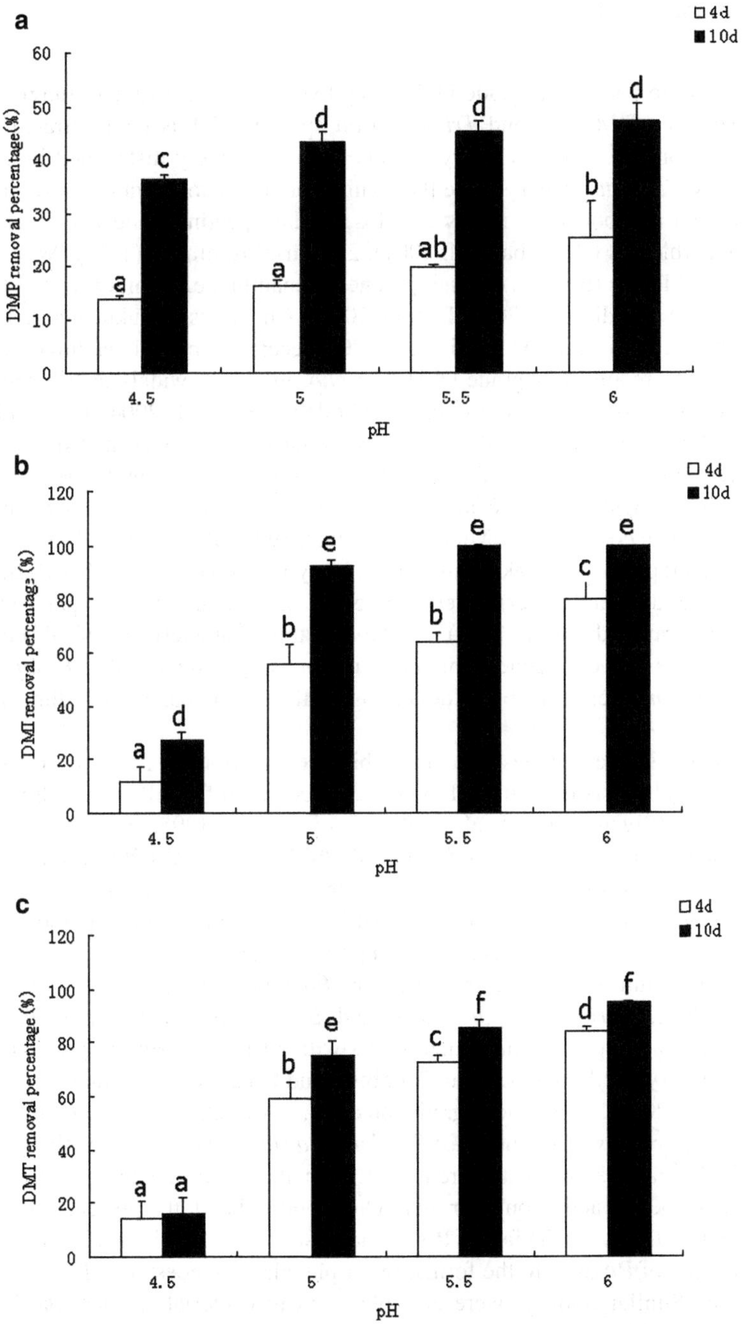

Fig. 15.11 Effect of pH levels on the degradation of DMPEs by *Trichosporon* sp. after 4 and 10 days of incubation: (**a**) DMP degradation, (**b**) DMI degradation, and (**c**) DMT degradation. *Bars* with different letters are significantly different at $p < 0.05$ (Tukey Test). *Error bars* show standard deviations among the triplicate samples

15.5 Discussion

The ability to partially degrade DMPEs by two mangrove sediment fungi, namely *Fusarium* sp. DMT-5-3 and *Trichosporon* sp. DMI-5-1 is demonstrated in this investigation. *Fusarium* is a large genus which is widely distributed in soil and on plants. *Fusarium* species have the significant ability to degrade a wide range of recalcitrant compounds, such as cellulose, xylan, pectin (Kang and Buchenauer 2000), herbicides (Bordjiba et al. 2001), 2,4,6-trinitrotoluene (TNT) (Weber et al. 2002), and PAHs (benzo(a)pyrene, pyrene, phenanthrene, anthracene, and benz(a) anthracene) (Verdin et al. 2004; Li et al. 2005c; Chulalaksananukul et al. 2006; Wu et al. 2010a, b). In addition, Chai et al. (2008) recently reported that three strains of *Fusarium* were able to degrade DEHP. *Trichosporon* is a widely distributed basidiomycetous yeast in the environment (Middelhoven et al. 2004; Kaszycki et al. 2006). *Trichosporon* species have the potential to metabolize a wide variety of organic compounds, including benzene compounds (phenol, benzoic acid, phenylacetic acid, cinnamic acid, and their derivatives as well as some other monoaromatic compounds) (Middelhoven et al. 1992; Middelhoven 1993; Sampaio 1999), cyclopentanol (Iwaki et al. 2004), biarylic compounds (biphenyl, dibenzofuran, and diphenyl ether) (Sietmann et al. 2002), and PAHs (phenanthrene) (MacGillivray and Shiaris 1993). As the structure characteristics of DMPEs were similar to the above benzene compounds, it is not surprising to isolate *Fusarium* and *Trichosporon* species capable of degrading DMPEs from mangrove sediment in this study.

Previous studies suggest that microbial degradation of PAEs is initiated by stepwise hydrolysis on both carboxylic ester bonds to form PA through phthalate monoester (Sivamurthy et al. 1991; Ganji et al. 1995; Pradeepkumar and Karegoudar 2000; Gu et al. 2005; Li et al. 2005b; Li and Gu 2006). *Fusarium* sp. as well as *Trichosporon* sp. isolated from the mangrove sediments in this investigation transformed three isomers of DMPEs to respective phthalate monoester or PA, which show a good agreement with previous reports.

A large number of bacteria such as *Pasturella multocida*, *Sphingomonas paucimobilis*, *Variovorax paradoxus*, and *Burkholderia cepacia* exhibited the capability to completely mineralize PAEs without leaving any by-products under the laboratory conditions (Li et al. 2005b; Li and Gu 2006; Wang and Gu 2006a, b; Wang et al. 2008). Complete degradation of PAEs was also observed in filamentous fungus *Aspergillus niger* and yeast *Rhodotorula rubra*. These observations indicate that PAEs can be completely removed from the environment by a number of microbial species acting on their own. On the other hand, the present study shows another possibility for the fate of PAEs under laboratory conditions, i.e., incomplete degradation of PAEs with the formation of phthalate monoester and PA as the end products. Similar findings were also observed in bacterial species *Rhodococcus ruber* and *Sphingonmonas yanoikuyae*, and fungal species *Sclerotium rolfsii* (Sivamurthy et al. 1991; Li et al. 2005a; Wang and Gu 2006a). In addition, there are reports describing how a bacterial consortia could completely mineralize PAEs, in which *Arthrobacter* sp. and *Klebsiella oxytoca* were only able to partially transform phthalate diesters to monoesters, and further metabolism of phthalate monoesters required the cooperation of *Sphingomonas paucimobilis* and *Methylobacterium mesophilicum* (Vega and Bastide 2003; Li and Gu 2007). These observations and our results suggest that utilization of reconstituted consortia, including both fungi and bacteria, may be more efficient to completely mineralize PAEs than single species of either group.

A series of enzymes were involved in the metabolism of DMPEs, including esterase, phthalate dioxygenase and protocatechuate dioxygenase (Ganji et al. 1995; Chang and Zylstra 1998; Stingley et al. 2004). Partial transformation of three isomers of DMPEs to respective phthalate monoesters or PAs by both test fungi in this study indicate that these two mangrove sediment fungi were only able to produce esterase but not other DMPE metabolism-related enzymes. Esterases are reported to show high regio- and stereo-specificity (Bornscheuer 2002). Previous reports for degradation of DMPEs by bacterial consortia had demonstrated that two different species were involved in hydrolysis of two identical ester bonds of DMPEs, suggesting that the esterases responsible of hydrolysis of the first and second ester linkages of DMPEs are different (Vega and Bastide 2003; Li and Gu 2007). It was demonstrated with *Micrococcus* sp. YGJ1 that there were two distinct esterases involved in the metabolism of dialkyl phthalates (DAPs), including DAP esterase and monoalkyl phthalate (MAP) esterase (Akita et al. 2001; Maruyama et al. 2005). DAP esterase was responsible for the hydrolysis of DAP to MAP, which was further transformed to PA by the action of MAP esterase. An esterase isolated from *Rhodococcus erythropolis* was reported to be only capable of hydrolyzing DMP and DMI, but not DMT (Kurane 1997). All these observations suggest that esterases are highly substrate specific and there may be different esterases responsible for hydrolysis of different isomers of DMPEs and monomethyl phthalate esters (MMPEs). The present study strongly supports this notion (Table 15.5). Partial transformation of DMI to MMI by both the test fungi in this study indicated that they lacked the capability to produce MMI esterase to further metabolize MMI to IA. Both test fungi were able to transform DMT to TA via MMT, indicating that they were able to produce MMT esterase. However, both the test fungi were not able to transform MMP and MMI to the respective PA, indicating that the MMT esterase produced by the test fungi lacked the ability to hydrolyze the second carboxylic ester-linked methyl group from MMP and MMI. Therefore, the difference in hydrolysis ability of test fungi for different isomers of DMPEs and MMPEs in this study demonstrated high substrate specificity of fungal esterases.

Table 15.5 Hydrolysis ability of test fungi on different isomers of DMPEs and MMPEs

Fungal strain	DMP	DMI	DMT	MMP	MMI	MMT
Fusarium sp. DMT-5-3	No	Yes	Yes	ND	No	Yes
Trichosporon sp. DMI-5-1	Yes	Yes	Yes	No	No	Yes

ND not determined

15.6 Conclusion and Future Studies

This is the first detailed study of degradation of DMPEs by mangrove sediment fungi. Two mangrove sediment fungi, *Fusarium* sp. DMT-5-3 and *Trichosporon* sp. DMI-5-1, could not completely mineralize DMPEs but transform them to respective MMP or PA. The biochemical degradation pathways for different DMPE isomers for both the fungi were different, indicating that the fungal esterases involved in the cleavage of two carboxylic ester linkages of DMPEs are highly substrate specific. Complete mineralization of DMPEs requires the cooperation of other microorganisms.

This study demonstrates the potential of mangrove sediment fungi in the removal of PAEs in the environment. However, this is just a case study. The degradation mechanisms of PAEs in fungal system as well as the role of mangrove sediment fungi in the fate of PAEs in the aquatic environment remain largely unknown. There are still some aspects worthy of further investigation: (1) diversity of PAE-degrading fungi in the mangrove environments, (2) identification and characterization of the key enzymes and genes involved in PAE degradation from mangrove sediment fungi, and (3) degradation of PAEs using a co-culture system of PAEs degrading fungi and bacteria isolated from mangrove sediments to explore the possible synergistic degradation in the natural environment.

Acknowledgments This work was substantially supported by grants from City University of Hong Kong (Project No. 7002220 and 9610037), the Research Grants Council of the Hong Kong Special Administrative Region, China (Project No. CA04/05. SC01), and National Natural Science Foundation of China (Project No. 41006099), which are gratefully acknowledged. The authors would also like to thank Ms Jessie Lai of The University of Hong Kong and Miss Alice Chan of City University of Hong Kong for technical support in HPLC analysis.

References

Ahn JY, Kim YH, Min J, Lee J (2006) Accelerated degradation of dipentyl phthalate by *Fusarium oxysporum* f. sp. *pisi* cutinase and toxicity evaluation of its degradation products using bioluminescent bacteria. Curr Microbiol 52:340–344

Akita K, Naitou C, Maruyama K (2001) Purification and characterization of an esterase from *Micrococcus* sp. YGJ1 hydrolyzing phthalate esters. Biosci Biotechnol Biochem 65:1680–1683

Baikova SV, Samsonova AS, Aleshchenkova ZM, Shcherbina AN (1999) The intensification of dimethylphthalate destruction in soil. Eurasian Soil Sci 32:701–704

Bauer MJ, Herrmann R (1997) Estimation of the environmental contamination by phthalic acid esters leaching from household wastes. Sci Total Environ 208:49–57

Begum A, Katsumata H, Kaneco S, Suzuki T, Ohta K (2003) Biodegradation of phthalic acid esters by bakery yeast *Saccharomyces cerevisiae*. Bull Environ Contam Toxicol 70:255–261

Bordjiba O, Steiman R, Kadri M, Semadi A, Guiraud G (2001) Removal of herbicides from liquid media by fungi isolated from a contaminated soil. J Environ Qual 30:418–426

Bornscheuer UT (2002) Microbial carboxyl esterases: classification, properties and application in biocatalysis. FEMS Microbiol Rev 26:73–81

Cartwright CD, Owen SA, Thompson IP, Burns RG (2000) Biodegradation of diethyl phthalate in soil by a novel pathway. FEMS Microbiol Lett 186:27–34

Chai W, Suzuki M, Handa Y, Murakami M, Utsukihara T, Honma Y, Nakajima K, Saito M, Horiuchi CA (2008) Biodegradation of di-(2ethylhexyl) phthalate by fungi. Rep Nat'l Food Res Inst 72:83–87

Chang HK, Zylstra GJ (1998) Novel organization of the genes for phthalate degradation from *Burkholderia cepacia* DBO1. J Bacteriol 180:6529–6537

Chang BV, Yang CM, Cheng CH, Yuan SY (2004) Biodegradation of phthalate esters by two bacteria strains. Chemosphere 55:533–538

Chatterjee S, Dutta TK (2003) Metabolism of butyl benzyl phthalate by *Gordonia* sp. strain MTCC 4818. Biochem Biophys Res Comm 309:36–43

Chulalaksananukul S, Gadd GM, Sangvanich P, Sihanonth P, Piapukiew J, Vangnai AS (2006) Biodegradation of benzo(a)pyrene by a newly isolated *Fusarium* sp. FEMS Microbiol Lett 262:99–106

Colón I, Caro D, Bourdony CJ, Rosario O (2000) Identification of phthalate esters in the serum of young Puerto Rican girls with premature breast development. Environ Health Perspect 108:895–900

David RM, Gans G (2003) Summary of mammalian toxicology and health effects of phthalate esters. In: Staples CA (ed) Phthalate Esters. The Handbook of Environmental Chemistry. Vol. 3, Part Q, 299–316, Springer-Verlag, Berlin, doi: 10.1007/b11470

Eaton RW (2001) Plasmid-encoded phthalate catabolic pathway in *Arthrobacter keyseri* 12B. J Bacteriol 183:3689–3703

Eaton RW, Ribbons DW (1982) Metabolism of dibutylphthalate and phthalate by *Micrococcus* sp. strain 12B. J Bacteriol 151:48–57

Fang HHP, Liang D, Zhang T (2007) Aerobic degradation of diethyl phthalate by *Sphingomonas* sp. Bioresour Technol 98:717–720

Fromme H, Küchler T, Otto T, Pilz K, Müller J, Wenzel A (2002) Occurrence of phthalates and bisphenol A and F in the environment. Water Res 36:1429–1438

Ganji SH, Karigar CS, Pujar BG (1995) Metabolism of dimethylterephthalate by *Aspergillus niger*. Biodegradation 6:61–66

Gartshore J, Cooper DG, Nicell JA (2003) Biodegradation of plasticizers by *Rhodotorula rubra*. Environ Toxicol Chem 22:1244–1251

Gu JD, Li J, Wang Y (2005) Biochemical pathway and degradation of phthalate ester isomers by bacteria. Water Sci Technol 52:241–248

Harris CA, Sumpter JP (2001) The endocrine disrupting potential of phthalates. In: Metzler M (ed) Endocrine Disruptors: Part 1. The Handbook of Environmental Chemistry. Vol. 3, Part L, 169-201, Springer-Verlag, Berlin, doi: 10.1007/10690734_9

Hartmans S, Smits JP, van der Werf MJ, Volkering F, de Bont JAM (1989) Metabolism of styrene oxide and 2-phenylethanol in the styrene-degrading *Xanthobacter* strain 124X. Appl Environ Microbiol 55:2850–2855

Harwood CS, Parales RE (1996) The β-ketoadipate pathway and the biology of self-identity. Annu Rev Microbiol 50:553–590

Hwang SS, Choi HT, Song HG (2008) Biodegradation of endocrine-disrupting phthalates by *Pleurotus ostreatus*. J Microbiol Biotechnol 18:767–772

Iwaki H, Saji H, Nakai E, Hasegawa Y (2004) Degradation of cyclopentanol by *Trichosporon cutaneum* strain KUY-6A. Microbes Environ 19:241–243

Jobling S, Reynods T, White R, Parker MG, Sumpter JP (1995) A variety of environmentally persistent chemicals, including some phthalate plasticizers, are weakly estrogenic. Environ Health Perspect 103:582–587

Jonsson S, Ejlertsson J, Ledin A, Mersiowsky I, Svensson BH (2003) Mono- and diesters from *o*-phthalic acid in leachates from different European landfills. Water Res 37:609–617

Kang Z, Buchenauer H (2000) Ultrastructural and cytochemical studies on cellulose, xylan and pectin degradation in wheat spikes infected by *Fusarium culmorum*. J Phytopathol 148:263–275

Kaszycki P, Czechowska K, Petryszak P, Miedzobrodzki J, Pawlik B, Koloczek H (2006) Methylotrophic extremophilic yeast *Trichosporon* sp.: a soil-derived isolate with potential applications in environmental biotechnology. Acta Biochim Pol 53:463–473

Keyser P, Pujar BG, Eaton RW, Ribbons DW (1976) Biodegradation of the phthalates and their esters by bacteria. Environ Health Perspect 18:159–166

Kim YH, Lee J (2005) Enzymatic degradation of dibutyl phthalate and toxicity of its degradation products. Biotechnol Lett 27:635–639

Kim YH, Lee J, Ahn JY, Gu MB, Moon SH (2002) Enhanced degradation of an endocrine-disrupting chemical, butyl benzyl phthalate, by *Fusarium oxysporum* f. sp. *pisi* cutinase. Appl Environ Microbiol 68:4684–4688

Kim YH, Lee J, Moon SH (2003) Degradation of an endocrine disrupting chemical, DEHP [di-(2-ethylhexyl)-phthalate], by *Fusarium oxysporum* f. sp. *pisi* cutinase. Appl Microbiol Biotechnol 63:75–80

Kim YH, Min J, Bae KD, Gu MB, Lee J (2005) Biodegradation of dipropyl phthalate and toxicity of its degradation products: a comparison of *Fusarium oxysporum* f. sp. *pisi* cutinase and *Candida cylindracea* esterase. Arch Microbiol 184:25–31

Kurane R (1997) Microbial degradation and treatment of polycyclic aromatic hydrocarbons and plasticizers. Ann N Y Acad Sci 829:118–134

Lee SM, Koo BW, Lee SS, Kim MK, Choi DH, Hong EJ, Jeung EB, Choi IG (2004) Biodegradation of dibutylphthalate by white rot fungi and evaluation on its estrogenic activity. Enzyme Microb Technol 35:417–423

Lee SM, Lee JW, Koo BW, Kim MK, Choi DH, Choi IG (2007) Dibutyl phthalate biodegradation by the white rot fungus, *Polyporus brumalis*. Biotechnol Bioeng 97:1516–1522

Li J, Gu JD (2006) Biodegradation of dimethyl terephthalate by *Pasteurella multocida* Sa follows an alternative biochemical pathway. Ecotoxicology 15:391–397

Li J, Gu JD (2007) Complete degradation of dimethyl isophthalate requires the biochemical cooperation between *Klebsiella oxytoca* Sc and *Methylobacterium mesophilicum* Sr isolated from Wetland sediment. Sci Total Environ 380:181–187

Li J, Gu JD, Pan L (2005a) Transformation of dimethyl phthalate, dimethyl isophthalate and dimethyl terephthalate by *Rhodococcus rubber* Sa and modeling the processes using the modified Gompertz model. Int Biodeterior Biodegrad 55:223–232

Li J, Gu JD, Yao JH (2005b) Degradation of dimethyl terephthalate by *Pasteurella multocida* Sa and *Sphingomonas paucimobilis* Sy isolated from mangrove sediment. Int Biodeterior Biodegrad 56:158–165

Li P, Li H, Stagnitti F, Wang X, Zhang H, Gong Z, Liu W, Xiong X, Li L, Austin C, Barry DA (2005c) Biodegradation of pyrene and phenanthrene in soil using immobilized fungi *Fusarium* sp. Bull Environ Contam Toxicol 75:443–450

Li R, Chen GZ, Tam NFY, Luan TG, Shin PKS, Cheung SG, Liu Y (2009) Toxicity of bisphenol A and its bioaccumulation and removal by a marine microalga *Stephanodiscus hantzschii*. Ecotox Environ Safe 72:321–328

Long JLA, House WA, Parker A, Rae JE (1998) Micro-organic compounds associated with sediments in the Humber rivers. Sci Total Environ 210–211:229–253

Luo ZH (2010) Degradation of three dimethyl phthalate isomer esters (DMPEs) by mangrove sediment fungi. PhD Thesis, City University of Hong Kong, Hong Kong

Luo ZH, Pang KL, Gu JD, Chow RKK, Vrijmoed LLP (2009) Degradability of the three dimethyl phthalate isomer esters (DMPEs) by a *Fusarium* species isolated from mangrove sediment. Mar Pollut Bull 58:765–768

Luo ZH, Wu YR, Pang KL, Gu JD, Vrijmoed LLP (2011) Comparison of initial hydrolysis of the three dimethyl phthalate esters (DMPEs) by a basidiomycetous yeast, *Trichosporon* DMI-5-1, from coastal sediment. Environ Sci Pollut Res. doi:doi:10.1007/s11356-011-0525-1

MacGillivray AR, Shiaris MP (1993) Biotransformation of polycyclic aromatic hydrocarbons by yeasts isolated from coastal sediments. Appl Environ Microbiol 59:1613–1618

Maruyama K, Akita K, Naitou C, Yoshida M, Kitamura T (2005) Purification and characterization of an esterase hydrolyzing monoalkyl phthalates from *Micrococcus sp.* YGJ1. J Biochem 137:27–32
Middelhoven WJ (1993) Catabolism of benzene compounds by ascomycetous and basidiomycetous yeasts and yeastlike fungi. A literature review and an experimental approach. Antonie Van Leeuwenhoek 63:125–144
Middelhoven WJ, Koorevaar M, Schuur GW (1992) Degradation of benzene compounds by yeasts in acidic soils. Plant Soil 145:37–43
Middelhoven WJ, Scorzetti G, Fell JW (2004) Systematics of the anamorphic basidiomycetous yeasts genus *Trichosporon Behrend* with the description of five novel species: *Trichosporon vadense, T. smithiae, T. dehoogii, T. scarabaeorum and T. gamsii.* Int J Syst Evol Microbiol 54:975–986
Nalli S, Cooper DG, Nicell JA (2005) Metabolites from the biodegradation of di-ester plasticizers by *Rhodococcus rhodochrous.* Sci Total Environ 366:286–294
Paxéus N (2000) Organic compounds in municipal landfill leachates. Water Sci Technol 42:323–333
Peterson JC, Freeman DH (1982) Phthalate ester concentration variations in dated sediment cores from Chesapeake Bay. Environ Sci Technol 16:464–469
Philips MW, Gordon GLR (1989) Growth characteristics on cellobiose of three different anaerobic fungi isolated from the ovine rumen. Appl Environ Microbiol 55:1695–1702
Pradeepkumar S, Karegoudar TB (2000) Metabolism of dimethylphthalate by *Aspergillus niger.* J Microbiol Biotechnol 10:518–521
Sampaio JP (1999) Utilization of low molecular weight aromatic compounds by heterobasidiomycetous yeasts: taxonomic implications. Can J Microbiol 45:491–512
Sietmann R, Hammer E, Schauer F (2002) Biotransformation of biarylic compounds by yeasts of the genus *trichosporon.* Syst Appl Microbiol 25:332–339
Sivamurthy K, Swamy BM, Pujar BG (1991) Transformation of dimethylterephthalate by the fungus *Sclerotium rolfsii.* FEMS Microbiol Lett 79:37–40
Staples CA, Peterson DR, Parkerton TF, Adams WJ (1997a) The environmental fate of phthalate esters: a literature review. Chemosphere 35:667–749
Staples CA, Adams WJ, Parkerton TF, Gorsuch JW, Biddinger GR, Reinert KH (1997b) Aquatic toxicity of eighteen phthalate esters. Environ Toxicol Chem 16:875–891
Stingley RL, Brezna B, Khan AA, Cerniglia CE (2004) Novel organization of genes in a phthalate degradation operon of *Mycobacterium vanbaalenii* PYR-1. Microbiology 150:2749–2761
Tam NFY (2006) Pollution studies on mangroves in Hong Kong and Mainland China. In: Wolanski E (ed) The environment in Asia Pacific harbors. Springer, Dordrecht
Tam NFY, Li SH, Lan CY, Chen GZ, Li MS, Wong YS (1995) Nutrients and heavy metal contamination of plants and sediments in Futian mangrove forest. Hydrobiologia 295:149–158
Tan GH (1995) Residue levels of phthalate esters in water and sediment samples from the Klang River basin. Bull Environ Contam Toxicol 54:171–176
Tortella GR, Diez MC, Duran N (2005) Fungal diversity and use in decomposition of environmental pollutants. Crit Rev Microbiol 31:197–212
Vega D, Bastide J (2003) Dimethylphthalate hydrolysis by specific microbial esterase. Chemosphere 51:663–668
Verdin A, Sahraoui AL, Durand R (2004) Degradation of benzo[a]pyrene by mitosporic fungi and extracellular oxidative enzymes. Int Biodeterior Biodegrad 53:65–70
Wang YP, Gu JD (2006a) Degradability of dimethyl terephthalate by *Variovorax paradoxus* T4 and *Sphingomonas yanoikuyae* DOS01 isolated from deep-ocean sediments. Ecotoxicology 15:549–557
Wang YP, Gu JD (2006b) Degradation of dimethyl isophthalate by *Viarovorax paradoxus* strain T4 isolated from deep-ocean sediment of the South China Sea. Hum Ecol Risk Assess 12:236–247
Wang Y, Fan Y, Gu JD (2003) Aerobic degradation of phthalic acid by *Comamonas acidovoran* Fy-1 and dimethyl phthalate ester by two reconstituted consortia from sewage sludge at high concentrations. World J Microbiol Biotechnol 19:811–815

Wang J, Zhao X, Wu W (2004) Biodegradation of phthalic acid esters (PAEs) in soil bioaugmented with acclimated activated sludge. Process Biochem 39:1837–1841

Wang Y, Yin B, Hong Y, Yan Y, Gu JD (2008) Degradation of dimethyl carboxylic phthalate ester by *Burkholderia cepacia* DA2 isolated from marine sediment of South China Sea. Ecotoxicology 17:845–852

Weber RWS, Ridderbusch DC, Anke H (2002) 2,4,6-Trinitrotoluene (TNT) tolerance and biotransformation potential of microfungi isolated from TNT-contaminated soil. Mycol Res 106:336–344

Williams SE, Woolridge EM, Ransom SC, Landro JA, Babbitt PC, Kozarich JW (1992) 3-Carboxy-cis, cis-muconate lactonizing enzyme from *Pseudomonas putida* is homologous to the class II fumarase family: a new reaction in the evolution of a mechanistic motif. Biochemistry 31:9768–9776

Wu YR, Luo ZH, Chow RKK, Vrijmoed LLP (2010a) Purification and characterization of an extracellular laccase from the anthracene-degrading fungus *Fusarium solani* MAS2. Bioresour Technol 101:9772–9777

Wu YR, Luo ZH, Vrijmoed LLP (2010b) Biodegradation of anthracene and benz[a]anthracene by two *Fusarium solani* strains isolated from mangrove sediments. Bioresour Technol 101:9666–9672

Xie Z, Ebinghaus R, Temme C, Caba A, Ruck W (2005) Atmospheric concentrations and air-sea exchanges of phthalates in the North Sea (German Bight). Atmos Environ 39:3209–3219

Xu XR, Gu JD, Li HB, Li XY (2005a) Kinetics of di-*n*-butyl phthalate degradation by a bacterium isolated from mangrove sediment. J Microbiol Biotechnol 15:946–951

Xu XR, Li HB, Gu JD (2005b) Biodegradation of an endocrine-disrupting chemical di-*n*-butyl phthalate ester by *Pseudomonas fluorescens* B-1. Int Biodeterior Biodegrad 55:9–15

Xu XR, Li HB, Gu JD (2006) Elucidation of *n*-butyl benzyl phthalate biodegradation using high-performance liquid chromatography and gas chromatography-mass spectrometry. Anal Bioanal Chem 386:370–375

Xu G, Li F, Wang Q (2008) Occurrence and degradation characteristics of dibutyl phthalate (DBP) and di-(2-ethylhexyl) phthalate (DEHP) in typical agricultural soils of China. Sci Total Environ 393:333–340

Zhang J, Cai L, Yuan D, Chen M (2004) Distribution and sources of polynuclear aromatic hydrocarbons in Mangrove surficial sediments of Deep Bay, China. Mar Pollut Bull 49:479–486

Index

C. Raghukumar (ed.), *Biology of Marine Fungi*,
Progress in Molecular and Subcellular Biology 53,
DOI 10.1007/978-3-642-23342-5, © Springer-Verlag Berlin Heidelberg 2012

9783642233418